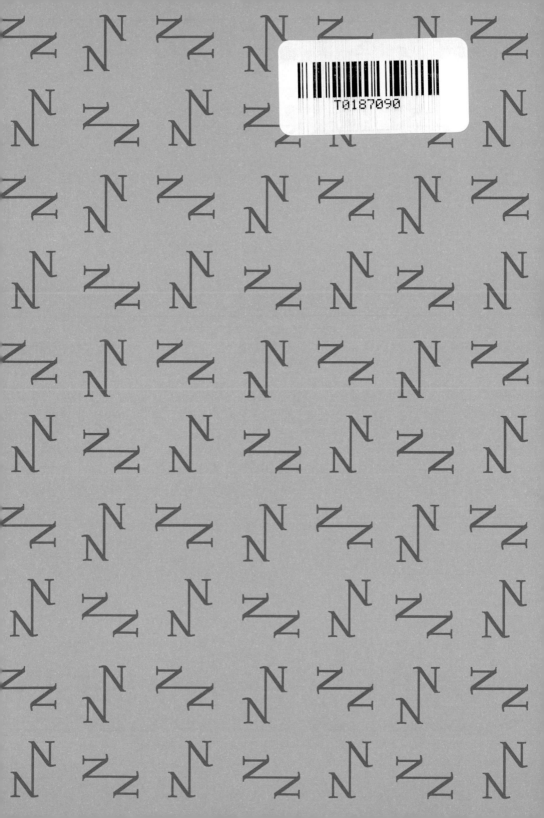

THE NEW NATURALIST LIBRARY

A SURVEY OF BRITISH NATURAL HISTORY

BEETLES

THE NEW NATURALIST LIBRARY

BEETLES

RICHARD JONES

WILLIAM
COLLINS

This edition published in 2018 by William Collins,
an imprint of HarperCollins Publishers

HarperCollins Publishers
1 London Bridge Street
London SE1 9GF

WilliamCollinsBooks.com

First published 2018

A CIP catalogue record for this book is available
from the British Library.

Set in FF Nexus

Edited and designed by
D & N Publishing
Baydon, Wiltshire

Printed in Hong Kong by Printing Express

Hardback
ISBN 978-0-00-814952-9

Paperback
ISBN 978-0-00-814950-5

Contents

Dedication

to Peter Hodge, coleopterist, mentor, friend

Editors' Preface

WHY HAS THE NEW NATURALIST LIBRARY waited so long to produce a book on beetles? We have long known that we needed one – beetles are ubiquitous, intriguing and much loved by many, from Charles Darwin to Christopher Robin, but they are surprisingly poorly covered in the literature. Comprehensive accounts of their natural history have been few, and until recently the keys needed by beginners to gain access to the group have been out of print, out of date or generally out of reach in terms of price. We knew that writing this book would be a challenge, requiring a huge fund of knowledge, boundless enthusiasm and an ability to do justice to a complex subject without flooding the reader with detail. It took time to find the right author. We are very glad that Richard Jones consented to meet the challenge, and we are delighted with the result. This book illustrates the pleasure that can be derived from the study of beetles, giving a conspectus of the group as a whole and delving into the quirks of natural history that make them so fascinating, and we expect that the key to beetle families and the listing of literature that enables one to move on to a species identification will help to remove some of the barriers that prevent more naturalists from indulging in coleopterology. Many established British naturalists found their early inspiration in a New Naturalist volume, and we expect this to inspire a new generation of beetle enthusiasts.

Author's Foreword and Acknowledgements

BEETLES ARE, ARGUABLY, the most important organisms on Earth. With nigh on half a million beetle species already described and catalogued in the world's museums, there are more types of beetle on this planet than any other group of creatures (quietly brushing aside nematodes for a moment, perhaps). They are certainly more diverse than any other category of living thing. Even so, we still know so little about them. Of those we do know, and have given names to, we hardly have a clue about their life histories, their larvae, their predators, or their interactions with other plants and animals. There are an estimated 1–3 million more beetle species out there waiting to be discovered, mostly in the deep, dark, dank rainforests of the tropics, although some would argue 30 million species are likely to exist. Even the experts can't agree.

This astonishing species diversity is matched by a similar diversity in shape, form, size, life history, ecology, physiology and behaviour. In effect, beetles occur everywhere, and do everything. And yet beetles form a clearly discrete insect group, typically characterised by their attractively compact form, with flight wings folded neatly under smooth, hard wing-cases. Almost anyone could recognise a beetle; indeed, many are intimately associated with human society. Groups like ladybirds are familiar to us from a very young age. Large stag beetles and handsome chafers are celebrated for their imposing size or bright colours. The sacred scarabs of the ancient Egyptians were given iconic, if not god-like, status, and even though the exact religious meanings may be fading after three millennia, the beetles' images remain in bewitching jewellery and monumental statuary.

Despite our ancient and easy familiarity with beetles, the order Coleoptera remains tainted by the notion that it is a 'difficult' group of insects to study. The traditional routes into studying British natural history, through birdwatching, butterfly collecting (back in the day) and pressing wildflowers, now extend to studying dragonflies, bumblebees, grasshoppers, moths, hoverflies and even shieldbugs. These are on the verge of becoming popular mainstream groups, but beetles remain the preserve of the expert, or so it seems. This is a shame, and something I'd like to set right with this book. Many British beetles are easy to find and easy to identify by the non-expert, but that bewildering background diversity, and the daunting numbers of species in the Coleoptera as a whole, have been enough to dissuade many a potential coleopterist from grasping the nettle and getting stuck in.

We are very lucky in Britain, in that we have a long and noble history of studying wildlife, call it what you will, from natural philosophy to nature conservation to citizen science. Insects, by virtue of being small and easily caught and examined, have led the way – beetles especially. It is no coincidence that Charles Darwin and Alfred Russel Wallace were, first and foremost, entomologists. Actually, no, any reading of their lives and works shows that they were, self-evidently, not just entomologists; they were coleopterists.

At school, at university and in my early professional life in medical publishing, my interest in insects was seen as a bit weird. But once any initial amusement at my curious demeanour had passed, I slightly relished the fact that I was the holder of a secret arcane knowledge, from which my contemporaries were self-excluded. That stag beetles grew from maggots, and not from smaller beetles, was a revelation to otherwise intelligent and well-adjusted friends. Visiting relatives in Florida 25 years ago, I found four species of beetle infesting a Tupperware box of flour in their larder; they were horrified, and not overly happy with my forensic dissection of the many hundreds of specimens in the congealed mess. I do, however, like to think that they were enlightened by my enthusiastic discussion of the beetles, and the fact that they represented four different beetle families, and had four different evolutionary routes, from bird nests, hollow trees, bumblebee colonies and leaf litter, into the twentieth-century cornucopia of the North American kitchen habitat. And, of course, every time a child asks 'Are ladybirds poisonous?', I can answer with a certainty that I know will startle them: 'Yes they are, but you have to eat quite a lot of them to do any real harm.' These simple basics of biology, ecology, physiology and evolution are a mysterious closed book to most people.

Yet one of the joys of entomology is that anyone can become an entomologist. With no need of expensive equipment, difficult techniques or specialist training, entomology has long been an egalitarian science, led by non-elitist, self-taught aficionados from all walks of life. When, as a teenage schoolboy, I attended the annual exhibitions of the various entomological societies and recording groups, I was a prole mixing with professors, but we spoke as equals when discussing obscure ground-dwelling *Trachyphloeus* weevils from the chalk hills behind my parents' South Downs house, or the abundance of *Brachinus crepitans* bombardier beetles on the undercliff of Newhaven's foreshore. Here also were dukes and dames, colonels and canons, engaged in a mutual, inclusive, friendly discourse.

That friendliness was extended to me, not because I had specialist information to give, or new data to be gleaned, but simply because I chose to engage. And engagement begins with picking up a hand lens and peering close at a beetle. Anyone can do this. In Britain, we have a perfectly manageable 4,118 species. This is *not* too daunting a number. It is enough that there will always be challenges to finding and identifying some of them. But with our rich history of beetle study, there are plenty of groups, like ground beetles, longhorns and leaf beetles, where relatively large and showy insects vie with each other for inclusion in picture books; and with the existence of numerous monographs there is a wealth of identification keys and research material already behind us. In the middle is a huge centre-ground, where finding and naming is doable, but where understanding is still waiting to be teased out. Beetle complexity in Britain is, in fact, perfectly aligned to encourage study, and nobody should be put off by a few of the more troublesome taxonomic knots.

I hope a New Naturalist volume on beetles will mark a watershed, by encouraging many who would otherwise be put off by the, to date, rather technical literature that has dominated this fascinating and beautiful group of insects. This will be my theme, throughout this book, as it has been throughout my life – that jargon can be mastered, that expertise can be learned, that arcane knowledge can be shared, that wonder at the natural world can be enlightening to all.

ACKNOWLEDGEMENTS

On 1 February 2014, Mark Telfer sidled up to me at the 11th British Entomological Society Coleopterists' Day at the Oxford University Museum of Natural History. He enquired, under his breath, whether I'd heard from Sarah Corbet yet, about the idea for a beetle volume in the New Naturalist series. I nearly dropped my cup of tea. It seemed that feelers had been put out for some victim to be approached. Mark, it transpired, had landed me in it. The thought of such a commission was daunting, to say the least, but I knew that if the offer came I'd hesitate for all of 2.5 seconds and then accept. When it came, I did. Producing this book has been even more daunting than I imagined at the time, and although it has been long and exhausting in the researching and writing, it has also been intellectually and emotionally rewarding beyond measure. My eternal thanks go to Mark, and I look forward to returning the favour with some equally monstrous task for him in the near future.

Lillian Ure-Jones was my first editor, and her reading through the initial typescript helped sort out some of the more garbled passages and highlighted the occasional lapse into unexplained jargon. She pulled me up when I kept repeating myself, or when I descended into the stickier depths of self-aggrandising pomposity. Later, it fell to Sarah Corbet to edit my words fully, and she did a masterful job. Despite her protestations that she had mostly marked lots of trivia and that I should ignore them if I disagreed, it was obvious that she knew what she was about, and I followed her on almost every point. In the final straight, Susi Bailey consummately copy-edited the text to the highest possible professional standard – it was at this point that all my embarrassing spelling mistakes were spotted, and schoolboy errors like when exactly Gondwanaland broke apart. And designers Namrita and David Price-Goodfellow brought order and style to my sometimes unhelpful and less than precise layout suggestions.

If the book has a retro feel about it this is because I have trawled through many old publications, reusing what I consider to be perfectly adequate diagrams and illustrations. Where I had gaps and could find nothing suitable elsewhere, Verity Ure-Jones provided me with several watercolours, pen-and-ink drawings and diagrams. My photographs are mostly from 35 mm transparencies, and although they might be brightly lit with flash, there is a certain dated celluloid look about them. Luckily, Peckham entomologist Penny Metal volunteered many of her superb digital images; the book is all the brighter for them.

During this book's construction, many people have also given me their support, their time or their knowledge. Andrew Duff saved me a tremendous amount of work by allowing me an early preview of the draft copy of the 2017 beetle checklist several months before publication; he cannot know how much effort he saved me in not having to go through making all the changes if I'd had to wait for a printed copy. Many thanks also go to the following for the help they have given me and for the information they have wittingly and unwittingly supplied: Nick Alexander, Rob Angus, Ralph Atherton, Tristan Bantock, Max Barclay, Charlie Barnes, Nick Bertrand, Steve Bolchover, Jeremy Bowdrey, John Bratton, David Buckingham, Andy Chick, Jon Cole, Martin Collier, Michael Darby, Tony Drane, Adrian Dutton, Brian Eversham, Beulah Garner, Stuart Hine, Peter Hodge, Trevor James, Martin Jenner, Jim Jobe, Mark Leppard, Johannes Lückmann, Richard Lyszkowski, Peter MacDonald, Darren Mann, Mike Morris, John Muggleton, Alex Ramsay, Harry Rutherford, Don Stenhouse, Malcolm Storey, Clive Turner and Richard Wright.

Finally, thank you Catrina Ure for being so patient and understanding of my long-standing fascination with beetles, and for your encouragement during the writing of this book.

CHAPTER 1

What is a Beetle?

HOW TO RECOGNISE A BEETLE

Everyone knows a beetle, surely? They're attractive, chunky, hard-shelled, long-legged insects that scuttle about in the garden, living under stones or running over the patio at top speed on a sunny summer's day. Winnie the Pooh's Christopher Robin had one (Alexander, he called it) in a matchbox – until Nanny let it out. Volkswagen used a sleek domed shape for their compact post-war family saloon car and it soon acquired the common name of its insect lookalike. The word beetle is so familiar to us that, at first, it seems unnecessary to question what it means. Mind you, what about those small brown cylindrical things under the cooker? Aren't they larder beetles? But they're a different shape. And carpet beetles are practically spherical. Things soon start to get a bit tricky.

Then there are ladybirds? In North America they're sometimes called ladybugs, but really they're beetles, not bugs. What about weevils? Are they different? No, they're beetles all right. Then it turns out that black-beetles are not beetles at all, but cockroaches, while similar-sounding cockchafers are chafers, which *are* beetles. Then there are the flies that are really beetles (fire-fly, Spanish Fly) and the worms that are beetles (woodworms, wireworms, glow-worms, mealworms). It's all very perplexing.

OPPOSITE PAGE: **FIG 1.** Living jewels: *upper left*: rosemary beetle (*Chrysolina americana*); *upper right*: bumblebee chafer (*Trichius gallicus*); *lower left*: black clock beetle (*Pterostichus madidus*); and *lower right*: bloody-nosed beetle (*Timarcha tenebricosa*). (Penny Metal)

There soon comes a point at which the apparent simplicity of naming and understanding these things descends into chaos and confusion. These familiar, simple common or folk names originated long ago with simple common folk for whom a detailed knowledge of insect classification was unavailable and unnecessary. Luckily, entomologists are an international bunch and are not fazed by the inadequacies of vernacular English (or any other language) to describe what is, in reality, a fairly cohesive and unified group of insects. They resort, instead, to Latin and Greek, the classical languages upon which modern scientific nomenclature stands, and all becomes clear.

Aristotle, in the fourth century BCE, is credited with first coining the term Coleoptera, the name of the insect order for beetles that we still use today. From the Greek words κολεος (*koleos*), meaning 'sheath', and πτερον (*pteron*), meaning 'wing', this is still the perfect basic description of a beetle, which, after all, keeps its delicate membranous flight wings folded underneath a tough sheath, made up of two (left and right) elytra (singular elytron).[1] These wing-cases, usually curved and moulded to cover the rear half or two-thirds of a beetle completely and tightly, are perhaps the most obvious and distinctive feature of the Coleoptera, but there are a few other key attributes that need to be mentioned for biological completeness. And there's a bit more entomological etymology on the way.

Beetles have chewing mouthparts, left- and right-hinged mandibles; these jaws are stout, roughly triangular cutting or biting plates that meet each other

FIG 2. The mandibles of the lesser stag beetle (*Dorcus parallelipipedus*) meet like the fingers of clasped hands.

---

1 This is where the jargon starts. There are a few new specialist terms to get used to with beetles, and elytra/elytron is the most important. Most technical terms are explained in the text, but there is also a glossary on pp. 409–15 detailing a whole host of words to be found in this and other beetle identification guides and keys.

FIG 3. Long, sharp, curved jaws are often a sign of a ferocious predator; this fits well with the aggressive hunting of tiger beetles, but the male also uses his mandibles to grip the waist of his mate during copulation.

like the blades of scissors, tongs, pliers or secateurs. The English word beetle is derived from various precursors in Old English and Anglo-Saxon: *betl, betel, bitel* and *bitula*, from the root *bitan*, meaning 'to bite'. Likewise, chafer (echoed in the modern German word for beetle, *Käfer*), derives from linguistic roots like *kaf, kef* and *ceafor*, linked to chaff (meaning 'to cut', 'chew' and 'gnaw'). For a time, especially during the nineteenth century, the term Mandibulata was widely used to include the beetles, and also other jawed insects like bees, wasps, ants, earwigs, grasshoppers and dragonflies.

The final defining feature of beetles is that they go through a complete metamorphosis – their development is said to be holometabolous. Hatching from an egg, the immature (and, more importantly, wingless), mobile, embryo-like maggot, grub, caterpillar or larva rearranges its body tissues in the pupa/chrysalis, and the final, complete, fully formed winged adult beetle emerges. They share this supposedly more advanced evolutionary trait with the other higher insects, such as Lepidoptera (butterflies and moths), Hymenoptera (bees, wasps, ants, etc.), and Diptera (flies). In contrast, groups like Hemiptera (leaf bugs and hoppers), Odonata (dragonflies), Mantodea (mantids) and Orthoptera (grasshoppers) pass through hemimetabolous development, with miniature nymphal versions of their respective adult forms gradually increasing in size, and with slowly enlarging wing buds, until the mature, fully grown winged adult finally appears.

These three physical characteristics (elytra, mandibles and holometabolism) serve to define beetles, the Coleoptera, in a biological sense, but only the first is easily observable, and this will be the most important part of recognising a beetle in the field. Well, recognising a typical beetle, that is. As is abundantly clear, beetles are a hugely diverse group of insects, and at the extremes of this diversity are long, thin beetles; short, fat beetles; domed beetles; flat beetles; shiny beetles; rough beetles; round beetles, square beetles even. There are also beetles that don't look anything like beetles.

This mesmerising and complex diversity makes beetles special; they are fascinating and varied, and there are so many new things to find out about them. However, this complexity has sometimes been enough to put people off studying the Coleoptera. There are just so many of them, and identifying some groups is a puzzle beyond patience – the hard work of many lifetimes. Oliver Wendell Holmes has his pompous entomologist in *The Poet at the Breakfast Table* (1872) decrying the childish simplicity of the other orders before the vast complexity and obviously serious importance of the beetles: 'Lepidoptera and Neuroptera for little folks; and Coleoptera for men, Sir!' But like all studies, it's easy enough to start simple. And for this the usual approach, one that I will follow too, is to start with a couple of 'typical' beetles.

HOW A BEETLE WORKS – GENERAL FORM, MORPHOLOGY AND STRUCTURE

As a teenager, one of my first major beetle book purchases was of the famous five-volume *Coleoptera of the British Islands*, by the Reverend Canon W.W. Fowler (published between 1887 and 1891), together with the supplementary sixth volume by Fowler and Donisthorpe (1913), which I bought from an antiquarian-book dealer in 1975. This was not the handsome (and very rare) large edition with 200 hand-coloured plates – that would have been well beyond my financial means – but the 'small paper' edition. Even so, £25, if memory serves, was then an eye-watering sum for a 16-year-old to rustle up. Despite missing the lovely coloured illustrations, this landmark monograph nevertheless came with a few black and white engraved plates at the beginning, and Fowler's choice of 'typical' beetles is still apt enough today to have me repeat them here.

A ground beetle (Carabidae, Fig. 4) is a good representative of the terrestrial beetle fauna, and conforms very well to that notional long-legged, hard-shelled beetle that Christopher Robin and others have found scuttling about in the garden. And a diving beetle (Dytiscidae, Fig. 5) serves to illustrate the differences (and yet similarities too) necessary for an aquatic beetle. The other labelled diagrams over the next few pages (Figs 6–8) are self-explanatory and, combined with the Glossary (pp. 409–15), should form the basic reference part of this book, enabling any reader to work through keys, or understand some of the sometimes obscure (but necessary) words that make up the coleopterist's lexicon.

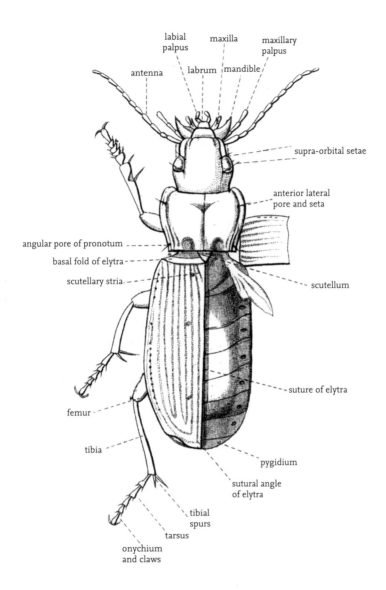

FIG 4. Generalised structure of a beetle, *Pterostichus melanarius*, upper side. From Fowler (1887).

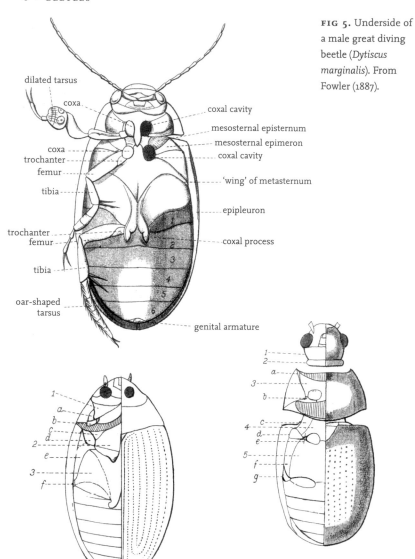

dilated tarsus

coxa

coxa
trochanter
femur

tibia

trochanter
femur

tibia

oar-shaped
tarsus

coxal cavity

mesosternal episternum
mesosternal epimeron
coxal cavity

'wing' of metasternum

epipleuron

coxal process

genital armature

FIG 5. Underside of a male great diving beetle (*Dytiscus marginalis*). From Fowler (1887).

FIG 6. Underskeleton of *Gyrinus*: 1, prosternum – (a) episternum, (b) epimeron; 2, mesosternum – (c) episternum, (d) epimeron; 3, metasternum – (e) episternum, (f) epimeron. From Fowler (1887).

FIG 7. Diagram of an imaginary beetle: 1, gular region; 2, collum; 3, prosternum – (a) episternum, (b) epimeron; 4, mesosternum – (c) episternum, (d) epimeron, (e) trochanter (shown with a fine line drawn through it); 5, metasternum – (f) episternum, (g) epimeron. From Fowler (1887).

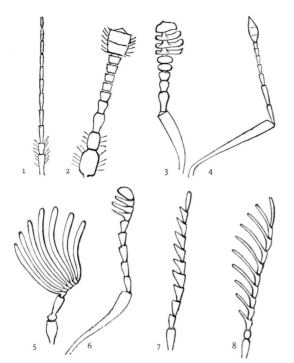

FIG 8. Antennal forms: 1, filiform (simple cylindrical segments); 2, capitate, clavate (tight, clubbed segments); 3, perfoliate club (with loose segments); 4, geniculate (elbowed, with a long basal scape segment); 5, lamellate (stacked plate-like segments); 6, fissate club (split with fissures); 7, serrate (toothed); 8, pectinate (comb-like). From Fowler (1887).

To start, though, it is important to have some understanding of the fact that beetles (indeed, all insects) are based on a series of segments along their bodies, and that the repetition of, or differentiation between, these segments is important in identifying and naming beetles, and understanding how they work. Research in analysing similarities between insects and other organisms, and the proposals of how they might all be related through evolutionary descent, is complex, and not without some mild-mannered controversy among biologists. However, a broad consensus of insect evolution is presented in the masterly volume by Grimaldi and Engel (2005), on which I rely heavily.

The very word insect means 'cut in', and refers to the fact that insects are divided into various segments by distinct boundaries that are visible as neat lines or grooves across their bodies, the most obvious ones dividing head, thorax and abdomen. Way back in distant geological time, well over half a billion years ago, a condition called metamerism evolved; it transformed long-extinct, probably soft-bodied creatures by allowing continued duplication (or self-budding perhaps) of a basic, simple, primitive body into a series of repeated similar segments. Whether this actually enhanced movement or burrowing through the primordial ocean bottom, or enabled different bits of the worm-like animal to concentrate on different physiological processes, is subject to much amicable debate among zoologists. Nevertheless, however or why ever it happened, the segmented worms were able to compete very effectively against the unsegmented animals, and they have more or less inherited the Earth. This relatively simple biological feature of multiple segmentation has since allowed the evolution of ridiculously complex organisms like obviously segmented insects and less obviously segmented humans (think repeated vertebrae in the backbone, the ribcage, a six-pack).

At the latest reckoning, the mostly terrestrial insects (along with fairly similar marine creatures, the Crustacea) are closely related to the Myriapoda (centipedes and millipedes), which together make up a significant part of the phylum Arthropoda (more Greek – αρθρον (*arthron*), meaning 'joint', and πους (*pous*), meaning 'foot', for their multi-jointed legs). The myriapods are obviously multi-segmented creatures, and even with their sometimes many hundreds of legs, evolution from some legless worm-like ancestor is an easily intuitive deduction. Our current understanding of insects is that they, too, all evolved from a multi-segmented, many-limbed animal that lived about 500 million years ago (mya). The peculiar many-legged velvet-worms (Onychophora, including members of the genus *Peripatus*) and the bizarre eight-legged microscopic water bears (Tardigrada) are somewhere close to the evolutionary nub where arthropods originated.

Having achieved the biological gains offered by metamerism, the next stage in evolutionary complexity arose by a proto-insect combining and consolidating multiple repeated segments into groups, each group having a specialist function. This process, called tagmosis, has contracted tubular velvet-worm-like, or millipede-like, repetition into the more compact insect body layout familiar to all as head, thorax and abdomen.

Modern beetles are far removed from their worm-like ancestors, although examining beetle larvae later on in the book will throw up some enlightening observations. Even so, under the microscope, that ancestral metamerism, followed by tagmosis, is still evident, and it's pretty clear that as far back as we can trace the beetles, they are founded on a body plan of 20 segments: six in the head, three in the thorax and eleven in the abdomen.

THE HEAD

A beetle's head is dominated by the sensory equipment of the eyes, antennae and palps, and the distinctive biting mandibles. The large compound eyes, sometimes made up of many hundreds of individual units (ommatidia), are usually positioned prominently at the sides of the head. They vary across the Coleoptera, from huge bulbous domes meeting in the centre of the frons (forehead), down to nothing at all in blind, eyeless cave-dwelling or subterranean species. In some species, the large eyes are kidney-shaped, partially divided by a lobe that is especially prominent in many dung beetles (Scarabaeidae), cardinal beetles (Pyrochroidae), bean weevils (Chrysomelidae, subfamily Bruchinae) and longhorns (Cerambycidae). In the surface-dwelling water beetles, the whirligigs (Gyrinidae), this division is complete, giving the insect what appears to be four separate eyes, two for peering down into the water and two for looking up into the air.

Insect vision is wholly different from that of vertebrates, and comparisons can only be analogous, at best. Each facet is its own light-focusing and light-sensing organ. The long, narrow, light-sensitive column beneath each facet, the rhabdom, triggers neurons to fire and these signals are then interpreted by the beetle's brain. Insects are very sensitive to movement, but the overall picture they have of the world, if indeed they have one at all, is probably extremely pixellated. Where beetle eye studies have been carried out, ommatidium surfaces tend to be small and flat in diurnal species active in full sunshine. Each flat facet lens normally focuses light onto a single rhabdom, and larger numbers of smaller ommatidia presumably offer a finer, higher-resolution analysis of the insect's surroundings. This type of structure is called an apposition eye, since the light points (analogous to pixels) are apposed, or next to each other. However, for those crepuscular or nocturnal beetles active in low light, the individual facets are generally large and coarsely domed, and several of them focus light onto a single rhabdom. These optical superposition eyes may give a coarser, more blurred interpretation of the world, but they make the beetle's sensitivity to light far greater, enabling it to function when its daytime competitors cannot. Many subterranean or cave-dwelling beetles lack eyes, but transcription studies on the apparently blind Kentucky cave beetle *Ptomophagus hirtus* show that its DNA still codes for light-processing proteins and it demonstrates light/dark sensitivity in laboratory studies (Friedrich *et al.*, 2011).

A few beetles also have small, primitive light-sensitive structures called ocelli (singular ocellus) on the top of the head – for example, there is a single ocellus in most larder and hide beetles (Dermestidae), and two ocelli in some rove beetles (Staphylinidae, subfamily Omaliinae) and the obscure family Derodontidae. In other groups (notably Diptera, Hymenoptera, Orthoptera and Odonata, where these structures are standard and often number three), the ocelli are thought to aid the flying insects with navigational horizon detection – as distinct from creating the more detailed, if pixellated, visual image of the world seen through the multi-faceted compound eyes.

Equally prominent on the head, the antennae can vary tremendously in size, shape and form between different families. In the majority of beetles, the antennae have 11 rather similar and roughly cylindrical units, historically always called joints or (more properly) segments, now frequently referred to as antennomeres. The first segment of the antenna, the one attached by a ball-and-socket pivot into the beetle's face, is called the scape; it is often larger than the others, either fatter, or long and stalk-like, and it contains its own independent musculature. The second segment, the pedicel, is usually small, but it also contains its own muscles and Johnston's organ, a sensory centre that detects and measures minuscule flexion and movement in the succeeding antennomeres – the flagellum. Johnston's organ is as near to hearing as beetles seem to get: the sound vibrations in the air are translated into minute mechanical movement of the flagellum, then into nerve impulses in the pedicel. The remaining segments (usually nine more in almost all beetles), all controlled by the same set of muscles, technically form the flagellum, but this is often masked by the enlargement of some of the terminal segments into a swelling or club, with the intermediate smaller segments sometimes forming the funiculus. The weevils (Curculionidae) ably demonstrate this basic antennal formula by having a long scape (sometimes as long as the remainder of the antenna), an abruptly elbowed (geniculate) angle at the pedicel, which with other segments form a funiculus, and finally a distinct terminal club.

In addition to being used as feelers, the antennae – and especially any enlarged finger-like protrusions, flattened plates or swollen segments – are covered with chemoreceptor sense organs of varying shapes and sizes; these

FIG 9. The extremely pectinate, verging on flabellate, antennae of this *Ptilinus pectinicornis* mark it out as a male. The female has much shorter lobes on her sometimes merely serrate or dentate antennal segments.

segments may appear dull or microscopically furry under the high-power microscope because of the densely packed sensory hairs (sensilla, singular sensillum). They detect airborne smells, for example the odours of foodstuffs, warning signals or chemical scents (pheromones) from potential mates. The sensitivity of beetle antennal chemoreceptors is astounding. Mate-finding pheromones tend to be small compounds (5–20 carbon atoms, molecular weights in the region 80–300), highly volatile and with frighteningly diverse chemical structures (Seybold & Vanderwel, 2003). Isolated test antennae in micro wind tunnels react to airborne species-specific scents at the level of the individual molecule. At the blunter end of the antennal structure range, some water beetles may simply use their antennae to break the surface to allow access to fresh air.

Antennal shapes and lengths vary greatly across the Coleoptera and are a basic feature in any identification guide. The antennae can be shortened to little more than finger-shaped blobs (e.g. Dryopidae), or extended to many times the length of the beetle's body (the aptly named longhorn beetles, Cerambycidae). Although an antenna of 11 segments is the standard for most British beetle species, those with 10 segments are fairly common, and in a few groups the number of segments is reduced by fusion down to five or six. Males and females sometimes have slightly different numbers. Very occasionally, an aberrant specimen may occur where antennal segments are missing, or duplicated, or where left and right antennae have different numbers of antennomeres, or are asymmetrical. In the most obscure book in my library, Mocquerys (1880) took great delight in illustrating various beetle abnormalities, including numerous 'monstrous' specimens with a branched antenna, supernumerary extensions or additional side segments. These were possibly the result of damage during metamorphosis, and analogous to the well-known extra limbs grown by amphibians. Up to about 20 segments appear in a few Australian species of the peculiar exotic family Rhipiceridae, and the antennae of some male longhorns (Cerambycidae, subfamily Prioninae) have up to 40 segments.

Between the eyes, the forehead, or frons, usually descends smoothly into the clypeus, a shielding plate that covers the upper limit of the mouthparts, and finally the labrum, the upper lip. Sometimes these are individually visible, separated by narrow grooves, but in many species they are welded together to form a single undifferentiated feature.

A beetle mandible is roughly triangular (tetrahedral in three dimensions), and hinged at the condyle (similar to a ball-and-socket joint) on the outer rear corner, with the sharp distal tips meeting or cutting past each other to bite and slice, and the stout inner corner (the mola) meeting its opposing mandible to grind and chew the food. Typical mandibles are short and stout, but they can be long and pointed (falcate) in predators such as tiger beetles (subfamily Cicindelinae), or hugely elaborate for wrestling, as in the famous stag beetles (family Lucanidae).

Next, just beneath the mandibles, come the maxillae, used for food manipulation during feeding. The basal parts move rather like a second pair of smaller, more delicate mandibles, while the terminal parts, called palps (or palpi, singular palpus), usually have three or four simple cylindrical segments, and resemble additional short antennae. Indeed, in some groups (e.g. Hydrophilidae), the maxillary palps take on the appearance (and some functions) of the antennae, and the true antennae are much shorter and insignificant looking.

The mouth cavity is bounded underneath by the labium (lower 'lip'). Attached at each side are the labial palps, again resembling small, short antennae, and usually with two visible segments in beetles.

THE THORAX

Immediately behind the beetle head is a large shield-like body plate. Called the thorax by the uninitiated, this is the pronotum, really just one of the upper plates of the thorax. The beetle thorax actually consists of three (pro-, meso- and meta-thoracic) ancestral metameric segments. Back in deep evolutionary time, some pre-beetle insect probably had rather box-like segments, with upper, lower and side plates hinged on at least slightly flexible membranous junctions; something of this arrangement is still visible in other insects like flies, but in the Coleoptera many of the head and thoracic plates have become welded immovably together for added strength, and the sutures are all but invisible on the outer surface. In some cases, individual portions of body segments are visible and can be identified; notionally, those of each thorax segment comprise the notum (upper, or dorsal plate), sternum (lower, or ventral) and pleura (singular pleuron, at the sides, lateral). Detailed knowledge of all these barely discernible body parts is useful at the more technical end of identification keys, but is unnecessary here. Despite their anatomical complexity, the three thoracic segments are still evident under a simple lens or microscope, without recourse to dissection – beneath each of them a pair of legs is attached.

The metathoracic (third) segment carries the delicate membranous flight wings, and the mesothoracic (second) segment carries the wing-cases, or elytra (we'll spend a lot more time on these later on in the chapter). In contrast, the upper surface of the prothoracic (first) segment is developed into a massive flat, arched or domed carapace – the pronotum – perhaps the single most important body plate on any beetle.

The pronotum acts as a tough armoured cover to protect the front half of the beetle; often the head can be withdrawn partly or completely underneath it. The structure's highly variable shape, combined with the shape of the elytra, is immensely useful in the initial identification of beetle groups. It may be broad and flat, curved, humped, domed, narrow, cylindrical, heart-shaped or extended out into broad flanges. By convention, the middle of the pronotum – indeed, the central area of any large flattish body part – is called the disc (and the edges are called margins). The pronotum can be decorated with nodules, spines, hairs or scales to achieve patterns or tufts, and its size and shape are important clues to the beetle's lifestyle.

The disc and margins of the pronotum are usually clearly dinted, like beaten metal, with a pattern or texture of microscopic indentations or pits. Traditionally called punctuation (sometimes now puncturation, to avoid confusion with commas and colons, or in the United States in particular, punctation), these tiny dents or dimples are very important for species identification. All parts of a beetle can have punctuation of varying types, from scattered pinpricks or isolated pore punctures, to large craters or broad dimples (foveae), to a more or less uniformly matt texture of minute semi-confluent pinpricks or massively corrugated, granular wrinkles. The problem has always been how to convey adequately in words what can be seen down the microscope.

Just behind the pronotum, between the two bases of the elytra, is a small, approximately triangular plate called the scutellum (or, technically, the mesoscutellum); this is the only part of the mesothorax that is visible from above, although it is sometimes so small as to be almost invisible, or is completely covered by the wing-cases. The scutellum acts as a bracing or alignment aid when the elytra are closed – a kind of door jamb for the wing-cases.

The scutellum also marks the nominal centre of a beetle. There is a strongly based convention in coleopterological literature for discussing bits of a beetle in terms of basal or apical parts (sometimes also proximal/distal, or basad/distad). The base of an antenna or the apex of a tibia are easily grasped concepts, but the base of the pronotum is the rearmost portion (nearest to the scutellum), whereas the base of the elytra is the foremost (again, nearest to the scutellum).

The rear half of most beetles is dominated by the elytra (see below), which cover both the hind two-thirds of the thorax, and the abdomen. In side view, it is easier to appreciate that the three structural body sections of an insect (usually head, thorax and abdomen) become, in beetles, (i) head, (ii) prothorax, and (iii) mesothorax, metathorax and abdomen combined under the elytra.

There is a simple and intuitively obvious reason why the thorax stretches down and under the abdomen like this: it gives stabilising centrality to the six leg bases. With two legs on each of the three thoracic segments, this central chassis allows precisely controlled and balanced walking and running in beetles, sometimes very speedily, rather than the dragging often seen in, say, beetle larvae (although the hugely disproportionate abdomens of female *Meloe* oil beetles form a fascinating exception).

Most beetle legs follow a standard pattern of segmentation, and are similarly articulated right across the Coleoptera. Starting from their bodily attachment, these segments are (single/plural): coxa/coxae, trochanter/trochanters, femur/femora, tibia/tibiae and tarsus/tarsi. The leg segments are variously decorated with hairs, spines, teeth, spurs and combs, or sometimes the occasional mucro (sharp apical point). Some legs are long and thin for fast running, others are short and broad for digging and burrowing, some are flattened for swimming and yet others are enlarged for jumping, but the basic structure remains constant. On the whole, they normally increase in size (or length) from front to rear, although the front legs can be modified out of this proportion since they are the first to meet prey to be clasped, or a substrate like dung or soil to be dug into. The front tibiae of a large portion of ground beetles (Carabidae) have a distinct notch on the inside, just up from the apex, complete with combs of fine hairs, through which the antennae are run during grooming.

The tarsus is analogous to the foot, and is the part of the leg that makes contact with the rock or leaf or stem or whatever the beetle is climbing up. It is divided into segments (joints in older literature), now generally referred to as tarsomeres, since, like the flagellum of the antenna, they are under the control of one set of musculature. The archetypal proto-beetle ancestor probably had five segments in each tarsus, and this is the default number across most extant modern beetle groups; however, there can be four or three segments in some legs, down to none in exceptional cases. Tarsomere number has been widely used as a classification and identification tool. A tarsal formula in the form 5–5–5 (five tarsomeres on front–middle–hind legs, respectively) would be the pentamerous norm; 4–4–4 (tetramerous) would mean four tarsomeres on each; while 5–5–4 (heteromerous) would mean only four tarsal segments on the back legs. Care is always needed when looking at tarsi; the formulae can sometimes vary between legs and between sexes. Individual tarsomeres are difficult enough to see on small beetles, even through a good microscope; the situation can be further clouded by some tarsomeres being so small that they are hidden and invisible. This leads to ladybirds, for example, being described as pseudotrimerous, because although they have four segments on each tarsus, the third one is so small that it is obscured by the fourth, and it looks as if there are just three.

The final tarsal segment usually has two hooked claws, or sometimes just a single claw, and in a very few species the claw joint is missing completely.

Clinging to the substrate is not just a grappling affair, and in some beetles (notably leaf beetles, Chrysomelidae) the tarsal segments are broad and flat. Some are divided into two pads (bilobed) and can be covered underneath by many branching, frond-shaped feathery hairs that allow the beetles to walk up smooth surfaces, just like geckos, harnessing the forces of molecular attraction at the sub-nanometre scale.

Beetle flight wings are relatively poorly studied and are easy to overlook, because they are so neatly tucked away under the elytra. For many people, it is a revelation that beetles can fly at all, so used are they to seeing them running around on the ground, apparently wingless. Beetle flight is covered in detail in Chapter 5, so it is enough just to have a cursory look at a typical beetle wing here.

The delicate flight wings of beetles arise on the hindmost section of the thorax, the metathorax, but they are mostly kept folded flat over the beetle's abdomen, covered by the elytra. Once a beetle wing is unfolded, its shape is remarkably similar to the wings of the Hymenoptera (bees, wasps, ants, etc.) and Diptera (flies). It is long-oval, leaf-shaped, slightly pointed at the end, membranous, and supported by a network of longitudinal veins that are usually reinforced by cross-veins to create enclosed clear, window-like cells. Where it differs is in the origami-like concertina folds by which it is scrunched up tight and stowed under the elytra. In most beetles, the structure of the wings, the alignment of the spar-like veins with their various breaks and hinges in suitable positions, the pleats of the membrane and perhaps even the minute hairs (microtrichia) that cover the cells, help hold the wing in stored readiness. The action of direct wing muscles, pulling the wing outwards and rotating it forwards, allows the spring-loaded structure of the aerofoil to unfurl, ready for flight. Likewise, in reverse, muscles pulling the wing back into stowage position trigger the folds and hinges to collapse the wing. Various analogies come to mind – folding soft-tops on cars, or pop-up gazebos. Beetles sometimes struggle to fold away their wings on landing, much like the newbie camper wrestling with a reluctant spring-loaded tent.

If beetle flight wings are easily neglected, the non-flight wings on the mesothorax – the elytra – are not. These rigid wing-cases are supremely important when it comes to beetle identification, and in understanding why beetles are so successful.

The combined shape and outline of the head, pronotum and elytra serve to give the first clue of a beetle's identity, and many families (or genera) can be recognised by silhouette alone. This will be the basis of Chapter 7, the catalogue of British beetle families. It is not unknown for a coleopterist to easily determine the remains of a beetle from just a single disarticulated elytron. A whole branch of archaeology is based on identifying loose beetle elytra found in subfossil deposits. The widespread and very common rove beetle genus *Tachyporus* (14 British species) abounds, and a key to loose elytra, using setae, is available (Booth, 1984). I was once handed a small glass tube of many dozens of shining green elytra found at the base of a Kent fence post where a hobby (*Falco subbuteo*) returned to perch and dismember its prey; they were immediately identifiable as those from 20 or 30 rose chafers (*Cetonia aurata*), which the bird had obviously been targeting as they flew about in the sunshine that day.

Elytra can vary from flat to domed, broad to long, and cylindrical to flanged, but this huge range of different shapes belies an underlying simplicity. Two elytra, usually exact mirror images of each other, meet in a straight line, the suture, down the middle of the beetle's back. In a very few species there is a small overlap. Though they appear to meet flush at the suture, there is usually a delicate tongue-and-groove joint here; most species are right-tongued and left-grooved, but there are others that are equally left- and right-tongued. In large flightless beetles the suture joint is fused, unable to open, and for some reason these are mainly right-tongued species. The front of the elytron, where it abuts the pronotum, is termed (as discussed above) the base, and the end at the tip of the body is the apex. The outer front corner is the humerus, humeral angle or shoulder. The long outer edge of the elytron is usually folded under slightly, sometimes with a sharp crease, to ensure a snug fit against the abdomen, and this narrow under-lip is the epipleuron. In the majority of beetles, the elytra completely cover the entire back half of the beetle, but in some groups they are shortened or cut off (truncated), leaving the tip of the abdomen exposed. In most of the rove beetles (family Staphylinidae), this shortening is extreme and the elytra appear little more than square flaps, leaving the abdominal segments clearly visible. Nevertheless, most rove beetles can fly actively and their flight wings are even more tightly and complexly packed away when folded.

Beetle elytra often appear ribbed, streaked or lined, the result of indented grooves or neat rows of punctures (or a combination) arranged in parallel lines – the striae (singular stria). These are numbered from 1 to whatever (usually between 8 and 10, but up to 25 in some Carabidae), from the innermost sutural stria outwards across the disc. A short (occasionally diagonal) stria at the base of the elytron, near the scutellum, is called, not surprisingly, the scutellar stria or scutellary striole. There may also be additional striae or bits of striae at the humerus or along the edge of the epipleuron.

The relatively smoother or convex regions between the striae are the elytral intervals, or interstices, and are also numbered from 1 (at the suture) outwards.

The standard model of an elytron represents the striae and intervals as a regularly undulating, corrugated structure, much like the galvanised corrugated-metal sheets used for lightweight roofing on temporary buildings, giving high strength and a certain degree of flexibility, but light weight. Primitive Archostemata beetle families show 10 striae representing 10 archetypal wing veins and two rows of punctures on each interstice.

The undersides of the elytra are usually smooth and unpunctured, lacking hairs or any significant surface sculpture. They are held in place by a locking mechanism – sometimes a flange along the sutural margins, which fits neatly into a median longitudinal slot in the last abdominal tergite, or a longitudinal ridge along the entire edge of each elytron from the shoulders to the tips, which fits snug against a matching depression that runs along the edges of the abdominal tergites – the insect equivalent of a Tupperware lid.

This is a useful starting point, but typical corrugated elytra are in the minority, and elytral diversity is huge. Both the striae and interstices vary from shallow to deep, flat to convex, narrow to broad. They are sometimes so faint as to be invisible, or they can be completely obscured by seemingly random (confused) punctures, or the elytra can seem brilliantly polished to an uninterrupted shine. Occasionally, areas of the interstices are raised into ridges, lozenges, spikes, nodules, knobs and other lumps and bumps, or the punctures across the elytra can be densely and massively compacted to create bizarre sculptured granulations. There can be shining, mirror areas or dull, matt areas. Underneath the microscope, a network of tiny criss-crossed lines, like the minute creases of human skin, can be revealed as mesh-like patterns (microsculpture, or reticulation), and even lace-like patterns within patterns (microreticulation). The varieties are endless, and beautiful. Like the head and pronotum, the elytra can be decorated with coloured hairs or scales. Or there can be any combination of these characters. It is impossible to overstate how varied and yet how important the elytra are to the beetle, and to the coleopterist.

THE ABDOMEN

By contrast, the abdomen of a beetle is little considered by most coleopterists. Indeed, not much of it is really visible, since it is almost entirely covered by the elytra above, and the underside is partly obscured at the front by the backward-jutting thoracic segments and the legs. Nevertheless, all beetles have abdomens, and they are usually made up of those 11 original metameric segments; typically, though, only five or six segments are visible.

The underside of the abdomen is protected by five (or up to eight) tough, sclerotised plates – the sternites. If the elytra and wings are lifted off, the upperside of the abdomen is similarly protected by five (or up to eight) dorsal plates – the tergites – although these are not so heavily sclerotised. Except in the rove beetles (and some oil beetles), with their shortened elytra, usually the only part of the beetle abdomen that is readily visible from above is the last segment, the pygidium, or sometimes the segment before it too, if the elytra are shortened slightly or rounded at the sutural angle.

The abdomen contains many of the beetle's vital organs – there will be a brief tour of internal anatomy later. Most of these will remain minor technical details to the beetle student, unless dissection or higher-level research is planned, but for species identification the internal genitalia are often of paramount importance. Effectively, the male genitalia are a telescoped or concertinaed series of two or three terminal body segments with at least some portion sclerotised – the aedeagus. This comes in various forms, from a curved comma shape in the carabids and staphylinids (colloquially known as the staphs), to long and thin in many leaf beetles and weevils, to bell-shaped. Through it passes a tube, the ductus ejaculatoris, which delivers the sperm. In many beetle groups, this hardened capsule differs distinctly between species that may appear visually identical in their external characters. The aedeagus can sometimes provide the only useful characters to distinguish otherwise apparently identical species clusters. Similarly, the reinforced helical/tubular sperm-storage organ of the female, the spermatheca, can sometimes be the only way to identify to species level. One notion was that male and female organs had co-evolved in isolated populations, such that a key-and-lock correlation had arisen. Intuitive though this may seem, there appears to be no fixed relationship, and in plenty of genera the males (usually) may be easily determined on subtle aedeagal shape differences, but the female spermathecae are all uniform. Details of genitalia dissection are given on p. 398.

THE ARMOUR AND ITS DECORATION

I have already slipped into the coleoptcrist's favourite analogy – that beetles are the knights in armour of the insect world. This is a rich and useful metaphor. Beetles are, without a doubt, armoured, but they owe this armour to a substance found throughout the insect class (and beyond) – chitin.

An insect wears its skeleton on the outside; it is this stiff exoskeleton (contrasted with the human endo-, or internal, bony skeleton) that gives insects (indeed, all arthropods) their stiff, inflexible, jointed appearance. This successful strategy evolved long before the land was colonised by living organisms and is shared by the vast assemblage of animals that make up the phylum Arthropoda. These include the now extinct trilobites, marine and terrestrial crustaceans (crabs, prawns, woodlice, etc.), millipedes, centipedes, spiders, scorpions, ticks, mites, and insects and a whole host of other less well known animals.

Ocean-dwelling crustacean exoskeletons are heavily calcified (those of barnacles, for example, are stone-hard), but insects have forgone brittle toughness for flexible strength, and although chitin occurs as the supporting element in all arthropod exoskeletons (and in fungi), it is in the insects that this substance has reached its pinnacle of development.

Chitin is a high-molecular-weight, unbranched amino-polysaccharide, mainly composed of β-(1-4)-linked units of N-acetyl-d-glucosamine. For the non-biochemically minded, this means countless billions of sugar-type molecules chemically linked into long polymer chains. Such polymers occur naturally in plants, where similar sugars are linked to produce cellulose, giving woody strength. They also occur unnaturally almost everywhere else, where short petroleum-derived molecules are chemically linked into man-made fibres like rayon and nylon, and in plastics (the poly- in polyester, polythene and polystyrene, for example). The immensely long chitin molecules are aligned into bundles (microfibrils) that are embedded in, and linked to, a protein matrix. The most usual form is for the bundles to be arranged in sheets, where the microfibrils are parallel, and for the sheets to be layered, each successive sheet with the parallel bundles at a slightly different angle. This lamination of sheets, with the parallel microfibrils arrayed at gradually changing angles, produces a versatile material that is flexible and yet has great tensile strength. Plywood, based on laminated thin veneers of wood in which the grain alternates in successive layers, achieves its flexibility and strength in the same way. In combination with lipid layers on its surface, the chitin helps create a mostly waterproof skin for the insect.

The layers of an insect cuticle are doubly strengthened by cross-linking hydrogen bonds between the chitin chains, and by phenolic bridges between the matrix of protein molecules in which the microfibrils are intimately embedded. This irreversible process, called sclerotisation, is perhaps analogous to tanning (leather production, not sunbathing); it darkens the cuticle and makes it rigid. The cuticle of insects is only 0.01–0.5 mm thick, and only the outer region of this (a quarter, say) is sclerotised, but even so this is enough to produce a structure of amazing strength and stiffness.

Sclerotisation darkens the outer cuticle, and this can sometimes be seen when an adult beetle emerges from its pupa. Initially, the insect is pale creamy

FIG 10. A teneral seven-spot ladybird (*Coccinella septempunctata*), recently emerged from its pupa; although the dark spots are visible, it will take many hours, and sometimes several days, for the final bright red and deep black markings to form fully.

yellow and soft (Fig. 10), but after a few hours, or at most a day or two, it darkens and hardens. In addition, coloured pigments are also laid down in the cuticle. Melanins are the almost ubiquitous pigment group, the same series of chemicals that darken all animals, including human hair and skin. Melanins, laid down at varying densities in the insect cuticle, produce colours ranging from yellow through red and brown to black.

These are the basic pigmented colours of most beetles, but these colour descriptions were seldom used in beetle books until fairly recently. Instead, a series of mysterious jargon terms, bordering on the arcane, were employed. Many years ago, a prominent and expert lepidopterist sidled up to me at an entomological meeting and asked in an undertone, so others would not overhear his ignorance, what 'testaceous' meant. This word, used frequently in old beetle books to mean a slight orange-yellow or reddish pink, was almost unknown to students of other orders. From the Latin *testa*, meaning 'tile' or 'shell', there was an implication of texture and light quality related to old clay roof tiles, snails and seashells, and it describes perfectly the translucent pinkish-yellow limbs of the 22-spot ladybird (*Psyllobora vigintiduopunctata*), where 'yellow' implies the colour of its much garishly brighter elytra. Other abstruse (but quaintly delightful) colour terms include: piceous (or pitchy), reddish black, the colour of pitch; castaneous, deep chestnut, reddish brown; fuscous, dull brown; fuliginous, the deep, opaque black of soot; rufous, pale red; luteous, egg-yolk yellow with a hint of red; sulphurous, brimstone yellow with a hint of green; and my personal favourite, ferruginous, a yellowy or brownish pink reminiscent of leaching rusty iron.

Many beetle colours, particularly the bright metallic sheens so distinctive of some groups, are not, however, the result of pigments, but are an artefact of submicroscopic structures on the cuticle. Parallel grooves and ridges across the hard cuticle of the head, pronotum and elytra, or indeed any of the armoured sclerites or segments of the legs or antennae, catch and reflect particular wavelengths of light while absorbing or scattering others. Just as sunlight passing through cut glass projects a spectrum of rainbow colours, refracting the slightly different wavelengths in slightly different directions, light is split when it hits the precisely aligned ridges on a beetle's cuticle, which form a diffraction grating. Depending on the periodicity, shape and size of the ridges, particular colours are preferentially reflected. This gives the glinting, reflectively metallic appearance so entrancing in many beetles, and accounts for the sometimes shifting colours as the animal alters its position in the sunlight, or the specimen is moved under the microscope lamp. These non-pigment structural colours are found throughout the insects, notably in blue morpho butterflies (*Morpho*), demoiselle damselflies (*Calopteryx*), blue- and green-bottle flies (Calliphoridae) and ruby-tailed wasps (Chrysididae), but nowhere are they more widespread or more resplendent than in the beetles.

A few glinting, metallic tropical species have become widely touted as living jewels (mainly cetoniine chafers, buprestids and leaf beetles), and their images regularly appear in books and magazines and on the Internet whenever the close-up wonders of a wildlife safari in miniature are required. However, even the commonest of British ground beetles are a marvel when seen in the right conditions. I've had many a squelching boot-full after sliding gently into the water when trying to grab at the tiny, glinting *Bembidion* and *Elaphrus* ground

FIG 11. The metallic gleam and rapid, erratic gait of ground beetles, like this *Notiophilus biguttatus*, make them difficult to follow as they dart across the ground.

beetles that haunt the damp herbage on stream and pond banks. Their shining bronze bodies, sparkling in the bright sun, are deceptively difficult to intercept with finger and thumb as they dash between drip-bespeckled stems and roots, and my concentration was stretched to the point of failing to keep a proper monitor on my overall balance, with the inevitable wet and muddy consequences.

Elsewhere, beetles get their colours and textures from various hair- or scale-like outgrowths on their bodies. They can be furry (the dung rove beetle *Emus hirtus*), prickly (the weevil *Sciaphilus asperatus*; Fig. 13), tufted (the longhorn *Pogonocherus hispidus*) or velvety (weevils in the genus *Cionus*). There is a spectrum of varieties to befuddle the beetle student, but a few simple definitions will make a start. The bodies of shining ground beetles are mostly smooth, without any hairs (glabrous), but even they have a few slim, whisker-like sensory hairs (bothria, or trichobothria, singular bothrium) usually sticking out of indented pore-like craters, and the precise positioning of these is useful in identification. Other beetles may have a sparse covering of regularly spaced hairs, short and even; this pubescence may be longer (pilose) in some species, or tufted, clumped or feathery (plumose), or contain a mixture of different-sized hairs, ranging up to the longer, taller, more upright spines called setae (singular seta).

FIG 12. The fluffy body of the summer chafer (*Amphimallon solsitiale*) gives this handsome beetle a cute, cuddly air.

FIG 13. Covered all over with round metallic scales, the ground-dwelling weevil *Sciaphilus asperatus* also has rows of raised, scale-like setae along the interstices of its elytra.

In the most extravagantly patterned or variegated beetles the setae are modified to form broad, flattened scales, and just as the same evolution produced the beautiful patterns of butterfly and moth wing scales, so the mosaic of colourful scales across the elytra and pronotum can produce veritable works of art. Some of the most delightful decorations occur in the weevils (Curculionidae and relatives) and hide beetles (Dermestidae), and the scales can be circular, shield-shaped, heart-shaped, triangular or feather-like. They can be flat and brilliantly metallic, but many are ribbed, although this is visible only under the very highest magnifications of top-end professional scanning electron microscopes. Under a hand lens or through a budget stereomicroscope, exactly where a seta becomes a scale is sometimes open to debate, and talk of scale-like setae and seta-like scales punctuates the coleopterological literature.

WHAT IS INSIDE A BEETLE?

Unless the coleopterist is undertaking detailed study, it is enough, here, to gloss over internal anatomy with just a whistle-stop tour. The major internal structures are the musculature, nerves, digestive tract, and respiratory, circulatory and reproductive systems. Whereas beetles are highly distinctive on the outside, their internal anatomy is very similar to that of other insects, and there are several fascinating general insect and beetle monographs where more detail can be found (Crowson, 1981; Beutel *et al.*, 2014; Gullan & Cranston, 2014).

Individual muscles are attached to the inside of the exoskeleton and usually work in pairs, one pulling an individual body segment or limb part in one direction, and an opposing muscle attached elsewhere pulling against it. The discovery that tarsal and some antennal segments share common muscles has led to a better understanding of limb genesis and evolution, but from our point of view it is enough simply to know that beetles have muscles, and use them to move their segmented bodies and appendages about.

Muscle movements are controlled by a nervous system, which in turn is informed by sensory nerves from the eyes, chemoreceptors on the antennae and palps, and touch sensors on the trichobothria dotted about the body. Nerves extend throughout each body segment, but are centralised in a nerve cord running through the underside of the insect, with coordinating nodules called ganglia. In more primitive insects, like cockroaches (Blattodea), each of the three thoracic segments and eight abdominal segments still contains its evolutionarily basic pair of ganglia, but in most higher insects (like beetles) these have become fused (combined and rationalised) into something approaching a single central nervous system. The ganglia in the head became combined into one unit back in the deep evolutionary past, and this is often called the brain, but comparison to the all-powerful and supremely controlling brains of vertebrates is unwise. To some extent, the ganglia continue to work independently within each segment, which explains why decapitated or dismembered insects continue to move for some time.

The digestive tract is at least analogous to our own, and it is easy to comprehend food being eaten at one end, passing through various digestive regions and being evacuated at the other. Typically, carnivorous species like tiger beetles (*Cicindela*) have short guts, just a single loop, whereas those of herbivores like cockchafers (*Melolontha*) are long and convoluted. This follows an ecological trend that also occurs in mammals, for example. The simplistic explanation for this is that protein-rich prey is easier to digest than the complex phytochemicals of plants, though larger species obviously have longer guts, and those feeding on protein-rich pollen (e.g. *Oedemera*) tend to have short guts. For some reason, the longest and most complicated guts are in the Scarabaeidae and Silphidae, but no explanation for this is as yet forthcoming. These scavengers/recyclers may need extra intestinal transit time to allow symbiotic bacterial action. Or they might not.

FIG 14. The devil's coach-horse (*Ocypus olens*), Britain's largest and most gothic rove beetle.

Food cut and chewed by the mandibles passes down the pharynx (throat) and oesophagus (gullet) into a crop. This is really a chewed-food storage area, allowing gentle trickle-down input to the mid-gut. This tubular region, with short blind-ending tunnel-like branches called caeca, is where digestive enzymes are released and the resulting broken-down nutrients absorbed. Eventually, food remains pass into the hind gut, where waste products from elsewhere in the insect's body are added and water is reabsorbed, before the faeces enter the rectum, in readiness for ejection through the anus. Generally, insect droppings are hard roundish pellets, called frass, but the exact texture of the output varies. In the devil's coach-horse (*Ocypus olens*), Britain's largest rove beetle, the flexible abdomen is waved over the back threateningly, to my mind a fell countenance unequalled in British beetles, but there is no venom or sting; instead, smelly liquid faeces are smeared on any would-be attacker, or entomologist.

All insects are air-breathing, at least as adults, but instead of having a single airway, from the mouth to the lungs, they still retain the two ancestral air openings (left and right) on each body segment. Air moving through the outer entrance hole, the spiracle, passes into a tube (trachea, plural tracheae), which branches into finer and finer tubules, ending at tiny fluid-filled cul-de-sacs called tracheoles. By diffusion, a very simple chemical transition, oxygen passes from air into the body cells of the insect around the tracheoles, and carbon dioxide passes out. Some ventilation by rhythmic muscular contraction and relaxation of the body segments enhances this passive process. Beetle spiracles are not usually visible because they occur on the upper edge of the abdominal segments, and are generally concealed by the elytra. Keeping a bubble of air underneath the elytra allows water beetles to breathe whilst they are submerged, but they must come back to the surface occasionally to replenish their scuba supply, as oxygen runs low and carbon dioxide builds up.

Haemolymph (the insect equivalent of blood) is not constrained by arteries and veins, but instead flows freely through the large body cavity of the insect, the haemocoele, bathing the internal organs. A watery liquid, usually colourless, but sometimes pigmented yellow, red, green or blue, it rarely contains respiratory compounds, and gas exchange occurs almost entirely through direct tissue contact with the tracheoles. The main function of the haemolymph is to transport nutrients, hormones and waste products about the body. Despite the absence of arteries and veins, there is a circulation system of sorts: a muscular tube running along the back axis of the insect pumps blood forwards. The main pumping area, sometimes inconsistently called the heart (towards the rear) or the aorta (in front), pushes blood through the tube to the head. After supplying the head organs, the haemolymph drains backwards into the body of the insect, through the thorax and abdomen, before returning to the pericardial sinus, the space around the heart, which is, in effect, a sump. Here, valved pores called ostia (singular ostium) allow one-way passage back into the dorsal pumping tube, to be recirculated.

The genitalia sit at the very tip of the beetle abdomen, but are usually housed internally and are invisible as the beetle goes about its general business. The final abdominal segments may appear tight-fitting and rigid, but there is some flexibility thanks to membranous hinges between the plates, and often a telescoping of tergites and sternites, allowing manoeuvrability during mating and egg-laying. Briefly, sperm created in the testes are stored in a bag-like seminal vesicle until mating. The male organ, the aedeagus, delivers sperm into the female genital tract, where it is stored in the spermatheca until needed for egg fertilisation. Eggs are laid through the short or longer telescoped ovipositor (egg-laying tube) in batches, or singly, depending upon the species. Mating and egg-laying behaviour is varied, fascinating and important, and will be revisited later on in the book.

WHAT IS NOT A BEETLE?

By way of a small diversion, it's worth considering what is not a beetle, to bolster an understanding of what beetles actually are and how they fit into the scheme of insects as a whole. As stated in the opening of this chapter, the widely known 'black-beetle' of popular pest-control handbooks is not a beetle at all, but the common (or oriental) cockroach (*Blatta orientalis*). It has similar biting mouthparts to a beetle and a broad pronotum, and its back is covered by leathery

wing-cases to protect the flight wings. These adaptations are partly responsible
for the cockroaches' own ecological success story, allowing them to eat diverse
foodstuffs, and giving them protection when crawling into tight spaces under
the fridge. Cockroaches (insect order Blattodea), though, are hemimetabolous:
they do not have larvae and do not go through complete metamorphosis. Instead,
their eggs give rise to nymphs, miniature versions of the adult, which develop
wing buds as they grow, increasing with each moult until the final, fully winged
state is achieved in adulthood. On a structural level, cockroach wing-cases are
much more wing-like than beetle elytra; they show a clear pattern of wing veins
(venation), are flat rather than domed, broadly overlap one another rather than
meeting at a suture down the middle, are less tight-fitting around the edges, and
are often short, barely covering the abdominal segments. Cockroaches also have
very long multisegmented antennae (100 segments are not uncommon), and a
pair of antenna-like sensory appendages called cerci at the tip of the abdomen.

Shieldbugs, stink-bugs and leaf-hoppers (and many other insects in the order
Hemiptera) are often mistaken for beetles, because they too have hardened and
coloured wing-cases folded over the body, protecting similar membranous flight

FIG 15. Not a beetle, but a tortoise bug, *Eurygaster testudinaria* (order Hemiptera, family
Scutelleridae). The scutellum covers almost the entire abdomen, making it look very like elytra.
I myself fell into the misidentification trap with a small, furry Greek scutellerid I exhibited
at an entomological society meeting thinking it to be a large nitidulid beetle. My error was
pointed out at the following meeting, leaving me red-faced but better informed. (Penny Metal)

wings underneath. The Hemiptera are a successful group, with tree-dwelling, herb-dwelling, ground-dwelling and aquatic species. Again, though, they are hemimetabolous, with nymphs rather than larvae. Some Hemiptera are plant-feeders, others are predators and a few are vertebrate-blood-suckers, but all have sucking, not biting, mouthparts. The antennae are short or long, but usually have only four to six segments, and sometimes only two, with a whisker-like appendage representing the flagellum. The wing-cases, often called hemelytra because they are hardened and coloured only at the basal half, and transparently membranous at the tips, overlap each other, often largely, but at least by a small flap. The small round wingless nymphs of some shieldbugs are brightly and warningly coloured and spotted, and often mistaken for ladybirds.

Crickets and grasshoppers (order Orthoptera) are fairly well known to the lay observer, and their large hopping or jumping hind legs are a bit of a giveaway. However, the occasional exotic species, broad, fat and brightly coloured, might be thought beetle-like, especially with the large saddle-shaped pronotum protecting the thorax behind the head. Orthopterans have biting mouthparts, but they are yet another hemimetabolous group. Generally, the front wings are toughened and coloured, but they are often narrow strips held ridge-like along the abdomen rather than flat like elytra. The net of veins is also pretty obvious, making the wings much more reminiscent of those of a dragonfly.

The systematic position of earwigs (order Dermaptera) often taxes the minds of expert entomologists as well as novice naturalists. Biting mouthparts, relatively

FIG 16. Despite its distinctive abdominal pincers, the common earwig (*Forficula auricularia*) is often mistaken for a rove beetle because of its short wing-cases and curled tail. (Penny Metal)

short antennae, prominent pronotum and short wing-cases (tegmina, singular tegmen) and exposed abdominal segments do give a rove-beetle appearance, but the terminal pincers (cerci) are so characteristic as to require no further description. Their exact use remains controversial and much discussed in entomological circles, but whether they are anti-predator deterrents, wing-folding tweezers, sexual ornaments or something else continues to fascinate us all. The Dermaptera are another hemimetabolous order, and are definitely not beetles.

Finally, there is an intriguing selection of odd small fry, which at first glance down the microscope might be mistaken for beetles – mostly because they seem to lack conventional wings. Some tiny (≈1 mm), hunched 'wasps' in the hymenopteran families Scelionidae and Encyrtidae, for example, either lack wings entirely, or the short, heavily clouded membranes are tucked in tight over the hind body in a passable representation of elytra. Several widespread flies have short, broad, darkened wings, which they hold beetle-like over their abdomens, including *Stegana coleoptrata* (Drosophilidae), *Phasia hemiptera* (Tachinidae), *Discomyza incurva* (Ephydridae) and *Camarota curvipennis* (Chloropidae); the subterfuge is enhanced by a somewhat scuttling gait. My favourites, though, are small, narrow hymenopteran parasitoids in the wasp family Bethylidae. These slim creatures run around in the net, not really flying and looking just like little rove beetles; they have the facies of something like *Astenus*, *Sunius* or *Rugilus*, complete with a broad triangular or pentagonal head and parallel-sided hind body (Fig. 17). I've got quite a collection now thanks to my confusion.

FIG 17. The small (3–4 mm), slim form of a bethylid wasp is very like a tiny rove beetle, especially with its shining head and thorax. (Penny Metal)

WHY ARE BEETLES SO SUCCESSFUL?

Beetles occur from seashore to mountain top, tropic to Arctic, forest depths to pond shallows, wilderness to car park, and in anthills, biscuit tins, mole nests, floorboards, cellars, cowpats and corpses. They are, without doubt, the most mind-bogglingly diverse group of terrestrial creatures on the planet. Experts may not agree about how many different species there are, but they all concur that beetles are supremely successful. To understand why they are so successful, we need look no further than the generalised form of a 'typical' beetle, and everything falls into place.

First, beetles have built on the success of insects in general. They are small, tough, active and mobile, and usually lay large numbers of eggs, and between them they have a broad palate, eating almost everything from basic plant material to each other. They have benefited from the standard insect trait of having a secretive but mobile embryo-like larval stage. This does most of the feeding and all of the growing in a safe (from predators), hidden location, but then gives rise to the winged and highly adventurous adult stage, which can disperse to find new localities to conquer. Then the beetles have gone a stage further.

It is generally acknowledged that the leap to hyperdiversity in the beetles is linked to the evolution of their body armour. It is perfectly possible to pick up a tiny beetle, little larger than the full stop at the end of this sentence, and hold it carefully between finger and thumb, before releasing it unharmed, if a little

FIG 18. Strength in the face of adversity. This vine weevil (*Otiorhynchus sulcatus*) was still alive and struggling its four-legged way along a garden path after breaking free from the spiderweb in which it was trussed. The silk wrap still coats its elytra and its back legs are hobbled. Forest Hill, 6 August 2002.

disgruntled. All humans are coleopterists by nature, otherwise what is the point of opposable digits? Take a fly or a wasp or a moth even remotely approaching this size, and it would be squashed to a pulp by such conduct, no matter how light-fingered one might be. Not only does the beetle's armour give strength, but the smooth lines of head, pronotum and elytra together are robust beyond the individual strengths of their constituent parts.

The elytral wing-cases in particular, one of the key features that most easily defines the Coleoptera, are usually regarded as conferring huge protective advantage onto beetles. However, it was only recently that an empirical study sought to quantify exactly how much protection they give. Linz *et al.* (2016) used cultures of the red flour beetle (*Tribolium castaneum*), a well-used experimental laboratory model, carefully cutting off most of the elytra from some specimens, then comparing survival to those of undamaged beetles under various environmental stresses. They looked at membranous flight wing damage, predation by *Pardosa* spiders, desiccation in varying dry atmospheres (0–100 per cent relative humidity, with or without food), and cold shock at –4 °C for 24 hours. In all groups, those undergoing elytral trimming showed greater morbidity and mortality, and lower survival rates, compared to the intact control beetles. The results were dramatic and obvious. The paper has become an instant classic of entomological literature.

Beetles, it seems, may be descended from some precursor insect that took to sheltering under the tough bark of tree ferns in the primordial antediluvian

FIG 19. Beetle success at tightly furling their wings is mirrored in other insects. This is a stonefly, *Leuctra geniculata* (Plecoptera), but its wrapped wings make it look very beetle-like. (Penny Metal)

forests of the Devonian era. Its ridged, corrugated wings were maybe tough enough to suffer the scraping in and out of its narrow hidey-holes, without compromising its aerial ability when it needed to shimmy out and fly off. This gave proto-beetle a great advantage over its less well protected neighbours, and it did well. Its uncountable, and mostly unknown, descendants today are testament to its success. In most modern Coleoptera, the wing-cases, so richly diverse across all of beetledom, are still, functionally, sheaths to cover the delicate flight wings folded away underneath; in large, heavy exotic species, they can be 1 mm thick. They protect against damage, against predators and against water loss, and act as thermal insulation. The elytra continue to give this tough protection, allowing beetles to push deep into leaf litter, root thatch, soil, dung, rotten wood or any other tight, safe haven, but at the twitch of a muscle or two they can be flexed open and the beetle can fly off unencumbered and unhindered by its magical, life-saving, evolutionarily sensational armour.

CHAPTER 2

Beetle Variety

HOW BIG IS A BEETLE ANYWAY?

The expression 'they come in all shapes and sizes' could have been coined just for beetles. More so than any other group of insects, beetles excel in diversity. Perhaps one of the most striking aspects of this variety is in their size. Insects are all pretty small anyway (the estimated modal length of all the world's beetles is 4–5 mm), but within these constraints beetles still manage to straddle the entire insect size range, from minuscule to massive. The Colombian ptiliid *Scydosella musawasensis* has a good claim (Polilov, 2015) to being the one of the smallest free-living insects in the world, at a fragile 0.325 mm long. Meanwhile, with a body length of up to 167 mm, the largest beetle is often reputed to be the obviously named *Titanus giganteus*, a massive longhorn from Brazil. Even using back-of-the-envelope calculations, and assuming a smooth, legless ellipsoid body shape, *Titanus* is a conservative 130 million times bigger than *Scydosella*.

British beetles don't quite reach these extremes, but our smallest, also a ptiliid, *Baranowskiella ehnstromi*, recently discovered in East Anglia (Duff *et al.*, 2015), can be 0.45 mm long, and our largest is either the stag beetle (*Lucanus cervus*) or the great silver water beetle (*Hydrophilus piceus*), both approaching 50 mm (not including the stag's giant mandibles) (Fig. 20), which still gives a size differential of something like 2 million. There can be a vast range within the same family. *Hydrophilus* is a truly impressive beast, and should be handled carefully since it has a large, sharp, backwards-pointing spine on the underside of the last thoracic segment, and short, high-leverage mandibles that can give

FIG 20. At up to 50 mm long, not including its giant antler-style jaws, the stag beetle (*Lucanus cervus*) is Britain's most massive beetle, although the great silver water beetle (*Hydrophilus piceus*) also makes a good claim to the title.

quite a nip. But its close relative, *Chaetarthria seminulum*, often found living in the same patch of water weeds, is only 1.3 mm long. Size disparities of almost equal magnitude also occur in the British fauna in the Carabidae (1.6–35 mm), Staphylinidae (0.7–28 mm), Dytiscidae (1.5–38 mm), Scarabaeidae (2–38 mm) and Cerambycidae (3–40 mm).

Arguments over biggest and smallest illustrate the difficulty, sometimes, of measuring insects. Several other beetles can make claims to be more massive than *Titanus giganteus*, including the elephant beetles (*Megasoma elephae* and *M. actaeon*) from South America and the African Goliath beetles (*Goliathus goliathus* and *G. regius*); although these genera in the Scarabaeidae may be heavier than *T. giganteus*, they are shorter and stouter, reaching about 137 mm and 112 mm, respectively. And while almost all of the Ptiliidae are under the 1 mm mark, with species often described in the scientific as well as the popular literature as being 'as large as the full stop at the end of this sentence', some are short and round, and others are long and thin.

For the modest British beetle fauna, it is usually sufficient just to give body length measurements in millimetres, but there are a couple of conventions. One is to ignore large, protruding horns or mandibles like those of the male stag beetle; the other is to exclude the variable and often sexually dimorphic snout, the rostrum, in weevils (Curculionidae, Apionidae, etc.), instead measuring these beetles to the tip of the elytra from an imaginary line drawn across the head at about the front edge of the eyes.

What ever became of the line?

Although the millimetre became the standard measure of insects well ahead of the adoption of the metric system in other fields (including in the notoriously inch-dominated United States), many old entomological books use another system of measurement. A line, $\frac{1}{12}$ inch (roughly 2.2558291 mm), was part of a now archaic duodecimal system once familiar throughout the world: 12 months in a year, 12 hours on the clock, 12 inches in a foot, 12 ounces in a troy apothecary pound, and goods sold by the dozen, or by a gross – a dozen dozens. Dictionaries are circumspect in the derivations of the word 'line', and although the idea of $\frac{1}{12}$ inch as a useful fraction ($\frac{1}{4}$) of a barleycorn, itself $\frac{1}{3}$ inch, regularly crops up, as do other suggestions that the line has, in the past, been used to denote $\frac{1}{10}$ inch, $\frac{1}{16}$ inch or $\frac{1}{40}$ inch. It has never been approved under any statute, and was always an unofficial measure. My personal suggestion is that it may have originally come from printing, where books (mostly Bibles and hymnals) typeset at 12 lines of text to the inch were once commonplace.

Stephens (1827–34), Spry & Shuckard (1840) and Cox (1874) all use lines, or fractions of a line, to measure beetles, but by the time Fowler started his six-volume monograph in 1887, he was using only millimetres. The transition is clearly discernible in British entomological journals of the time. The millimetre started to appear during the 1860s, mostly by corresponding European authors. During the 1870s, lines and millimetres overlapped; indeed, both were often given together in the same articles. By the 1880s, the line had all but disappeared, but with a few hard-core old-timers (the ptiliid expert A. Matthews for one) clinging onto the old measure until about 1893 (Jones, 1994).

The line has now gone the way of the ell (a double-arm stretch of cloth or rope) and the cubit (a forearm measure for masonry and bricklaying). But such was its monopoly at a time of great entomological advancement that we should at least respect its memory and understand its translation into modern units.

Size can be a frustratingly subtle identification tool. Many's the time an important monograph will present a key with one of the couplets stating 'larger', then giving size limits, and contrasting this with 'smaller' but giving a size range partly overlapping the first. To the museum curator with a large number of specimens available, this can be a useful distinction, but to the field naturalist peering at a singleton down the microscope, it can be worse than useless. Nevertheless, size is an important guide to distinguishing a beetle, and sometimes a specimen crawling in the hand can be accurately identified at a glance. The two brilliantly metallic green leaf beetles *Cryptocephalus aureolus* (larger, 5.5–7.8 mm) and *C. hypochaeridis* (smaller, 4.0–5.7 mm) obviously differ in size when compared sitting in neighbouring yellow hawkbit flowers, but under the lens any structural differences in terms of shape, punctuation, leg and antennal length are almost impossible to describe accurately. And it's worth remembering that size cues are lost or obscured in photographs, where even a brightly crisp and clear image of a single beetle may not be nameable, because it is not quite obvious whether it is the larger or smaller of the two.

SILHOUETTES AND CONTOURS

Most beetle diversity can be explained in terms of ecology, life history and habits. The ground beetles (Carabidae) and water beetles (Dytiscidae), which so easily lent themselves to the introductory chapter because of their typicalness, are general in their body forms, the one for running fast over the soil and the other for swimming fast through water. From each of these, the variety spreading out into other more specialist body structures is endless, from the sylph-like elegance of the tiny, pale latridiid *Adistemia watsoni* crawling up a mouldy museum storeroom wall, to the globular, gleaming metallic bauble of the golden leaf beetle *Chrysolina banksi* (Fig. 21) ponderously heaving its way over a black horehound (*Ballota nigra*) leaf.

FIG 21. Seemingly made from beaten and polished metal, the glorious golden globe of *Chrysolina banksi*. (Penny Metal)

FIG 22. The typical beetlish form of *Carabus problematicus*, a large and widespread carabid ground beetle.

Beetle shape varies tremendously, but as a general rule, the sleeker the beetle and the smoother its body outline, the better it is able to shimmy down into that crevice or push through water, mud, soil, dung, herbage, or other liquid or flexible medium. Although they are ground beetles, two British carabid species illustrate a nice intermediate silhouette zone. *Omophron limbatum* is smoothly rounded and globular, rather like a *Haliplus* water beetle; its mottled straw and dark markings are also very similar. It lives in the loose sand at the edge of flooded gravel pits in south and east England, but is truly terrestrial. The best way to find it is to splash the banks with water to frighten the beetles into scurrying to the surface. *Oodes helopioides*, which is also smoothly rounded at the sides but rather flat, spends much of its time underwater as it crawls through the flooded roots, vegetation and leaf litter at the edge of lakes, fens and slow rivers.

By contrast, a globular or uneven beetle is usually one that simply walks about on the substrate, much as do the leaf beetles (Chrysomelidae), with their shining, rounded and domed forms. *Cryptocephalus* are about the most convex members of this large family, their body underside almost as convex as the upper surface. The reed beetles (subfamily Donaciinae), though, are more conventionally elongate, with a narrow head and pronotum contrasting against the well-shouldered elytra, and with a slightly flattened body. This may reflect their more agile activity through waterside vegetation, and their potential to push down into the reed, sedge and bur-reed leaf curls and sheaths on which the majority feed. On another tangent, the distinctive ground beetle *Cychrus caraboides* shows a combination of silhouette forms in the same body. Its elytra-covered abdomen is round and convex, but its head and thorax are contrastingly narrow. As a long-legged predator, it can chase around in the open or push through light leaf litter, but when it finds its chosen prey, a snail, it can get its front end, complete with long, curved jaws, deep into the snail shell as its victim tries to retreat inside.

At the extreme, the tortoise beetles (Chrysomelidae, subfamily Cassidinae) are flattened and flanged around their nearly circular or smoothly oval edges. They can withdraw legs and antennae in, completely hidden beneath an impregnable carapace, as they clamp themselves down hard onto a leaf surface.

Also at the more depressed end of the beetle convexity spectrum are the thoroughly flattened species that occur under decaying bark (notably members of Cucujidae, Silvanidae, Laemophloeidae and Colydiidae), which seem able to slide themselves into the minute gaps at the heartwood–cambium and cambium–phloem boundaries as fungal rot sets in. There are other good examples among the more generally shaped beetles. The ground beetle *Dromius quadrimaculatus* is a decidedly subcortical species (*D. meridionalis* and *D. agilis* sometimes also occur under loose bark), whereas equally flattened close relatives occur in the loose soil (*Syntomus, Microlestes*), cracked ground (*Drypta dentata, Polistichus connexus*), grass and reed stems (*Demetrias, Odacantha melanura*) and leaf litter (*Dromius linearis*). Another very flattened carabid genus, *Aepus*, dwells in tiny silt-filled cracks in seashore rocks; so tight are these crevices that they maintain air-filled spaces when they are submerged at every high tide. Generally, the Carabidae are experts at pushing into tight spaces, if not under bark or stones, then into the root thatch and leaf litter. Hugely expanded muscular trochanters articulating on the coxae of the back legs give terrific push-up as well as push-forward power, allowing these beetles to 'wedge-push' as they shimmy their typically wedge-shaped bodies deeper and deeper into the substrate (Evans, 1977; Forsythe, 1983). In South-east Asia, the aberrant ground beetles *Mormolyce* are not only flattened, but have elytra that are expanded well beyond the edges of the abdomen into broad rounded plates. The beetles find small prey between the layers of bracket fungi, but their ungainly elytral adornments must offer some mysterious advantage other than simple flatness.

The Histeridae are mostly convex, lentil-shaped beetles that occur in dung, carrion or other putrescent decay, but a few occur under rotten bark, including the bizarre *Hololepta plana*, first found in Britain in 2009. It must rank as one of our flattest insects, being more than three times as broad as it is deep (Allen & Hance, 2009). The darkling beetles (Tenebrionidae) are also mostly chunky, relatively convex, long-legged species, but a few, like *Crypticus quisquilius* and *Phaleria cadaverina*, are smoothly oval and slightly flattened (resembling carabids) to ease their movement through roots and decaying plant matter in the loose sand of coastal dunes.

One of the oddest flat beetles in Britain is the parasitic beaver beetle (*Platypsyllus castoris*) (Duff *et al.* 2013), whose compressed form presumably eases its passage down into the thick waterproof pelt of its hosts. It has, like all flat beetles, a dorsoventral compression, but in this case there is a clear analogy to the lateral compression seen in fleas, which also need to push deep into the fur or feather coating of their chosen victims. The adult beaver beetle and its larvae spend almost their entire lives on living beavers, and are true parasites, feeding on skin particles, fatty secretions and possibly blood. About 60 per cent of beavers are host to the beetle, but, unlike fleas and lice, no serious detrimental effect has ever been demonstrated on the hosts. Similar flattened, smooth-lined rove beetles, *Amblyopinus* and *Chilamblyopinus* species, live in the fur of Central American water rats, but are predators of the rats' fleas (Ashe & Timm, 1988, 1995).

Boring through dead wood is at its most efficient for a cylindrical beetle, with obvious parallels in the civil engineering world of mining and tunnel-building. Larval tunnelling may be different (see Chapter 3), but if a beetle is chewing into a tough woody substrate, a circular cross section to its bore is an optimal solution. Most wood-chewers have small, tough mouthparts on the front underside of the head, but by shuffling a gentle spiral in its tunnel, the beetle can address

FIG 23. The aptly named *Platypus cylindrus*, showing its elongate cylindrical form, almost perfectly circular in cross section and ideally suited to boring through wood.

FIG 24. Once considered extremely rare, *Agrilus biguttatus* has either spread, or awareness of its distinctive D-shaped exit holes in rough oak bark has allowed its true distribution to be discerned, even though the adults remain elusive and rarely seen. Honor Oak Park, 4 September 1995.

its jaws to the full circumference of the drill hole, exactly as a drill bit chews its way through a plank. To keep the boring (and therefore energy expenditure) to a minimum, it pays off to have as narrow a cylindrical body form as possible, as perfectly exemplified by the chief wood-borers: woodworm and deathwatch (Ptinidae, subfamily Anobiinae), jewel beetles (Buprestidae), powder-post beetles (Bostrichidae), longhorns (Cerambycidae), bark beetles (Curculionidae, subfamily Scolytinae) and pinhole borers (Curculionidae, subfamily Platypodinae). Even if much of their burrowing is done as a larva, the adults still have to tunnel out from their pupal chambers, or they need to dig maternal burrows in which to lay their eggs. The necessity for completely circular holes is not altogether fixed though; adults of *Agrilus* (Buprestidae) chew characteristic D-shaped exit holes through sometimes very thick, gnarled bark (Fig. 24). Predators of wood-boring beetles also tend to be cylindrical, typical examples being the Cleridae, including *Opilo mollis*, *Tillus elongatus* and *Korynetes caeruleus*.

Halfway between being flat and being cylindrical is a versatile middle ground dominated by the rove beetles (Staphylinidae). Characterised by shortened elytra, they combine the benefits of protected flight wings, extra-closely folded, with the worm-like flexibility of an exposed hind body. These mostly short-legged beetles are renowned for their ability to disappear into the tiniest of crannies: broad, flat species like *Micralymma marinum* in seashore rock cracks; *Platystethus*, *Megarthrus* and *Anotylus* in dung; *Achenium depressum* in leaf litter; and the more cylindrical *Bledius* burrowing in sand. Elsewhere, they can also be shining sub-cylindrical (*Tachyporus*, *Bolitobius*) or almost hemispherical (*Cypha*).

One of the flattest British staphylinids, *Encephalus complicans*, is most peculiar because its flatness seems not connected with any ability to insinuate itself into tight places, but to enable it to curl its tail right up and over the rest of its body so that it resembles a globular seed. This extreme contortion merely emphasises a tail-curling behaviour common to many staphs, including the small but chunky *Aleochara* species, which raise their tails up as they scatter from carrion beaten over a collecting sheet, and that large Gothic monstrosity, the devil's coach-horse, (*Ocypus olens*).

Whether flat, globose or cylindrical, short and stout or long and thin, broadly rounded or constricted and broad-shouldered, the outline of a beetle is also the first clue to its identity. Almost any British beetle can, with a little practice, be quickly allocated to one of the 102 families of beetles currently on the British checklist. Most families have a 'look' to them – what birdwatchers have long known as the jizz (possibly a corruption of gist). This is sometimes called the facies in more technical monographs and is illustrated by a habitus figure – a whole-insect illustration, either of a typical member of the group, or slightly stylised, but still demonstrating the quintessential physical features of that particular family. Although colour, sculpturing, legginess and antennal form might be useful in the initial identification, the outline and general body shape is without doubt the most important feature.

This essence of beetleness reflects a family's evolutionary history. In some groups it is more cohesive than in others. Among the most obviously shaped beetles are the elongate and sub-cylindrical clicks (Elateridae), with their smartly parallel-sided elytra and slightly rounded pronotum, swept back into pointed hind corners. Ladybirds (Coccinellidae) are all fairly ladybird-shaped, some much more rounded than others, but all smoothly ovate, with a short, broad pronotum and domed elytra. And the lovely skipping Mordellidae are distinctive by virtue of their keel-shaped bodies, slightly hunched form and long, pointed abdomen tip, called the style. Sometimes this essence is more subtle and difficult to put into words. There is something evidently recognisable in the bullet-shaped elytral tapering that immediately identifies a buprestid, even though these species range from elegantly long and narrow (*Agrilus*) to short, fat seed mimics like *Trachys*. The latridiids are all minute, but the proportions of the relatively large head, narrowed pronotum and round-shouldered elytra have a uniformity to the coleopterist's eyes, despite wide differences in colour, punctuation and hairiness.

That this jizz works is easily demonstrated by a visit to another country. Even on the other side of the planet, it's possible to find a beetle unlike anything known from home and be able to identify it at least to family level on the basis of some vague notion about its shape, its body proportions and its overall appearance – the cut of its jib.

LIFE AND LIMB – LEG FORM AND FUNCTION

Limb diversity in beetles goes hand in hand (or tarsus in tarsus perhaps) with body shape, the one obviously having co-evolved with the other. A typical beetle shape (Carabidae, of course) is usually accompanied by typical elegant beetle legs. These propel the beetle over what is in effect a highly granular and uneven world for a tiny animal. Six legs is the minimum requirement for negotiating an uneven surface by any stable gait; this the beetle achieves by an alternating tripod walk. At any one time, a minimum of three legs make contact with the surface – front and back on one side of the body, and the middle leg on the other side – while the other three legs move forward for the next step. Ordinarily, in most beetles this movement is too quick to see with the naked eye, but it can be observed in the slow, clockwork gait of large, cumbersome species like the bloody-nosed beetle (*Timarcha tenebricosa*), the cellar beetle (*Blaps mucronata*) or the vine weevil (*Otiorhynchus sulcatus*). Incidentally, these make easy, charming and low-maintenance pets.

The long, slim legs of the Carabidae are perfect for running fast, and it is sometimes quite a challenge to grab an *Amara* or *Pterostichus* once they're off. For some time, the American cockroach (*Periplaneta americana*), very beetle-like in its form, was celebrated as the fastest-running insect, clocking up speeds of 1.5 m/sec (5.4 kilometres per hour). But this world title has now been taken by a flightless Australian tiger beetle, *Rivacicindela hudsoni*, with bursts of 2.5 m/sec (9 kilometres per hour) (Kamoun & Hogenhout, 1996). Sadly, no speed trials have been conducted on our native tiger beetles, but they are genuinely very fast, and there ought to be a prize for catching a green tiger beetle (*Cicindela campestris*) with your bare hands.

One of the easiest structural specialities to understand is the broad shovel-like legs of digging beetles. Nowhere are these more obvious than in the stout soil-moving limbs of the dumbledor (or simply dor) beetles (Geotrupidae). These large domed black beetles have heavy, muscular bodies, smoothly rounded between thorax and abdomen to give at least some ease of movement through the earth, and their legs are incredibly powerful. I am not alone in recounting my first childhood encounter with dor beetles as one of the most memorable and exhilarating discoveries in the beetle world. Many other coleopterists also tell tales of their own private wonder at the power of some large scarab or other, as these magnificent animals heaved their way out through the clenched fingers of an awe-inspired child, to fly off into the dusk like a miniature helicopter.

The legs of geotrupids, especially the front tibiae, are broadly flattened, with large teeth jutting from the outer edges. The dors are predominantly dung beetles in Britain, digging a burrow under or near the pat and burying a bolus of faeces 10–50 cm underground, on which an egg is laid. This heavy earth-moving requires all legs to be so armed, and the tibiae are prominently ridged for optimal bulldozing. So good are dors at digging, that one species, the deep-digger scarab *Peltotrupes profundus*, is a putative depth record-holder, with one tunnel reported to descend more than 2.3 m into the sandy soil of Putnam County, central Florida (Howden, 1952). Many other dung beetles, notably *Colobopterus* and *Aphodius* species (Scarabaeidae), also have all six legs strongly ridged and toothed, although they spend more time in the dung itself rather than in the soil. Like the Geotrupidae, they have smooth body lines, but are more cylindrical in body form, again an aid to better movement through the clagging mire in which they live.

Elsewhere, soil-, decay- or dung-burrowing (fossorial) limbs, at least in the front legs, are found in *Clivina* and *Dyschirius* (Carabidae), Histeridae, *Leiodes* (Leiodidae), *Bledius* (Staphylinidae), *Melanimon* (Tenebrionidae), *Trox* (Trogidae), *Hoplia* and *Cetonia* (Scarabaeidae), Lucanidae, Scolytidae and *Sphaeridium* (Hydrophilidae).

This last genus is a nice example of a seeming divergence, a cross-over from one adaptation to another. *Sphaeridium* lands on very fresh, semi-liquid cow dung, quickly folds away its flight wings and pushes its way down into the fresh dropping. It is not so much burrowing into the dung as swimming, much as its close relatives *Coelostoma* and *Chaetarthria* swim in mud and pond-edge vegetation.

FIG 25. A female *Copris lunaris* uses her rounded, shovel-shaped head and broad toothed legs for digging a deep burrow in the soil and stocking it with dung on which to lay her eggs. Possibly extinct in Britain, the species remains widespread on mainland Europe. Saint-Amand-en-Puisaye, Nièvre, France, 24 August 2000.

Several very convex species, like the pill beetles (Byrrhidae), also have broad, flattened legs, but rather than for digging, they are shaped for hiding. If disturbed, these species draw their legs (and antennae) in tight against the body, where they are often received into grooved recesses on the underside. This makes them look like seeds or tiny pebbles, and also prevents any aggressor grabbing hold of a loose limb and dragging the insect off to be dismembered. They occur at the mixed soil–grass-root thatch boundary and can be easily overlooked. Perhaps the most extreme example is the American pill scarab *Ceratocanthus aeneus* (Hybosoridae), which curls its entire body, including its broad, flat tibiae, into an almost perfect, shining green sphere, reminiscent of the pill woodlouse (*Armadillidium vulgare*) and pill millipede (*Glomeris marginata*), offering nowhere for an attacker to gain purchase.

According to beetle anatomy, digging is done with the front legs, while swimming is done with the back legs. Nowhere is this more obvious than in the large hind limbs of the true water beetles, the Dytiscidae. Here, the third pair of legs is the longest, with broadened and flattened tibiae and tarsal segments. These are usually fringed with a fine skirt of long hairs, which expand to catch the water on the backwards thrust, but collapse to reduce turbulence and friction on the forwards recovery stroke. For the dytiscids, a sculling action with the legs, similar to the frog-kicking of a swimmer doing breaststroke, propels the beetles with startling agility and speed (leg strokes registered at sixteen per second in small *Hydroporus*, seven per second in larger *Ilybius* and two per second for the giant *Dytiscus*), a technique also adopted in water boatmen bugs (*Corixa*, *Notonecta*), which have even more disproportionately long rear legs.

The other main water beetle group, the Haliplidae, also have long back legs, but with their stouter, deeper, globular and less hydrodynamic body shapes, they are less rapid in the water. They use alternate, rather than simultaneous, leg movements, appearing to run through the water rather than swim.

Whirligigs (Gyrinidae) adopt a slightly different approach. They also use rear-leg propulsion, but in their case all four back legs are reduced to short, stout, flat paddles, each pair beating simultaneously, but the hind legs being out of phase with the middle legs. The hind legs produce most of the forward thrust, but the middle legs are important for stable trajectory and steering. Curved trajectories appear to be more energy efficient than straight lines through the water, which partially explains the wildly curved spirals these beetle execute so well. As gyrinids live a dual life, both on and below the surface, the plane through which their four legs beat is altered when they switch from normal surface mode to underwater diving (Xu *et al.*, 2012). The whirligig's front legs are, by contrast, long and well articulated; they are used for grappling and manipulating prey.

The back legs of various jumping beetles have also become the primary power source for movement, in this case hopping. The appropriately named flea beetles (Chrysomelidae, tribe Alticini) are all very small (the largest British species measures about 5 mm) and have grossly enlarged hind femora containing a structure called the metafemoral spring. This central scroll-shaped chitinous organ stores the energy created by the powerful femoral muscles. Distorted by muscular exertion, the curled spring tissue snaps back into position with a sudden movement when tension is released, exerting a force that allows the beetle to spring 70 cm in the blink of an eye (Furth, 1988; Nadein & Betz, 2016). By that time, of course, they hope to have escaped from the eye of any potential predator.

Conversely, the enlarged legs of some Oedemeridae remain completely mysterious. Although swollen, perhaps even more so than in hopping flea beetles, the hind femora of the very common *Oedemera nobilis* and rather scarce *O. femoralis* males (Fig. 26) contain no springing mechanism and the beetles are unable to jump. Instead, they fly and crawl about in exactly the same manner as the slim-legged females. There is some suggestion that they may use their large hind femora and thick, curved hind tibiae to cling onto the female during mating

FIG 26. The swollen femora of *Oedemera nobilis* males remain an enigma. The beetle does not jump or kick, or burrow with them.

(Furth & Suzuki, 2008), but the closely related British *O. lurida* and many other species worldwide do not possess this peculiar adaptation. In tropical South-east Asia, the monstrous swollen hind legs of the bizarre frog beetle *Sagra buqueti* (Chrysomelidae, subfamily Sagrinae), are thought to be weapons. The powerfully muscular femur, especially pronounced in the males, articulates like a nutcracker against the reinforced, curved tibiae (Emlen, 2008). Although the implication is that males joust for females or territory, or both, this behaviour has yet to be observed, or at least reported. Obviously, much more research is needed.

The tarsi, nominally five-segmented but variously numbered across the Coleoptera, and offering useful identification pointers, are more than just the pads on which the beetle rests its weight when walking. Like antennomeres, tarsomeres can vary from short and broad to nearly as long as the tibiae, especially on the hind legs. On the whole, though, the major variation (apart from number) is in whether the individual segments are bilobed (divided into two flat plates) or conventionally cylindrical. At the greatest extreme – in many of the Chrysomelidae, for example – the underside of the broadly bilobed tarsal segments are covered with sub-microscopic feathery hairs, allowing the beetles to climb up shiny surfaces like glass. In the well-known North American palmetto tortoise beetle (*Hemisphaerota cyanea*), powerful adhesive power is obtained by secretion of an oil into the tarsal pads, effectively creating a stickiness in which the divided and subdivided hairs can latch onto the substrate (Eisner & Aneshansley, 2000). It seems that this adhesion is not so much for climbing up glass, or grass, but for males especially to grip onto the shining, domed backs of the females during copulation (Gloyna *et al.*, 2014). The front tarsi are usually larger and broader in male leaf beetles, especially in the handsome members of the genera *Chrysolina* and *Timarcha*. There may be a similar mate-clasping advantage at work in the swimming dung beetles, *Sphaeridium* (Hydrophilidae), where the males have a large final tarsal segment and the claw enlarged into an almost prehensile grappling hook.

An even more extreme sexual dimorphism occurs in a few of the larger dytiscid water beetles. Again, it is the front tarsi of the males that are grossly enlarged into broad round wads, covered in adhesive setae and sucker pads. The evolution of these broad front tarsi in the males is often accompanied by the appearance of ribbed, ridged or heavily granulated elytra in the females. The two sets of structures are thought to have evolved because the female sculpturing interferes with the tiresome attentions of amorous males, allowing the females some choice when selecting a mate, while the males' foot pads have enlarged to try to overcome this (see Miller, 2003, and Karlsson *et al.*, 2013, for good overviews).

THE ANTENNA – JUST ANOTHER APPENDAGE?

A beetle's antennae are, arguably, its most important sense organs. Although eyesight is undoubtedly useful to many species, there are a few beetles that lack eyes completely – none, though, lacks antennae. Normally, 11 antennomeres make up one beetle antenna; this is quite an unusual number among insects, most of which have far fewer (true bugs, most flies) or many more (cockroaches, grasshoppers, crickets, butterflies, moths, many other flies), and suggests that 11-segmentedness was an early evolutionary adaptation. At least the distal antennomeres carry chemosensors and the majority of them also have tactile hairs. Popularly referred to as 'feelers', this is obviously one of the primary purposes of antennae, and some beetles bat the ground alternately with them as they walk along. Others hold them out erect in the air ahead of them.

At their most simplistic (yes, Carabidae again), the 11 antennal segments are a more or less uniform necklace sequence of rounded beads (moniliform), or short or long cylinders (filiform). This belies an underlying structure that is not uniform. As discussed in Chapter 1, the first segment (scape) and the second (pedicel) contain their own individual muscles and proprioceptor systems, but the remaining antennomeres combine into a composite strand, the flagellum. The differences in antennal shape and structure are useful identification characters, and as with the legs there is a concomitance between antennal form and habit or lifestyle.

One of the commonest antennal forms is the clavate club, where anything from one to all of the flagellar segments become thickened towards the ends of the antennae. Clubbed (clavate) antennae occur widely across different beetle families, but often intermittently, even within closely related genera, suggesting that this structural form has evolved on numerous different occasions. Unique to beetles (occurring in Leiodidae and some Staphylinidae, Cucujidae and Chrysomelidae, among others) is the interrupted club, where segments 7, 9, 10 and 11 are large, but segment 8 is small. The significance of this unusual development is undocumented. Intuitively, enlarged club segments have a greater surface area, to increase the number of chemoreceptors, and there is some evidence that the larger the segments, the more a species relies on chemical detection to find a mate, food supplies, or both. A good example is given in the dung beetles, scarabs and chafers. The plate-like (lamellate) formation of the club gave this group its former suborder name Lamellicornia. Members of this obviously natural grouping of beetles have the final few antennal segments flattened into broad, plate-like leaves, which lie together at rest or can be splayed apart – the beetle equivalent to flaring one's nostrils.

It has long been known that dung beetles find faeces by smell – they invariably fly upwind to the source, and if they lose the scent for a moment, they stop flying forwards and start weaving increasingly left and right until they find themselves back in the scent plume again. Experiments using detached antennae, rigged with electrodes, can identify which particular chemicals are stimulating the chemoreceptors. The response to various test substances wafted over the antennae in micro wind tunnels can be measured by electrically monitoring the nerves. It is no surprise that dung beetle antennae react strongly to excremental substances like 2-butanone, phenol, indole and especially skatole, a heterocyclic organic compound, a bacterial decay product and a powerfully odorous substance that gives dung its characteristic smell (Tribe & Burger, 2011).

In dung beetles and chafers, the final three segments of the antennae are normally lamellate, but in some genera segment 4 and up to 7 may also be plate-like. At the extreme, the doubtfully British (or occasionally vagrant) *Polyphylla fullo* has seven extravagant broad, curved plates for a club. Dung beetles fly to dung and here, coincidentally, they find their mates. Other insects use chemical lures to attract each other. Whether *Polyphylla* is using its grossly enlarged antennae to detect subtle sex pheromones on the wind is, as yet, unclear, but in its close relative, the common cockchafer (*Melolontha melolontha*), the club of the male has seven large plates, while the antennae of the female have only six small club segments. The implication might be that although both sexes need to find some particular effluvium given off by the soil, or roots in which to lay eggs, the male also uses his extra olfactory ability to locate a female.

FIG 27. The seven wonderfully layered, lamellate plates of the antennal club of the non-British ten-lined chafer (*Polyphylla decemlineata*) identify this as a male. Central Guatemalan highlands, 22 October 1992.

Pheromone chemistry is relatively well known in some moths, many of which have strongly feathery antennae in the males. Caged virgin females, reared from caterpillars by Victorian lepidopterists, were used to lure attentive males. Today, synthetic pheromones are used in sticky traps in orchards to detect the presence of pest moth species so that precisely timed insecticide spraying can be targeted against mass emergences. Chemical studies in beetles lag behind rather, but a project to monitor the large red click beetle *Elater ferrugineus* using female pheromones showed that this handsome species was far more widespread than previously thought in Europe (Andersson, 2012) and Britain (Harvey *et al.*, 2017). In 2013, one flew down to land on Peter Hodge at a garden picnic party near Lewes, the first time this beetle had ever been seen in Sussex. He had put a small vial of the test pheromone in his shirt pocket a few days earlier. For the chemically minded, the female pheromone, 7-methyloctyl-(Z)-4-decenoate, is based on a series of 6-, 8- and 10-carbon methylated fatty acid alcohols (Tolasch *et al.*, 2007). The larvae of this large click beetle are predatory on other beetle larvae in rotten wood, and in at least one study (Svensson *et al.*, 2004), females are attracted to sex chemicals (a decalactone) given off by adult males of their prey (*Osmoderma eremita*, a central European scarab that does not occur in Britain). The antennae of male and female *Elater* are serrate, but not very different from each other, suggesting that they have similar olfactory acuity, even if this is for slightly different chemical substances.[2]

The diminutive male of *Drilus flavescens* (Drilidae) has flabellate antennae, another means by which the surface area of the individual antennomeres can be greatly increased. The bizarre wingless female (which has short, squat, rather redundant antennae) is rare and secretive, but it is clear that the males must find their mates by tracing a subtle pheromone scent across the dry open grassland where this species occurs, and over which it flies freely.

Similar, sometimes extreme, differences occur throughout the Coleoptera. Enlarged or dilated antennae occur in the males but not females of (among others): *Attagenus pellio* (Dermestidae), *Malachius bipustulatus* (Malachiidae), *Ptilinus pectinicornis* (Ptinidae), *Agrilus laticornis* (Buprestidae), *Ctenicera pectinicornis* and *Megapenthes lugens* (Elateridae), *Melasis buprestoides* and *Microrhagus pygmaeus* (Eucnemidae), *Metoecus paradoxus* (Ripiphoridae) and *Callosobruchus chinensis* and *Luperus flavipes* (Chrysomelidae). The male timberman (*Acanthocinus aedilis*) has resplendent wiry antennae that are twice as long as the female's, and more

---

2 In a strange but enlightening departure from normal behaviour, males of the primitive *Priacma serrata* (Archostemata, Cupedidae) are attracted to washing hanging on lines throughout their geographic range in the western United States. Apparently, the species-specific pheromone is very similar to a chemical used in a laundry soap powder containing bleach (Atkins, 1957).

than four times its body length. When it flies, these trail out around it into the air, presumably picking up airborne chemicals released by the female, which seemingly needs only to find dead pine logs with her rather inferior organs. The longest antennae of any beetle are reputed to be the trailing streamers of a large longhorn, *Batocera kibleri*, from the Solomon Islands, which can reach 230 mm, although this is 'only' about 2.5 times its body length of about 90 mm.

In a curious departure from most other beetles, male *Lymexylon navale* and *Hylecoetus dermestoides* (Lymexlylonidae) have maxillary palps that are intricately feathery and complexly divided, and are thought to be responsive to female pheromones. Likewise, the snail-predator ground beetle *Cychrus caraboides* (Fig. 28) has strangely enlarged securiform final segments of both maxilliary and labial palpi; these are packed along their final apical surface with chemoreceptors. When this narrow-headed beetle attacks a snail, its head and pronotum are inserted deep into the shell opening, but the antennae remain curled back outside; it is possible it may be using the palpi to detect subtle changes to the scent of the snail's mucus while feeding that its long, ungainly antennae cannot pick up. Similar enlarged final palp segments occur in mollusc-attacking *Carabus* species (Symondson, 2004).

In the oil beetles, the two most widespread in Britain (I hesitate to say commonest, for these are sporadic insects), *Meloe violaceus* and *M. proscarabaeus*, have kinked antennae. In the male, this kink is extreme – a crook-like hook at the flattened and curled antennomeres 6 and 7 – then is followed by four more or less straight, moniliform segments, almost as an afterthought. The male uses this curl to grip a slightly less extreme kink on the female antenna during

FIG 28. The narrow head and thorax of *Cychrus caraboides* are adapted for pursuing a snail as it retreats back into its spiral shell.

FIG 29. The swollen, pale basal segments of the antennae of the male malachite beetle (*Malachius bipustulatus*) can clearly be seen as the female goes in to lick them. Shooter's Hill, Greenwich, 13 May 1998.

mating, accompanied by a caressing of her body. Exactly what is going on is not clear, but mutual exchange of pheromone scents seems plausible. In some exotic laemophloeids, the enlarged, curved basal segment of each male antenna looks like a mandible, leading some (e.g. Lefkovitch, 1958) to speculate that they might be used in combat.

In the smaller and much commoner malachite beetle (*Malachius bipustulatus*, Malachiidae), the four basal segments of the male's antennae are enlarged; the second has a large, flat, pale protrusion, and the third and fourth have slightly smaller, hooked, pale teeth. During courtship, the pair of beetles meet head to head and the female nibbles excitedly at the male's enlarged antennomeres (Fig. 29), which appear to be bathed in a viscous liquid secreted from 'excitator' glands on the front of his head. Again, exchange of pheromones appears to be the best explanation.

Another common antennal form in the Coleoptera is the elbowed (geniculate) antenna, found in Sphaeritidae, Histeridae, Lucanidae, some Staphylinidae and especially in the majority of weevils (Curculionoidea). In the curculionids, the long, narrow basal segment, the scape, is sometimes as long as the remainder of the antennomeres together – usually six or seven segments of the funiculus, and three forming a club. The purpose of this antennal arrangement fits neatly with the egg-laying behaviour of the adult weevil. As it chews a drill hole into plant material using the small but sharp jaws at the tip of the rostrum, the antennae, which are usually inserted near the tip of the snout, might be expected to get in the way. However, by flexing back the long scapes and tucking them into grooves (scrobes) alongside the rostrum, the weevil effectively repositions its antennae further back, allowing it to keep drilling beyond the antennal insertion and yet continue to monitor the substrate with its flexible funiculus and sensory club segments. Once the extra-deep hole is completed the beetle turns round and deposits an egg in it.

Less well understood mechanisms have given rise to some other odd antennae. In the aquatic or semi-aquatic Gyrinidae, Dryopidae and Heteroceridae, the antennae are short and stout, with most or all of the terminal segments forming a compact blade-like club, of unknown purpose. I'm minded to posit that the structure may have something to do with a physical interaction at the water surface for these boundary-zone beetles. The bizarre ant-nest beetle *Claviger testaceus* (Staphylinidae), a permanent 'guest' of the yellow meadow ant (*Lasius flavus*), has only six short, stout antennomeres, so perhaps its cosseted underground life means it does not need to feel very much (it also lacks eyes).

In aquatic members of the Hydrophilidae and Helophoridae, the clubbed antennae are usually very short, and are exceeded in length by the long, slim maxillary palpi, whereas terrestrial species have more conventional antennae and shorter palps. This may reflect the lesser importance of olfactory cues in water, but the increased significance of tactile stimuli in the murk where these

FIG 30. A female (with two small warts in the middle of the frons) lesser stag beetle (*Dorcus parallelipipedus*), showing the long scape and angled flagellum of the geniculate antennae. The punctuation of the pronotum is also shown to good effect with the electronic flash.

beetles live. The great silver water beetle (*Hydrophilus piceus*) has a large hastate, four-segmented club on its antenna, three broad cup-shaped antennomeres sitting inside each other, and terminates with an asymmetrical pear-shaped segment. Together, these form a funnel, which, as the beetle rises head first to the water's surface, breaks the meniscus, allowing air to flow down and replenish the breathing bubble held on the beetle's underside, where it is trapped by a multitude of minute hydrofuge hairs. This thin bubble, the plastron, appears pale and shimmering, giving the beetle its common name.

The long setae on its four basal antennomeres give the common ground beetle *Loricera pilicornis* its name (*pilus* is Latin for 'hair'). They are much longer and stronger than normal, and specially settled into deep sockets so that they do not bend outwards or deform if pressure is applied to them. Such pressure is applied by struggling springtail prey, which try to escape, but which *Loricera* pins down in the basket of criss-crossed bristles just in front of its jaws (Hintpinzer & Bauer, 1986). A similar basket has recently been found in the now extinct scydmaenid *Cascomastigus monstrabilis*, described from Burmese amber dated to about 100 mya (Yin *et al.*, 2017).

Leaving our shores to look at exotic species reveals a bewildering series of other bizarre antennal shapes, from the huge, flattened, sometimes dish-shaped clubs of the aberrant ground beetles (Carabidae, subfamily Paussinae) that live in ant nests, to the extra-feathery antennae of the Australian *Rhipicera femorata* (thought to be a cicada parasitoid), with, uniquely, more than 20 super-flabellate antennomeres. One of the most extraordinary, however, is the South American longhorn beetle *Onychocerus albitarsis*, which has a sharply pointed hook on its bulbous 11th antennal segment, and can deliver a painful pinprick if handled incautiously. Although this hook was first observed more than 130 years ago (Smith, 1884), recent work shows that it is not merely a mechanical defence. The finely pointed spike at the tip has two minute channels fed by pores, through which venom is injected; the whole set-up is reminiscent of the scorpion sting (Berkov *et al.*, 2007). Remarkable.

BEETLE DRESS CODES – COLOUR, CRYPSIS AND COMMUNICATION

Contrary to literary misconceptions, the beetle default colour is not black, but brown, and not a very dark brown at that. Take any one of the myriad tiny rove beetles from a dung pat or a dead fox, look at the click beetles, chafers or weevils crawling over the herbage, or the countless tiny water beetles paddling about in the dip-net, and almost all of them are dull brown. Look in any identification guide, though, and you might be surprised by this fact, since such books are full of technicoloured jewels and trinkets. That's because, consciously or unconsciously on the part of the compilers, pretty, brightly coloured and distinctive species are more easily identified, look better on the page, and are much more attractive to the buyer and reader of the guide. There are, obviously, gems among the brown, but brown is the baseline.

As mentioned earlier, the main pigment in beetles – indeed, in all insects and most other animals – is melanin, a light-absorbing polymer containing heterocyclic indole components derived from the oxidation of the amino acid tyrosine (4-hydroxyphenylalanine) by the enzyme tyrosinase. Its near ubiquity across the animal kingdom reflects its very early evolution in living organisms, but also its versatility. This single chemical can produce a spectrum of colours, from pale beige to pitch black; on the way, it passes through yellow, orange and, of course, brown. Perhaps it isn't the most challenging colour palette, but reds are in there too, provided by carotenoids. These are the same red pigments found in carrots, and although they tend to fade after death, together they are more than enough to create a dazzling variety in beetle life

And if single-colour beetles are also the baseline, those colour identification guides are quick to point out some of the more striking patterns across the Coleoptera. Camouflage has an obvious advantage to any small creature, and perhaps some of the best examples occur in the longhorns, the Cerambycidae. A tiny *Pogonocherus hispidus* (Fig. 31), half resembling a bird dropping and half a lichen excrescence, is all but invisible on a tree trunk, and similar mottles

FIG 31. On a leaf, *Pogonocherus hispidus* resembles a bird dropping; on lichenose tree bark it is all but invisible. (Penny Metal)

occur in *Leiopus, Acanthocinus, Rhagium* and *Mesosa*. Weevils, too, offer splendid camouflage patterns, notably *Polydrusus cervinus, Liophloeus tessulatus*, various *Hypera, Ceutorhynchus* and *Cionus* species, *Cryptorhynchus lapathi, Grypus equiseti* and my own personal favourite, *Platyrhinus resinosus* (Fig. 32). This last weevil, which breeds in dusty black cramp ball fungi, (*Daldinia concentrica*) on old ash logs, is sub-cylindrical, gnarled in texture, and topped and tailed with pale blotches to so closely resemble a broken twig that I have never succeeded in finding a specimen, even though I know the species occurs less than a mile from my house.

At the other end of the beetle pattern book are brightly stark beetles like ladybirds, with their striking red or yellow colours and crisp dot decorations.

FIG 33. Although often hidden on sycamore (*Acer pseudoplatanus*) foliage high in the trees, the orange ladybird (*Halyzia sedecimguttata*) is one of our most brightly (warningly) coloured beetles.

The meaning is clear – do not eat. Ladybirds owe their distastefulness to a series of alkaloids, including coccinelline, hippodamine and myrrhine – tricyclic amines first isolated from *Coccinella, Hippodamia* and *Myrrha* ladybirds, respectively. In a recent study measuring toxicity by using crushed ladybirds to kill test *Daphnia* (Arenas *et al.*, 2015), the brighter the ladybird, the more poisonous it was. At the top end of the scale was the bright orange ladybird (*Halyzia sedecimguttata*), which sits brazenly on sycamore (*Acer pseudoplatanus*) and other leaves grazing mildew (Fig. 33), while the inconspicuous and secretive brown larch ladybird (*Aphidecta obliterata*), which is relatively camouflaged, was the least toxic. This feature, satisfyingly called signal honesty, presumably contrasts with the signal dishonesty of play-acting mimics.

That ladybirds taste foul is well known and the subject of many experiments, usually with unfortunate bird predators tricked into trying to eat them (e.g. Marples *et al.*, 1989, 1994; Marples, 1993). Birds soon learn to avoid the brightly coloured insects, and these warning (aposematic) colours are reinforced at each tasting error the birds make. Ladybirds also taste bad to humans. You would be ill-advised to pop a live specimen into your mouth, but rough handling will induce one to reflex-bleed (see p. 122) onto your fingers, and a quick lick will detect a peppery acridity in the liquid. Normally, ladybird–human tongue interactions are uncommon, but when the Asian harlequin ladybird (*Harmonia axyridis*) arrived in California (after being released as a biocontrol agent against exotic aphids), new opportunities arose. The Californians were not eating the ladybirds, but a propensity in the beetles to seek feeding or roosting sites on damaged grapes led to them being crushed along with the fruit during wine-making. Just one or two ladybirds per kilogram of fruit was enough to taint the wine, making it unpalatable, or at least unsaleable (Pickering *et al.*, 2004; Linder *et al.*, 2009). Vintners throughout the world now have to consider this beetle when it comes to harvest time.

Ladybird-like spot themes also occur in *Endomychus coccineus* (Endomychidae), *Mycetophagus quadripustulatus* (Mycetophagidae), *Scaphidium quadrimaculatum* (Staphylinidae) and *Clytra quadripunctata* (Chrysomelidae). The first three of these are subcortical species, feeding out of sight under fungoid bark, so what benefit they gain from aposematic colours is uncertain. Many exotic endoymchids are also very brightly coloured, so may be genuinely distasteful to predators (as are many chrysomelid leaf beetles), but most mycetophagids and staphs are demure, cryptic and presumably rely on hiding rather than warning to avoid being eaten. There is a treasure trove of study still waiting to be explored in the world of beetle palatability.

The strong contrast of black spots on a background of red or yellow is not the only prominent warning colour theme to permeate the Coleoptera. One particular combination that struck me long ago was the frequent pairing of a pinkish-orange head and pronotum against bright metallic blue elytra. This eye-catching contrast occurs throughout many beetle genera, including *Lebia* and *Brachinus* (Carabidae), *Paederus* (Staphylinidae), *Tillus* (Cleridae), *Triplax* (Erotylidae), *Tetratoma* (Tetratomidae), *Vincenzellus* and *Salpingus* (Salpinigidae), *Acmaeops* and *Phymatodes* (Cerambycidae), *Zeugophora* (Megalopodidae), and *Oulema*, *Gastrophysa*, *Derocrepis* and *Podagrica* (Chrysomelidae), to name but a few.

The first three genera are well known to be noxious, and all of those listed are likely to be also.

There is also no doubting the urgency of the warning given by a series of bee- and wasp-mimicking beetles. Black and yellow bars, stripes or chevrons are a danger signal the world over, and enough of their owners have a powerful sting at the tail to make everyone sit up and pay attention. Several longhorns (e.g. *Clytus arietis*, *Pachytodes cerambyciformis*, *Rutpela maculata* (Fig. 35), *Leptura quadrifasciata* and *L. aurulenta*) are beautifully wasp-marked, and their eager, rapid flight over flowers is remarkably deceptive. The bee-chafer (*Trichius fasciatus*) easily fools bumblebee-watchers, its size, colours and bobbing flight perfectly matching any accompanying bumbles. The pride of Kent, a rove beetle (*Emus hirtus*), is reputed to be a bumblebee mimic, but its habit of visiting very fresh cow dung is completely un-bee-like; it is very rare in Britain (I've certainly never seen it), and observations of it in flight are scant. Again, whether these beetle mimics are all bluff and bluster, or whether any of them reinforce the warning by being distasteful, is not widely reported.

Metallic colours are a strong feature in many beetle families, in Britain notably the ground beetles (Carabidae), jewel beetles (Buprestidae), dor beetles (Geotrupidae), rose chafer (*Cetonia aurata*, Scarabaeidae), leaf beetles (Chrysomelidae) and some weevils (*Byctiscus* etc., Rhynchitidae). Caused by optical refraction rather than by pigments, these brilliant shining rainbow colours decorate some of our most attractive beetles. Multilayer reflectors caused by

FIG 35. The wasp longhorn (*Rutpela maculata*) is well named, being remarkably wasp-like when flitting about the flowers.

the laminated chitin layers of the cuticle, sub-microscopic parallel grooves on the cuticle surface creating diffraction gratings, or repeated three-dimensional crystalline structures all cause white light to be prismatically broken and selectively reflected. The spacing and regularity of the layers, grooves or crystals determine the wavelength (and therefore the colour) of the refracted and reflected light, and this can change depending on the incidence angle of the light, which itself changes as the beetle moves. A good review of the mechanisms by which metallic and iridescent beetles shine is given by Seago *et al.* (2009).

That these striking beetles attract our notice might seem a little counterproductive, because this would imply that they also attract the unwelcome attentions of predators. However, the metallic glints may offer a variety of defensive and avoidance strategies. The Spanish fly (*Lytta vesicatoria*, Meloidae), a rare vagrant in Britain, is a deep metallic green, but is packed with the chemical cantharidin, a powerful blistering agent that would soon put predators off shiny things for life. The two metallic green leaf beetles mentioned earlier, *Cryptocephalus hypochaeridis* and *C. aureolus* (Fig. 36), sit prominently in yellow flowers and must present obvious targets to predators, but they appear to remain unmolested. The bodies of chrysomelids are well known for containing chemical defences (Deroe & Pasteels, 1982; Pasteels *et al.*, 1989); several other British members of the genus are brightly (and therefore warningly) coloured red, yellow and black, and many other chrysomelids are strongly coloured or garishly patterned.

In some circumstances, shining metallic glints are excellent predator- and entomologist-avoidance devices. The many metallic *Bembidion*, *Elaphrus* and *Agonum* species to be found running across wet mud and through the waterside herbage of ponds and ditches are remarkably hard to follow. As they zigzag about, accelerate, then stop dead, the sun catches their shiny bodies but confuses the eye among the similar reflected glints of water droplets and shining wet leaves, making them very difficult to intercept. The whirligigs, *Gyrinus*, are likewise devilish tricky to catch, and increasingly desperate snatches with the water net sometimes end in damp disaster. Similar avoidance benefits have been mooted for metallic tiger beetles occurring in tropical forests (Schultz, 1986), and some work has been carried out experimentally using domestic quails as predators, which take more pecks to capture metallic virtual beetles than targets without these interference colours (Pike, 2015).

Throughout the world there are plenty of garish, strikingly patterned beetles, notably the chafers, longhorns and jewel beetles, which sport bizarre patterns of

FIG 36. The striking green gem that is *Cryptocephalus aureolus*, in its typical yellow flower backdrop. (Penny Metal)

bright, strong colours, often in geometric or seemingly artistic designs. Although it is easy to claim these as aposematic warnings, there may be a counter-intuitive camouflage going on here. Low down in the herbage, vertebrate predators (mostly birds) do most of their hunting by silhouette outline, effectively looking for insect-shaped insects. Counter-shading, where the underside of an animal is paler than the upper surface, to counter under-body shadows, is widespread and effective, from carp to caterpillars. Bright colours can break up an outline, such that although a hunter may see the beetle, it may not immediately recognise it as such, giving its potential victim vital moments to escape. Such was the rationale behind the stark geometric 'dazzle' (also 'razzle-dazzle') paint jobs given to naval ships during the First World War. In Britain, the asparagus beetle (*Crioceris asparagi*) might just fit this model (Fig. 37).

Oddly, it is the most sombre of beetles, the all blacks, that are among the most obvious – walking brazenly across bare ground or short grass, or through the herbage. These are also some of the species best avoided if you are a potential predator. Top of this list are the bizarre but wonderful oil beetles, *Meloe* species. These huge inky-black insects, parasitoids in solitary bee nests (see p. 74), are flightless and cumbersome, the females usually bloated with thousands of eggs. They are, though, among the most unpalatable of potential prey items, containing the same chemical cantharidin as the Spanish Fly (see p. 121).

The similarly sized and coloured bloody-nosed beetle (*Timarcha tenebricosa*, Chrysomelidae) is equally nonchalant as it trundles through the herbage. Like oil beetles, it is flightless, large, heavy and unpalatable to predators. Its haemolymph is rich in anthraquinones (Jolivet & Verma, 2002), which it readily gives up by

FIG 37. The striking geometric pattern of the asparagus beetle (*Crioceris asparagi*) may help break up its silhouette outline. Here, a mating pair demonstrate the usual beetle copulation arrangement – male perched on top.

FIG 38. The cellar beetle (*Blaps mucronata*), a large, waddling tenebrionid, sadly is now declining as cellars become less inviting to beetle life, so entomologists are less able to experience its foul-smelling chemical defence.

reflex bleeding (auto-haemorrhaging) from its mouthparts – hence its highly apt common name.

The once widespread cellar beetle (*Blaps mucronata*, Tenebrionidae) is another quintessential clockwork waddler, and whenever it is found it is clear that it is in no hurry to escape; it doesn't need to rush, as it is also immune to attack by virtue of its chemical defences. Many of the tenebrionids exude (some spray) a defensive exudate from the tail end, as they adopt a headstand position (Tschinkel, 1975). This has been known since antiquity, and although he thought he was describing a type of cockroach, Pliny the Elder, writing in the first century CE, was obviously describing *Blaps* when he wrote '*blatta odoris taedio invisum, exacuta clune*' ('having an unpleasant smell and being pointed at the tail end') (Beavis, 1988). Again, the major component of its chemical defence is a cocktail of quinones (Peschke & Eisner, 1987). There are also black ladybirds (*Exochomus*), black rove beetles (devil's coach-horse, *Ocypus olens*), black click beetles, black ground beetles and black carrion beetles; the message is clearly delivered – danger, keep back!

On the other hand, no such warning is given by tortoise beetles (*Cassida*); these generally leaf-green beetles are carefully camouflaged against green foliage. Their ability to hide in the herbage is enhanced by their flattened and flanged body shape, discussed earlier. Some species are marked with brown blemishes to add further to the cryptic effect. Despite the seemingly obvious advantages in terms of camouflage, green pigments are relatively rare in adult insects. Orthopteroid groups like grasshoppers and bush-crickets, and leaf insects in Phasmatodea, are often green, and there are some heteropteran 'true' bugs, but green is decidedly scarce, for example, among the pigment-rich butterflies and moths (despite an abundance of green caterpillars), and all but unknown in

Hymenoptera (bees, wasps, ants, etc.). In beetles, any greens are usually structural metallic (refraction or crystalline interference) colours, and true green pigments are rare. Those that occur in tortoise beetles are apt to denature, and beetles, bright green in life, can often fade to dull straw or brown on death (similar fading can occur in Orthoptera and Phasmatodea). Even here, though, among the modestly secretive tortoise beetles, there is the occasional bright shocker – *Cassida murraea* is sometimes green, but changes to a deep scarlet, flecked with black spots at sexual maturity, and *C. nobilis* has two metallic silver or golden stripes down its back. *Cassida murraea* feeds on fleabane (*Pulicaria dysenterica*), and one assumption is that it may sequester the toxic, or at least highly aromatic, chemicals that formerly made this plant a bane to fleas and a treatment for dysentery. *Cassida nobilis* feeds on goosefoots (*Chenopodium*) and glassworts (*Salicornia*), and although both of these groups of plants are edible to humans, they should be cooked to remove the distasteful saponins and other less-than-palatable components they contain. Whether the golden insignia on *C. nobilis* is an aposematic warning or an eye-confusing glint defence remains to be tested on the usual selection of willing or unwilling bird predators.

It seems that every group of beetles has both brightly coloured baubles and discreetly demure colour forms among its complex diversity. This is part of the overall attraction of the Coleoptera. Diversity is ever present, to give something more to titillate the eye.

FIG 39. The flattened and flanged form of an obviously named tortoise beetle, *Cassida rubiginosa*, on bramble (*Rubus fruticosus*).

CHAPTER 3

Life Histories – Larval Forms and Behaviour

GLORIOUS GRUBS

There is an easy misconception that the adult stage is the most important in an insect's life. It is the striking, strange, colourful or bizarre adults, after all, that are featured in identification guides; they are photographed on flowers or under rocks, form the basis for museum collections and receive formal scientific names. There was always a tendency in older books to describe the adult as the 'perfect' state, and the imago (literally the 'image' of perfection) is still a standard, albeit slightly archaic, term for an adult insect. But from almost every biological point of view, it is really the larva that is the dominant stage.[3] To some extent, the adult is merely a transient, short-lived but highly mobile dispersal phase. It is the larva that does most of the feeding and certainly all of the growing, from egg to pupa. It is the larval stage that dictates how and where beetles occur, and the larva that takes centre stage in the beetle's ecological interactions with the rest of the world. No analysis of beetles can be complete without constant reference to their life history and how the immature stages succeed (or not) in their monumental odyssey through larvahood.

Larvae (or alternatively nymphs in hemimetabolous groups) offer an organism a huge benefit in terms of dividing up and sharing out the primary biological

---

3 The word *larva* is originally Latin, meaning 'ghost' or 'mask', implying a disembodied spirit or otherworldly shape, echoing, but different from, the true, 'perfect' corporeal form of the adult.

imperatives of feeding, growing, mating, dispersing and reproducing. Having a larval stage allows immatures and adults to occupy non-competing niches. The larva – an extended, sedentary but semi-mobile self-nourishing embryo (metaphorically at least) – often utilises different foodstuffs from the adult, in different microhabitats, against an independent time frame, and usually with complementary seasonal requirements. However, when it achieves nutritive and developmental completion, it can magically transfigure itself into a highly mobile, exploratory, gene-exchanging, colonising, egg-laying adult beetle. This is a true miracle of life.

Adults and larvae sometimes do occupy similar habitats – the same foodplant, for example – but whereas a vine weevil (*Otiorhynchus sulcatus*) nibbles leaf edges, the grubs feed underground at the plant roots (Fig. 40). Dung beetle adults effectively take the dung soup course, ingesting only the smallest particles (8–50 µm in diameter) in the more liquid fraction, whereas the larvae have tougher grinding mandibles to deal with a slightly wider mix of more coarse, fibrous material. More often, adults occur well away from larval habitats, although they obviously visit to lay eggs. Thus the longhorns *Rutpela maculata* and *Clytus arietis* are flower visitors as adults, while the larvae feed in dead timber or woody plant roots. Stag beetle larvae (*Lucanus*) also feed in mouldering logs, but the adults do not feed, beyond sucking up a bit of moisture to stave off dehydration. This functional split between immatures and adults gives rise to all the bizarre life histories of the Coleoptera, but our knowledge of the early stages of the order remains fragmentary and obscure. What follows is a simplified précis of knowledge to date. And I repeat earlier pronouncements from every other beetle book ever written – there is still much to uncover, and entomologists will find endless new stories to tell if they spend as much time looking at larvae as they do poring over the adults.

Starting at the beginning, beetle eggs are all very much of a muchness. They tend to be rather egg-shaped, smoothly membranous and have soft, relatively flexible shells, lacking any of the surface sculpture, ribbing or hatch-opening mechanisms exhibited by those of other insects. Only a few species of the primitive family Cupedidae (not occurring in Britain) show such architectural details. Size varies commensurate with adult size, thus the large violet ground beetle (*Carabus violaceus*) lays eggs 5 mm long, while ptiliid eggs, produced and laid just one at a time by the minute beetle, are like grains of dust. Eggs may be laid individually or in batches, inserted into potential foodstuff or deposited on the surface (attached by a sticky secretion), depending on the ecology of the individual species. Most are hidden away secretively, but gardeners may be familiar with the clusters of 10–50 yellow or orange ladybird eggs, each about 0.5–

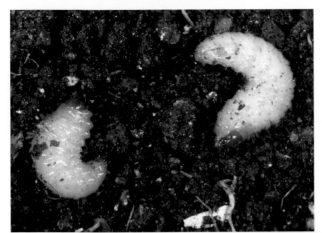

FIG 40. C-shaped grubs of the vine weevil (*Otiorhynchus sulcatus*) are notorious root-feeders, but the adults chew characteristic perforations around leaf edges (below).

0.7 mm high, laid on tree trunks or leaves (Fig. 41). The great silver water beetle (*Hydrophilus piceus*) is famous for laying its 50 or so eggs in a special air-filled silk cocoon, measuring roughly 20 mm by 10 mm and attached to the underside of a floating leaf. At one end this bears a tall chimney, which reaches up and out of the water to supply oxygen to the developing embryos. A few other hydrophilid water beetles make similar, sometimes portable, egg sacs. On the whole, though, beetle eggs are seldom seen and thus poorly studied. At least the larvae are easier to come across.

FIG 41. A clutch of standard ladybird eggs.

FIG 42. This lily beetle (*Lilioceris lilii*) laid her eggs in convenient half-dozen batches.

If adult beetles seem a relatively cohesive group – moderately recognisable, and all structurally united by their sheath-winged, smooth-lined, leggy, beetlish body forms – their larvae show a contemptuous disregard for such conventions. If anything, beetle larval diversity is even greater than that shown by the adults. Indeed, most in-depth Coleoptera monographs bemoan the fact that there is no simple diagnostic feature by which beetle larvae can be exclusively separated from other immature insects. In debates on beetle evolution, this is taken as good evidence that beetles are, genuinely, part of the holometabolous insect line. It does mean, however, that identifying a beetle larva is fraught with even more difficulties than identifying an adult. This confusion is compounded by the unfamiliarity of most beetle larvae to the general public, itself a feature of the creatures' secretive and extraordinarily diverse life histories. There are not many easily available beetle larva keys, but those of Böving & Craighead (1931), van Emden (1939–49), Chu (1949) and Klausnitzer (1991–2001) may suffice.

FIG 43. Female *Galerucella calmariensis* laying eggs. Each is topped with what looks like a morsel of excrement – possibly a way of passing on symbiotic gut bacteria to her offspring. Brede, East Sussex, 23 June 1992.

At best, a nominally 'typical' beetle larva, if there were such a thing, might be described as follows. The head capsule lacks compound eyes, but there can be one to six stemmata (sometimes called larval ocelli) on each side; these are simple clear or crystalline domes, with a cluster of light-sensitive cells beneath. They may be closely grouped – an incipient compound eye perhaps – or widely spaced, often in genus- or family-distinctive constellations. Many subterranean species are blind, lacking all light-detecting organs. Loss of stemmata through evolution appears to be irreversible, so the precise numbers are often good indicators of phylogenetic lineage. Thus larvae of darkling beetles (Tenebrionidae) and their relatives never have more than five stemmata on each side, rove beetles (Staphylinidae) no more than four, ladybirds (Coccinellidae) three, weevils (Curculionoidea) two, and scarabs (Scarabaeidae), click beetles (Elateridae) and soldier beetles (Cantharidae) have only one on each side. The two short, often squat, antennae are generally three-segmented, comprising a scape, pedicel and undivided flagellum (but between two and five segments are frequently found).[4] The penultimate segment usually has a very small additional side segment, called a sensorium, the exact purpose of which is not known. Antennae are particularly

---

4 The larvae of the marsh beetles, Helodidae, have thin, whip-like antennae with up to 150 minute ring-like segments reported, the numbers of these increasing at each moult.

short or abbreviated in boring or internally feeding larvae. The two mandibles, usually described as orthopteroid for their stout triangular secateur biting form, are used for crushing, grinding and tearing, although some predatory species have sharp, pointed, sickle-shaped jaws, complete with grooves to deliver digestive juices and drain partly digested victims. The thorax usually has three rather similar segments, although the first may be enlarged or have a broad, thick plate-like shell – this will eventually develop into the all-powerful beetle pronotum. Each thoracic segment has two legs with gross segmentation approximating that of the adults: six segments (coxa, trochanter, femur, tibia, tarsus and pretarsus with one or two claws) in the Adephaga families, but five (fused tibiotarsus) in Myxophaga and Polyphaga. Legs may vary from rather long in active predatory species, to very short or nearly absent in species that develop hidden inside their foodstuff – mostly plant-feeders, but also parasitoids. The abdomen is usually made up of nine or ten very similar segments, each soft and pliable to allow some expansion after feeding. Occasionally, the abdominal segments may have fleshy wart-like processes, but these do not achieve the same powerfully prehensile abilities as the muscular abdominal prolegs and anal claspers, armed with rings of crochet hooks, seen in butterfly and moth caterpillars.

FIG 44. The familiar flat, pale larva of one of the cardinal beetles (*Pyrochroa*) under fungoid bark, clearly showing off its urogomphi tail spikes.

The nearest we get to a distinctive identifying beetle larval trait is on the ninth abdominal segment, where most species have fixed or articulated projections called urogomphi. These are, perhaps, analogous to the segmented sensory appendages, cerci, found in adults and nymphs of Orthoptera (grasshoppers and crickets), Dictyoptera (cockroaches) and other insects, but as entomologists are pleased to point out, true cerci are always attached to the tenth abdominal segment, not the ninth. Attachments of urogomphal musculature and other internal details also support the idea that urogomphi have evolved quite separately from cerci, and only in the Coleoptera.

At a more technical level, no beetle larva has silk-producing labial glands (which are widespread in Lepidoptera and Hymenoptera larvae, for example, and used for spinning a silken cocoon in which to pupate), the third thoracic segment lacks spiracles (present in Diptera, for example), and they possess a maximum of only six gut filtration extensions, the Malpighian tubules (often multitudinous in the larvae of other insect groups). Variations on the beetle larval head capsule are very diverse, but a primitive feature may be the distinctive U- or Y-shaped frontal sutures on the top of the head, and a groove running from this towards the back of the head, where the sclerotised left, right and front protective plates fuse together. This is particularly emphasised in the primitive family Cupedidae, and in groups where burrowing or boring larvae have powerful muscles, anchored to an internal sclerotised keel, corresponding with the external sutures.

VARIATIONS ON A THEME

Despite this lack of overall unity, beetle larvae can be identified and studied in all their glorious, if slightly confusing, diversity. Just as with adult beetles, a feel or impression – the jizz – of a larva is often enough to start the identification process. Helpfully, beetle larvae can be broadly classified into about eight types, based on their body forms, a result of their ecologies and life cycles, and although not immutable, they are ultimately a reflection of ancestral phylogeny (Lawrence *et al.*, 2011).

Eruciform – slow and caterpillar-like (*eruca* is Latin for 'caterpillar'), with well-developed but relatively short legs and usually with fleshy protuberances along the sides of the abdomen. These include the larvae of ladybirds (Coccinellidae) and leaf beetles (Chrysomelidae, Fig. 45), which feed openly on plant surfaces. Just like true caterpillars, the body form is simply a more or less cylindrical food-digesting bag, with legs enough to give some mobility or grip onto the substrate.

FIG 45. The short, fat but nevertheless clearly eruciform larva of a leaf beetle, the uncommon *Chrysolina oricalcia*, feeding on hedge parsley (*Anthriscus sylvestris*). Nunhead Cemetery, south-east London, 20 April 1993.

FIG 46. The prepupa of the byony ladybird (*Henosepilachna argus*), just as spiny as any exotic moth caterpillar. Wisley, Surrey, 17 July 2000.

FIG 47. Typical curved, fat-bodied, short-legged scarabaeiform grub of the stag beetle (*Lucanus cervus*). Dulwich, south-east London, 8 July 1997.

Scarabaeiform – sluggish, fat C-shaped grubs, just like those characterised in the Scarabaeoidea (Fig. 47). Although the legs are well developed, they are claw-like and able to do little more than hoick the large and often bloated body of the grub around. Such grubs often feed inside their foodstuff, hollowing out voids or shuffling through a friable substrate like dung (scarabaeid dung beetles), leaf litter (chafers) or wood mould (stag beetles). This larval form is very rare outside the Coleoptera, in which it appears to have evolved several times.

Elateriform – long, slender, toughened cylindrical larvae with short legs, exemplified by the wireworms, the larvae of click beetles (Elateridae, Fig. 48), and mealworms (Tenebrionidae). These are often species adapted to burrowing in confined spaces, feeding under bark, boring in rotten wood or fungi, or in the soil at plant roots.

FIG 48. Typical annular cylindrical elateriform larvae of click beetles.

Vermiform – thick, legless (or nearly so) grubs (Fig. 49), like those of many weevils (Curculionidae etc.). Not all are very worm-like (although *vermis* is Latin for 'worm'), but in common with the archaic term wyrm (meaning any legless or reptilian creature, including serpents, snakes and dragons), applied to the legless maggot-like larvae which, like those of flies, feed inside their foodstuff, they do not require locomotory powers provided by limbs. This form is sometimes called apodous (lacking legs).

Campodeiform – flattened, elongate, leggy, generally active larvae, with a well-articulated head, distinct antennae and usually long, segmented urogomphi; typical of hunting predatory larvae of ground beetles (Carabidae), water beetles (Dytiscidae and Gyrinidae, which also have 10 pairs of fluffy-looking external gills down the abdomen), rove beetles (Staphylinidae, Fig. 50) and the like. Named after *Campodea* (Fig. 51), a small, pale bristletail (hexapod non-insect order Diplura), which is very similar in form.

FIG 49. The apodous vermiform larva of the jewel beetle *Agrilus biguttatus*. The swollen thoracic segments and narrow 'tail' are characteristic of this family. Sydenham Hill Woods, south-east London, 8 July 1997.

FIG 50. Typical campodeiform larva of a rove beetle.

FIG 51. *Campodea*, the small two-tailed bristletail after which campodeiform larvae are named.

Cheloniform – broad and oval, with flattened and flanged body segments, and a concealed head, somewhat resembling a compressed turtle (*khelone* is Greek for 'tortoise'). Examples include water pennies, the flat, almost round aquatic larvae of Psephenidae, whose smooth, streamlined shape allows them to cling to the underside of stones in fast-flowing rivers and streams.

FIG 52. Onisciform larva of the common glow-worm (*Lampyris noctiluca*).

Onisciform – broad, or domed, with all body segments fairly heavily sclerotised and appearing relatively uniform (Fig. 52), sometimes with flattening into a flange-like edge, a bit like a woodlouse (*Oniscus*). The carrion beetles (Silphidae) are typical of this group.

Fusiform – broadest in the middle, but narrowed to head and tail ends (*fusus* is Latin for 'spindle'). This descriptor can also be used to modify many of the previous terms. Many ladybird (Coccinellidae) larvae are this shape, especially the orange ladybird (*Halyzia sedecimguttata*) and 22-spot ladybird (*Psyllobora vigintiduopunctata*), which seem more active and less dumpy than some of the commoner garden species.

Other authors have chosen to claim five, or just four, main larval forms; as with adults, this confirms the complexity and diversity of body structures achieved in the Coleoptera. Such divisions may be helpful in describing beetle larvae, but they are not rigidly defined, and contrasts within families and genera are frequent. Variations within individual species are known, and the oil beetles (Meloidae) exhibit one particularly strange example.

In a process called hypermetamorphosis, the larva of an oil beetle passes through several utterly different body forms during its complex development. Hatching from the tiny egg (often laid in their thousands in a small depression in the soil), the first stage is a minute but highly active, long-legged campodeiform larva – a triungulin, named for the three (rather than the usual two) tarsal claws possessed by many species.[5] It scampers up the vegetation and waits, usually with a gang of its siblings, on a flower, and attempts to grip onto the body of a passing solitary bee to be taken back to the host burrow (a transport option termed phoresy). Not just any bee will do; most oil beetles have species- or sometimes genus-specific host requirements. Hitching a lift with the wrong bee, or on an entirely different insect altogether, is fatal, and mortality at this point is huge. But, for the lucky larva that achieves its goal in reaching the right nest, it changes into a fusiform larva (reminiscent of many carabid species) to locate the bee's brood and food store. Next, it transforms into a fat, short-legged scarabaeiform grub, scrabbling around and feeding in the bee's brood cells, until it is ready to pupate. Eventually, a new adult emerges and climbs out of the nest remains. Similar gross hypermetamorphic disparities of form occur in other parasitoid groups, such as the wasp-nest beetles (Ripiphoridae).

Oil beetle larvae are also unusual because, contrary to the usual larval/adult behaviour dichotomy, many of the adult beetles (*Meloe* species especially) are huge, lumbering, wingless (hence flightless) and cumbersome (Fig. 53), unable to move far from the host bee nest from which they have emerged. Instead, it is up to the triungulins to become the insect's dispersal phase, hijacking the highly mobile bee to, hopefully, establish new outlying oil beetle colonies. I write this

FIG 53. The huge, bloated form of an egg-laden oil beetle, *Meloe proscarabaeus*, only the second example I have found in 50 years of serious Coleoptera study. St Lawrence, Isle of Wight, 17 April 2014.

5 In truth, triungulins have but a single claw (actually a large spatulate, claw-shaped tarsal segment), but with a large claw-like seta each side of it, giving the trident appearance.

part of the chapter in late 2016, having recently watched ivy bees (*Colletes hederae*) on the flowers of their favoured foodplant in my parents' garden at Newhaven, Sussex. This bee is a recent coloniser to Britain (it arrived in 2001), but it has now spread widely over southern England, and as I watched, I wondered how long it would be before the bee's well-known European oil beetle parasite, *Stenoria analis*, might arrive here too. For larvae of this beetle, 'hijack' is exactly the right word. Clusters of *Stenoria* triungulins, huddling on a flower or tall stem, actively attract males of their host bee species. The arrival of the male bees, in a manner exactly reminiscent of the pheromone attraction by which they detect newly emerged females, suggests a chemical lure, with long-chain alkenes the likely candidates (Saul-Gerschenz & Millar, 2006). The male bees attempt to mate (pseudocopulation) with the triungulin cluster, easily allowing the *Stenoria* larvae to clamber aboard. These are later delivered to female bees during genuine copulation, and end up back in the nests being excavated by these females as they unload pollen masses into the brood cells (Vereecken & Mahé, 2007). The *Stenoria* larvae then go through similar hypermetamorphosic costume changes. It seems too easy. In this species, it is both the triungulin larvae and the adult flying beetles that can spread to find new colonies of the bee, so I anticipate *Stenoria* arriving in Britain shortly.

A beetle I do not anticipate arriving any time soon in Britain (although it has reached Austria, Gibraltar, South Africa and Hong Kong, in imported timber; Beutel & Hörnschemeyer, 2002) is the North American telephone-pole beetle (*Micromalthus debilis*). This is a great shame, because it is one of the most unusual organisms ever discovered. Exhibiting the only known instance of larviform reproductives in the Coleoptera, otherwise known as paedogenesis, its larvae can lay viable eggs or live larvae, without metamorphosis or mating. This is a peculiarity otherwise known only in a few flies and aphids; *M. debilis* has one of the most complex life histories of any creature.

Normal female *Micromalthus* beetles – tidy, cylindrical, shining black, and with slightly shortened wing-cases and an exposed hind body – lay eggs that hatch into active triungulins, sometimes called the 'caraboid' stage for their carabid-like form. These climb into the rotten wood substrate in which the beetles breed, where they moult into a legless 'cerambycoid' form, reminiscent of a longhorn grub. They retain this shape for three more moults, feeding for several months, and then after a short resting stage they make a strange choice unparalleled elsewhere in beetledom. The next moult will take the larva down one of four possible developmental pathways. It may produce a conventional pupa and then a new adult female beetle; this is the least unusual formula. Alternatively, it moults to produce a new larval form, in which the ovaries are already active and working, but in which no sexual fertilisation is required for more eggs or larvae to emanate – the paedogenetic larval stage. This type of larva has ovaries, but no equivalent of the vagina, uterus or spermatheca found in adult beetles.

This larva can now take one of the remaining three life tracks. First, it can produce more triungulin larvae – usually about 10 – born live (viviparity) a couple of weeks later, without any egg stage. These then go around the initial life cycle option again. Second, it can lay a single large egg, which remains attached to the mother larva. After about 10 days, the egg hatches into another type of legless larva, this time curculionoid (resembling a weevil grub), which eats its parent (matriphagy), chewing its way through the vent that gave it birth. This larva later pupates, and eventually a tidy, cylindrical, shining black, winged adult male beetle emerges. Third, it can produce an intermediate paedogenetic larva capable of delivering either female triungulins, a male egg or sometimes both.

Since this peculiar life cycle was first examined (Barber, 1913a,b), re-examined (Pringle, 1938) and re-re-examined (Scott, 1938, 1941; Smith, 1971; White, 1973; Pollock & Normark, 2002), our understanding, along with our wonder at this beetle, has increased. The males have been reported to be 'non-functional' – i.e. they do not copulate with adult females – and indeed males are very rare, and appear to be completely absent from South African colonies. This seems unlikely, as even rare genetic exchange is very important to maintain some minor diversity in populations, and the high cost of producing males would otherwise be counterproductive.

The males, it appears, are haploid, with just one of each of the 10 standard *Micromalthus* chromosomes (n = 10 in the formal notation of cytogeneticists) in each body cell. Females are diploid (with 2n = 20), a matching pair of each chromosome.[6] The retention or reduction in chromosome number occurring in the ovaries of the *Micromalthus* paedogenetic larva is accompanied by spontaneous abortion of some embryos; this accounts for the male/female (or both) possibilities during experimental breeding. Haplodiploidy (also called arrhenotoky), where chromosome numbers dictate male or female offspring, also occurs in a few bark beetles (Scolytinae, tribe Xyleborini), but is especially well known in the Hymenoptera (bees, wasps, ants, etc.), where its influence on the evolution of social behaviour and caste formation is still being vociferously debated by biologists. Thelytoky, in which unmated females create viable eggs

---

6 Twenty chromosomes (n = 9 + X/Y) is common across much of the Polyphaga, but many groups have lost a chromosome, and numbers vary from four (n = 1 + X/Y) in the South American click beetle genus Chalcolepidius, to 69 (n = 34 + X/O) in the central European ground beetle genus *Ditomus*. Both standard insect sex-determination XO and XY systems are in play through the Coleoptera.

or embryos and produce only female offspring, is also known in many insect orders – in the Coleoptera, it occurs in about 50 species of weevils (Curculionidae) and at least seven other families, including Ptiniidae, Ciidae and Chrysomelidae (Smith & Virkki, 1978).[7] But the occurrence of these life history strategies together in *Micromalthus* paedogensis is unique. Work continues, and if ever this beetle becomes established in Britain, I'm going to get my own breeding colony to worship.

Hypermetamorphosis and paedogenesis are at one extreme of the larval form and behaviour spectrum, but whether active, immobile, chewing, grinding or stealing, the body structure of the larva, and its habits, are everywhere inextricably linked to the adult beetle's ecology and evolution. This is immediately apparent when considering the length of time it takes a beetle to grow.

GROWTH AND DEVELOPMENT

The larval stage is often (in beetles) the longest of the life cycle, ignoring any dormant overwintering of the egg where no development is actually taking place. Most species (in Britain at least) have a one-year life cycle, each stage adapted to some seasonal advantage. For leaf-feeding larvae, synchronisation to leaf availability during spring and summer, after bud-burst, is a distinct bonus, whereas those feeding in fungal fruiting bodies tend to occur later into autumn and through winter, as toadstools and brackets sprout and then senesce. Even if foodstuffs are not limited by seasonal plant growth cycles, seasonal temperature cycles have profound effects on emerging adults, which benefit from warm, dry weather to find mates, new oviposition sites or overwintering roosts. If burgeoning food supplies are sufficient and clement weather allows, multiple generations can overlap, thus ladybird population explosions occur as densities of their aphid prey increase.

On the other hand, life cycles extending over more than one year are also common, especially in large species, where the gradual accumulation of body mass, from sometimes nutrient-poor food, is a slow process. In Britain, the larva of the great silver water beetle (*Hydrophilus piceus*), a wrinkled grey science fiction horror, usually spends two to three years eating protein-rich snail prey before transforming into an adult, whereas that of the similar-sized stag beetle

---

7 Another curious strategy, pseudogamy, is demonstrated by the rare spider beetle *Ptinus clavipes*, which reproduces females parthenogenetically, but only after insemination, even though no male genetic material is incorporated into the developing embryo. It has a bisexual form, which is diploid (2n = 18), but also a female form, *mobilis*, which is triploid (3n = 27). Bizarre.

(*Lucanus cervus*), a pale, flaccid, glutinous handful of a maggot, takes three to seven years of grinding protein-poor wood mulch before it is ready to pupate. Extended larval times are difficult to measure in rearing studies, but occasional anecdotal evidence suggests variation can be extreme. The house longhorn (*Hylotrupes bajulus*) normally takes about three years to develop into an adult, but development times of up to 30 years have been recorded (Hickin, 1975). The North American jewel beetle *Buprestis aurulenta* regularly emerges from furniture crafted from the Douglas-fir (*Pseudotsuga menziesii*) heartwood timber in which its larvae (pale and worm-like, but with a bloated thoracic region) have been boring deep for several years. There used to be a display in London's Natural History Museum showing a thick-soled wooden sandal from which a large specimen had emerged, much to the consternation of its owner; and during refurbishment of a building in British Columbia, two *B. aurulenta* larvae were exposed as the original floorboards were being sanded, where they had reputedly been nibbling away, very slowly indeed, for 51 years without quite achieving adulthood (Smith, 1962).

Growth in beetle larvae, in common with all insects, is not a continuous, gradual process, but is stepped, in sudden, discrete incremental stages – instars. Although best known in lepidopterous caterpillars, growth patterns in beetle larvae are under similar hormonal control. At a certain weight or size trigger, a burst of moulting hormones, ecdysone and ecdysterone, stimulates the larva to form a new epidermis and cuticle inside the old one; it then splits and sheds its old stretched, tight-fitting skin to reveal a new, temporarily expandable cuticle, which soon hardens to a larger, tough new shell. All the usual analogies about putting on weight inside your suit of armour work here. Meanwhile,

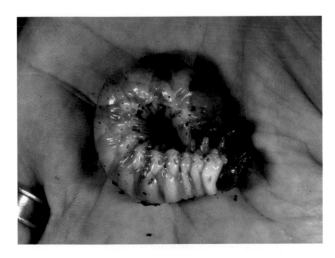

FIG 54. Although not quite fully grown, a stag beetle grub (*Lucanus cervus*) is still an imposing animal.

juvenile hormones (a series of acyclic sesquiterpenoids) prevent the larva from metamorphosing into a pupa (and then to an adult) until a final weight, size or time barrier has been broken.

Moulting occurs to a standard pattern in most beetles. The skin splits lengthways, starting at the head between the eyes (that distinctive U- or Y-shaped frontal suture), and continuing along the thorax and some way along the abdomen. As the larva pulls itself free, it leaves behind the linings of the gut and of the tracheal breathing tubes. There is some ability to regenerate lost limbs, or bits of limbs, which can be replaced at a moult.

The number of moults through which Coleoptera larvae must pass before a final transformation into the pupa is not uniform through the order. Although between three and five instars are common development schedules, instar numbers can range from two (some Histeridae) to a hugely variable 10–24 (*Tenebrio molitor*, Tenebrionidae).[8] Although each individual species will have a relatively fixed number of instars, there can be considerable variation, even within closely related populations of the same species. The hide beetles, Dermestidae, are well known for highly inconsistent instar numbers (precise instar numbers can easily be counted in laboratory cultures). It seems that instar plasticity here allows larvae to continue developing during unfavourable conditions, when moults occur even though little or no feeding or growth has taken place; larval sizes remain the same, or actually decrease. Additional moults, without growth, can be induced experimentally by administration of juvenile hormone towards the end of what might be an otherwise normal larvahood. The regular shedding of the integument may be a necessary developmental imperative; where the irreversible curing of sclerotisation continues, whether the larva is feeding or not, physiological stasis may not be an option – perhaps rusty armour might be a useful analogy here? However and why ever instar numbers deviate from a norm, extreme variability appears to be one of the adaptations that has allowed some domestic pest species (e.g. Dermestidae, Anobiidae and Tenebrionidae) to succeed in harsh, dry, low-nutrient environments, where they are able to eke out a meagre living until the times of plenty return. A useful review of larval growth and instar numbers, covering all insects, not just beetles, is given by Esperk *et al.* (2007).

Larval integument stiffness can affect instar numbers, with the soft, folded, more flexible skin of scarabaeid larvae (generally needing only three instars) able to expand much more between moults than the tough, constraining cylindrical

---

8 A strange exception to this occurs in the Leptodirini, a cave-adapted tribe of the Leiodidae, where the non-feeding larva obtains all its nourishment in a uterus-like organ inside the adult female beetle's abdomen until large enough to be born. It has but a single instar, which does not feed, or moult, but which soon pupates to change into an adult.

segments in the Elateridae (in which 10–14 instars are regularly recorded). Larvae feeding internally inside their foodstuff, or boring through it, are often softer, wrinkled and more maggoty-grubby than the rather heavily sclerotised campodeiform or onisciform larvae that have to push about through the root thatch or against other obstacles, but all beetle larvae have toughened heads, to house the jaws and anchor the mandibular musculature. Measuring the width of the well-sclerotised, inflexible head capsule is a useful determinant of instar number, with each incremental jump approximating to about 1.5 times the previous head width – known as Dyar's law (Dyar, 1890).

Away from precise rearing experiments, it is sometimes difficult to gauge a beetle instar number, other than by approximation to some final estimate of a fully grown larva. Leaving aside the obviously extreme form of triungulins, newly emerged from the average beetle egg, the first instar (called the *larvule* in French, a word surely worthy of adoption into English) is often slim and ill-defined; almost inevitably, larval illustrations in identification guides focus on the later instars, usually the largest and most distinctive forms.

The field coleopterist seldom comes across a beetle pupa;[9] it seems reasonable to expect a final instar larva to take itself off into even more secretive corners to pass this highly vulnerable stage. Perhaps the most frequently found are ladybird pupae, attached to leaves or tree trunks. These closely resemble hunched, hardened versions of the larvae, which indeed they are. Having attached itself to the leaf surface by a knot of proteinaceous silk (the cremaster), the larva bows its back and its body becomes stiff; this is the prepupa stage, effectively the last larval instar preparing for its final moult. Sure enough, the skin down the back of the prepupa splits and is shuffled down, much like someone removing overalls until they are scrunched up on the floor. In most species, the final larval skin is concertinaed up to a mere crumpled belt at the tail end, but in the Chilocorinae (in Britain, three black species with red spots), the final larval skin remains as a split inflated shell around the pupa (Fig. 55). The plump pupa is now revealed, with a vague outline of wing buds exposed, and legs, antennae, compound eyes and other fully adult features just about visible, although covered in the outer layer of the crisp pupal shell, as if wrapped in a layer of pupal cling film. This is the obtect pupa that characterises some beetle families, notably the Chrysomelidae and Staphylinidae.

Another pupa that is sometimes uncovered is that of a longhorn beetle, *Rhagium* or *Leptura*, under the bark of a rotten log or tree stump. This time, the

---

9 The word pupa is from the Latin for 'doll' (or 'little girl'), presumably for its crisp, hard, immobile but accurately mannequin-like resemblance to the final adult.

FIG 55. Pupa of *Exochomus quadripustulatus* (Coccinellidae, subfamily Chilocorinae), in which the final larval skin is not completely shed, but remains as a split shroud over the chrysalis within.

legs and antennae are free of the pupal body, and the insect looks more like a pale, shrivelled, mummified version of the adult beetle. This is the exarate pupa, characteristic of most beetle families. A more or less standard feature of many exarate pupae is a series of bristle brushes arranged down the back. Since an exarate pupa is often curled up, resting on its back in the hollowed pupal chamber, these tufts serve to keep it up and away from the surface of the chamber, preventing damage from moisture or mould.

A curious anatomical feature occurs in some beetle pupae, notably the Coccinellidae and Dermestidae; what have been termed gin-traps appear down the back of the abdominal segments. Sharply raised ridges bordering the tergites, with gaping clefts between them, are held open like a series of, usually four, mouths at rest, but if the pupa is tickled with a bristle, the abdomen flexes, snapping shut the ridged trap edges together in a rapid biting movement. When this was first observed (Hinton, 1946), it was suggested that it was a defence mechanism against predators or parasitoids. Little work has been done on examining this, but Eisner & Eisner (1992) tested pupae of the North American ladybird *Cycloneda sanguinea* using bristles and live fire ants (*Solenopsis invicta*), and managed to get scanning electron micrographs of the gins closed tight around the ants' probing antennae.

FOODSTUFFS — EVERYTHING AND ANYTHING

Beetle diversity is reflected in beetle (mainly larval) foodstuffs. There seems to be virtually no natural product on the planet that is not eaten by them. They also show the greatest diversity of appetites, from omnivores that will scavenge almost any organic material, to the highest food specificity imaginable. The cellar beetle (*Blaps mucronata*, Tenebrionidae) will eat anything from stored grain to carrion, but the rare *Meligethes atramentarius* (Nitidulidae) will feed only in the flowers of yellow archangel (*Lamiastrum galeobdolon*) in ancient woodlands.

Much to the consternation of farmers, gardeners and foresters, many beetles eat plants; indeed, there is barely a plant anywhere in the world that is not eaten by some beetle or other. Phytophagy, the eating of plants, has allowed some groups of beetles – notably Chrysomelidae (leaf beetles), Cerambycidae (longhorns) and Curculionoidea (weevils) – to become hyperdiverse. Arguably, there is a link between plant diversity and insect diversity, as the two kingdoms have evolved either in a battle as the devoured attempt to outmanoeuvre their devourers, or through the mutual benefits of nectar production and pollination. About half the world's beetle species are phytophagous (Farrell, 1998), and this link between plant diversity and beetle diversity has been used to try to estimate just how many insect species we have here on planet Earth (see p. 327).

Fairly obviously, leaf beetles eat leaves, and many well-known pest species have acquired common names because of the extensive damage they do to crops or prize blooms. The asparagus beetle (*Crioceris asparagi*), with its Mondrianesque juxtaposition of red border, cream squares and inky blue-black outlining, feeds

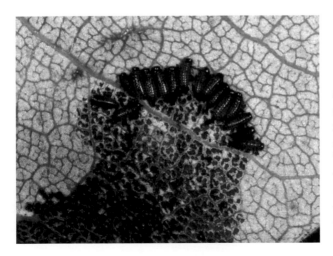

FIG 56. Safety in numbers: larvae of many newly hatched chrysomelid leaf beetles (*Pyrrhalta viburni* here) feed gregariously, grazing the leaf surface together rather than individually.

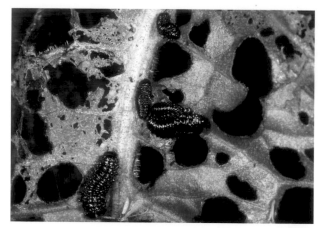

FIG 57. By the time they are nearly fully grown, beetle larvae can do serious damage to ornamental plants and farm crops. Those of the viburnum leaf beetle (*Pyrrhalta viburni*) shred the leaves of their host plant, much to the disconsolation of gardeners.

FIG 58. Not all ladybirds are predators – larvae of the 24-spot ladybird (*Subcoccinella vigintiquattuorpunctata*) graze the leaf surface of false oat-grass (*Arrhenatherum elatius*), leaving characteristic zigzagging scars behind them.

on asparagus (*Asparagus officinalis*) fronds, or rather its pale grey eruciform grub does (Fig. 59). Likewise, the rosemary beetle (*Chrysolina americana*) chews on rosemary (*Rosmarinus officinalis*). In a strange departure from orthodox anatomy, larvae of the lily beetle (*Lilioceris lilii*, feeding on lilies, obviously) are still vaguely eruciform, but have the anus situated halfway down the back. Through this they cover themselves in their own copious semi-liquid red/brown faeces (Fig. 60), presumably as a deterrent to predators, either through disguise or distaste. The flattened green larvae of tortoise beetles (*Cassida*) use a similar but less messy technique, attaching dried excrement to the faecal fork at their tail, to accumulate a nodule or tangle of dark frass under which to hide (Fig. 61). British species mostly raise a blob-like parasol, but in the southern United States, the palmetto tortoise beetle (*Hemisphaerota cyanea*) collects a lifetime's thatch of extruded excrement over its back, and looks more like a disembowelled ball of string than a living creature (Eisner & Eisner 2000).

FIG 59. The caterpillar-like eruciform larvae (inset) of *Crioceris asparagi* do enough damage to asparagus (*Asparagus officinalis*) crops to give this pretty and distinctive species its common name of asparagus beetle.

FIG 60. The slug-like
larvae of the lily beetle
(*Lilioceris lilii*) are all but
invisible under their
shrouds of glistening
excrement.

Alternatively, larvae may burrow through the leaf, leaving upper and lower
epidermis layers intact, but excavating a pale, air-filled squiggle or blister in
their wake. Blotch mines created by the caterpillar-like grubs of the tiny squat
jewel beetle *Trachys scrobiculatus* on ground-ivy (*Glechoma hederacea*) are easier to
find than the secretive seed-mimicking adult insects. The vermiform maggots
of the hopping weevils *Isochnus* and *Orchestes* are also leaf-miners, in a variety of
foodplants, mostly trees. Food-plant fidelity, and the precise patterns of mine
architecture, can very often be used to identify the species without either larva
or adult being present, but care should be exercised. What were once thought
to be the distinctive blister mines of *Orchestes erythropus* (a beetle then unknown
in Britain) in leaves of its unique foodplant, holm oak (*Quercus ilex*), were
later suggested to be simply *O. hortorum*, which attacks numerous native and
introduced oak species (Thompson, 1994).

Larvae that feed inside a leaf, rather than openly on its surface, have the
advantage of remaining hidden from potential predators and parasitoids, but
only very small larvae can burrow between the sandwiching epidermal layers.
Larger larvae can overcome this, though, if their parents are able to construct
a leaf home, in which they can feed. The birch leaf-roller (*Deporaus betulae*,
Rhynchitidae) lays an egg on a leaf, then chews sinuous cuts at either side from
the leaf edges but stopping at the mid-rib, causing the leaf blade to curl over into
an untidy roll. The short, squat, eruciform larva can develop hidden from view,
feeding on the plant material, which although damaged is still supplied with
some nutrients through the central uncut vein. Similar leaf-cutting by the oak
leaf-roller (*Attelabus nitens*) and hazel leaf-roller (*Apoderus coryli*) (both Attelabidae)
creates slightly tidier packages resembling short, squat cigar stubs still attached
to their respective foodplants (Fig. 62). The larva pupates in the roll, which
eventually falls with all the other leaves in autumn.

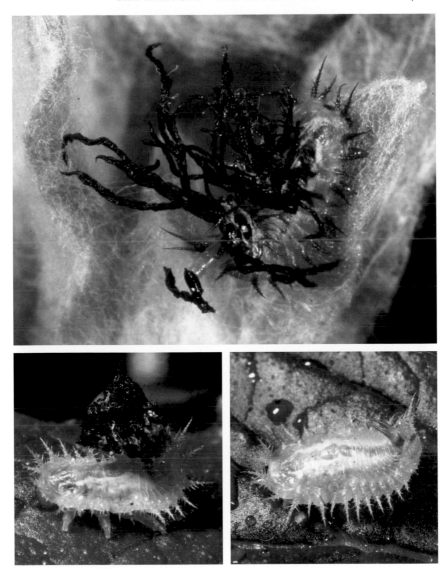

FIG 61. The well-disguised spiny larva of the red tortoise beetle *Cassida murraea* (top) on fleabane (*Pulicaria dysenterica*) is a mass of tangled black spires — strands of frass with which the larva has decorated its body. Meanwhile the larva of *C. viridis* (above left) on water mint (*Mentha aquatica*) uses frass and moulted larval skins to build a knob-like parasol. If this is removed by an inquisitive but inconsiderate coleopterist, the larva is revealed as rather soft and vulnerable (above right).

FIG 62. The bizarre and handsome hazel leaf-roller weevil (*Apoderus coryli*), and the neat leaf roll in which it has laid an egg.

Other portions of the vegetation are eaten too, and the weevils (families Apionidae, Curculionidae, etc.) are prime herbivores the world over, with species feeding in all parts of living (and dead) plants. Using its long snout, with tiny jaws at the tip, a female drills a hole into the foodplant and lays an egg inside; by utilising different parts of the same herbage, several beetle species can occupy the same plant without competition. In Britain, mallows are host to *Malvapion malvae* in the fruits, *Aspidapion radiolus*, *A. soror* and *A. aeneum* in the stems, and *Pseudapion rufirostre* in the seedpods. The length of the female weevil's snout is often an indicator of how deep into the plant tissue the grub will develop. On closely related hollyhock plants (*Alcea rosea*), the stem-feeding *Aspidapion* species are still able to cut into the tall, stout stalks for their grubs, but the huge flower buds are now drilled into by *Rhopalapion longirostre*, the female of which has a nearly straight rostrum that is almost as long as the rest of her body.

Together with the ovules, styles, pollen masses and other structures that develop at one of the most important growing points on the plant, petals represent a highly nutritious target for many groups, often lumped together in the vague and unhelpful term 'flower' or 'pollen' beetles. Untold millions of *Meligethes* (Nitidulidae) and *Olibrus* (Phalacridae) occur in wildflowers, and in those of cultivated crops like rape (*Brassica napus*). The apple blossom weevil (*Anthonomus pomorum*) was once a scourge of commercial apple and pear orchards, the larvae developing in the flower buds and preventing them from opening or being pollinated, with consequent loss from missing fruit later in the year. Insecticidal spraying has now reduced this beetle to the point of being uncommon in Britain. Luckily, the common *A. rubi*, sometimes an equal menace to strawberry and raspberry farmers, has a natural reservoir in bramble (*Rubus fruticosus*) flowers, so is common and widespread.

Seeds and pods are also invaded. The aptly named seed beetles (*Bruchus, Bruchidius*, etc., Chrysomelidae) lay their eggs in the flowers, or nascent pods, of mainly legumes, but the grubs really only start to feed when the seeds within the pods are beginning to mature. They hollow out a seed to leave an empty shell, and emerge as adult beetles when the pods have fallen or have split open. The human habit of storing food for later consumption has brought bruchids (together with many other beetles) into sharp focus. Harvested beans, peas and lentils are frequently contaminated by the larvae, pupae or adult beetles. Similar destructive infestations occur with *Sitophilus* weevils in stored grains of wheat (*S. granarius*), rice (*S. oryzae*) or maize (*S. zeamis*).

The acorn weevils *Curculio glandium* and *Curculio venosus* develop, not surprisingly, in acorns – the adult female has a terrifically long, thin snout to dig deep into the huge seed. A single egg gives rise to a single grub, which feeds until the acorn drops; it then chews a circular exit hole and pupates in the soil. The similar *Curculio nucum* develops in a comparable way in hazel (*Corylus avellana*) nuts. In an interesting deviation from this seed-based life history, larvae of the very closely related *Curculio villosus* develop in oak apples – large fleshy, vegetative growths caused by the presence of eggs, then larvae, of the gall wasp *Biorhiza pallida* in the leaf buds of oak trees. But whereas acorns (and hazel nuts) are large enough for just a single *Curculio* larva, oak apples can contain as many as four to six weevil grubs, together with sometimes hundreds of the original tiny gall wasps, and various other parasitoids, inquilines and lodgers.

Stems, stalks and twigs are bored into by a succession of beetle grubs. Several *Thryogenes* species (Erirhinidae) develop in waterside reeds and rushes, while the chunky *Grypus equiseti* (Erirhinidae) bores into the thick stems of horsetails – quite a sensible strategy given the narrow, tough, needle-like quality of the leaves. Several *Mordellistena* (Mordellidae) larvae feed in plant stems, causing the plant to create a swelling (a gall) around them. Such gall-making is rare in beetles, and seems to be more connected to physical damage caused by the larvae feeding inside the plant rather than by the complex cocktails of growth-usurping chemicals distinctive of galls caused by, say, Hymenoptera (Cynipidae) or Diptera (Cecidomyiidae). Larvae of the reed leaf beetles (*Donacia*) live in various aquatic plants, as well as reeds, often beneath the waterline and obtaining oxygen from the airways in their hosts' stems and roots. The common 'flower' beetles *Oedemera nobilis* and *O. lurida* occur in profusion on flowers, but their larvae develop in plant stems.

The distinction between roots and stems is often hazy in the thick thatch of herb growth, so many borers choose either location, based on stem diameter and moisture content, rather than botanical diagnosis. Longhorns (Cerambycidae) like *Phytoecia cylindrica* and *Agapanthia villosoviridescens* live in the stems of umbels and thistles, respectively; again, the adults are often found on flowers. Many click beetles (e.g. *Agriotes*, Elateridae), chafers (*Melolontha*, *Cetonia*, etc., Scarabaeidae) and weevils (*Otiorhynchus*, *Cathormiocerus*, *Strophosoma*, etc., Curculionidae) are also root-feeders, living in the soil and grazing roots and rootlets just as other beetle larvae might graze on leaves above ground. A similar situation occurs with algal growth in streams, so the elateriform or sometimes onisciform larvae of riffle beetles (Elmidae) simply graze the green encrustations on the underside of stones in the water flow.

Many beetles develop in tree trunks, branches, logs and stumps, feeding in or under the bark or deep in the heartwood. Weevils like *Pissodes* and *Magdalis*, feeding under the bark of small branches or boring in terminal twigs, respectively, have simply adapted typical weevil plant-feeding to a slightly

FIG 63. A typical wood-boring weevil grub (Scolytinae), leaving frass-filled burrows under the bark.

FIG 64. Muddled burrows of *Hylesinus varius* under ash bark. The axis of the trunk runs left to right.

woodier level. Bark beetles (Curculionidae, subfamily Scolytinae) are perhaps the extreme in this family. Typically, a female chews her way into the bark and then excavates a long burrow at the junction of the heartwood and cambium layer. Eggs are laid at regular points along this maternal gallery. Hatching grubs then chew away from the birth gallery, at right angles to it. Depending on the beetle species, the maternal burrow may be horizontal across it (e.g. the ash borer, *Hylesinus varius*) with larvae then burrowing up and down the line of the bole (Fig. 64), or vertical, up the axis of the trunk (e.g. the elm bark beetles *Scolytus scolytus* and *S. multistriatus*), with larval tunnels radiating out horizontally left and right (Fig. 65). Later, when the tree is dead and the bark peeling off, the beautifully intricate engravings left by the mother beetle and her grub offspring are uncovered, and can often be identified to species level because of their distinctive

FIG 65. Vertical maternal burrow of one of the elm bark beetles (*Scolytus*), showing the radiating juvenile tunnels chewed outwards, from left and right, by the developing grubs.

orientation signatures. Under spruce or pine bark, the adult female of *Pityogenes bidentatus* makes a six-pronged star gallery and the radiating larval tunnels make an elaborate snowflake woodcut.

Several bark beetles in the families Platypodidae and Curculionidae (Scolytinae) are collectively known as ambrosia beetles, because they brew their own special food by encouraging fungal decay in the dead wood. As the beetles mine the phloem layer in the bark, they inoculate the plant tissue with symbiotic fungi, the spores of which they carry in small depressions on various parts of their bodies, called mycangia. In *Platypus cylindrus*, the spores are carried in prothoracic depressions in the female; in *Xyleborus*, they are held in paired organs near the mandibles. The fungus then attacks the host tree, and the beetles and their offspring feed on the fungal decay zones. In North America, the Asian *Xyleborus glabratus* has become a major pest of native redbay (*Persea borbonia*) and avocado trees (*Persea americana*), which are soon killed by the fungus transported by the beetle (Fraedrich *et al.*, 2008).

Across the world, many jewel beetles (Buprestidae), bark beetles (Scolytinae), longhorns (Cerambycidae) and other pinhole borers (Anobiinae, Platypodidae, Bostrichidae, Lyctidae) are well known for causing damage to forestry trees and the harvested timber they produce, or for spreading diseases that prematurely kill the trees. The domestic woodworm *Anobium punctatum* is notorious enough to have produced a common English name for the adult beetle from its wyrm-like (in this case actually scarabaeiform) larval stage. Its unwelcome chewings through construction timber, furniture and fittings can probably be found

FIG 66. Cascades of fresh sawdust down a tree trunk show where the bark beetle *Platypus cylindrus* is mounting a concerted attack. These are the activities of females digging egg-laying burrows. Downham Woodland Walk, Bromley, 25 August 1999.

in every street in the country. Similar destruction to buildings and property, especially old and prestigious edifices, has been wrought by the deathwatch beetle (*Xestobium rufovillosum*). The large *Xestobium* holes (3–4 mm in diameter, compared to 1 mm for *Anobium*) can commonly be found in the exposed heartwood of broken old oak, elm and willow trees growing in the countryside; no wonder the larvae can do so much damage in construction timbers. Unlike *Anobium*, though, the deathwatch does not readily fly into houses, infesting new bits of furniture; indeed, it has often – wrongly – been reported as being incapable of flight. Instead, colonies of the beetle were brought into the buildings when they were constructed, the larvae secretly gnawing away in small pockets of fungal decay inside the felled and seasoned timber, which was then used for the large supporting joists and beams. Deathwatch growth may be slow – up to 12 years as a larva are quoted (Hickin, 1975) – but it is insidious. Through ensuing centuries, generations of the beetle continue to gnaw away at the wood, eventually creating sponge-like cavities and hollow voids enough to significantly endanger the structural integrity of the supportive beams, to the point of the building's collapse. *Anobium* and *Xestobium* develop in dry, seasoned timber, but even regularly inundated wharf and groyne timbers are prone to attack from larvae of *Nacerdes melanura* (Oedemeridae) and *Pselactus spadix* (Curculionidae).

Plenty of other beetle groups also eat the wood, particularly as fungal rot and decay set in. Arguably, the sound timber of growing trees is a tough diet, comprising mainly celluloses and lignins, but with fungal mycelia actively digesting these structural chemicals, the job is made easier. A whole host of beetles can then take advantage of the tastier conditions. The distinctive flat larvae of *Pyrochroa*, the cardinal beetles, are often found under loose bark, usually in a sinuous or circular void that they have hollowed out, and which is littered or bounded by a wall of round frass particles.

At this fungal level, it is unclear what many beetle larvae (e.g. those of Cucujidae, Bitomidae, Melandryidae and Mordellidae) are feeding on – fungal hyphae, fungal wood-digestion products, or both. At least within the large fruiting body of a bracket fungus, puffball or subterranean truffle, it is fairly clear that the larvae of Erotylidae, Ciidae, Endomychidae and Leiodidae, are devouring the fruiting body itself. Fungi and fungal associations are poorly studied, but their relevance is becoming increasingly clear. The widespread and often abundant small black and brown longhorn *Pseudovadonia* (*Leptura*) *livida* is a frequent flower visitor in spring (Fig. 67), and although widely reported as breeding in rotten logs, its larvae are now known to feed in the soil, on humus particles and roots infected with the hyphae of *Marasmius oreades*, a common grassland toadstool that forms fairy rings (Svacha & Danilevsky, 1986).

FIG 67. The larvae of *Pseudovadonia (Leptura) livida* feed in the mycelia of toadstool-forming fungi in the soil and leaf-litter layer, earning it the common name fairy-ring longhorn beetle.

Dead and decaying organic matter is a lovely catch-all concept, frequently used for describing the habits of beetles we don't really understand very well. A pile of leaf litter, a compost heap and manure piles all have their scavengers, from tiny latridiids, probably eating moulds, to dumbledors (Geotrupidae) and rove beetles, feeding in the rancid bacterial decay. Some species, though, are more specific in their decaying organic matter requirements. At one end of the scale, leaves recently fallen from the plant are food for the larvae of *Cryptocephalus* leaf beetles, which retain a certain amount of host-specificity even though the leaves are dead. These curious larvae create a bag-like case around their bodies, using their own excrement, with just head and front legs poking out. This process begins as the eggs are laid, the adult female beetle covering each with a layer of waxy secretion and her own frass – a process known as scatoshelling. Using her rather long front two pairs of legs to grip a leaf or stem, she manipulates the egg in a dimple recess at the tip of her abdomen and uses her hind legs to mould the coating, before letting the egg drop to the ground. As they grow, the larvae increase the size of their frass capsule to form a broad rimmed vessel, earning them the name pot beetles.[10] Again, adults can sometimes be beaten from host

---

10 There may be something that facilitates case creation in detritus-feeders, since this same behaviour is adopted by tineid and psychid moth larvae and terrestrial caddisflies.

trees, or found sitting in flowers, but the larvae occupy a wholly different part of the landscape.

Mammalian dung, sometimes little more than processed plant material, is used by a wide variety of scavengers, but especially the scarabaeoid dung beetles – *Aphodius, Onthophagus, Copris* and Geotrupidae. The fat grey or white grubs are a familiar feature in mouldering pats littering a grazing meadow. Although fresh dung attracts plenty of adult beetles, they have all gone a week or two later, leaving their larvae to feed peacefully, and removing themselves from competing with their own offspring. Dung beetle grubs certainly break up and help break down the dung, but they are probably taking as much nutrition from the bacterial soup as they are from the remnants of the plant material.

The dung beetles are a particularly well-studied group of insects, not just for their helpful recycling of smelly droppings, but for their nesting and brood-care behaviours. Excellent reviews of dung beetle (adult and larval) ecology are presented by Hanski & Cambefort (1991), Scholtz *et al.* (2009) and Simmons & Ridsdill-Smith (2011a). Although we have none of the exotic sacred dung-roller scarabs in Britain, many species (*Onthophagus, Copris, Typhaeus,* etc.) dig tunnels into the soil and relocate boluses of dung down into roughly accumulated masses, or separately constructed cells, in which to lay their eggs. In temperate northern Europe, pressures from large beetle numbers and rapid drying out of the damp droppings are much reduced, so it is easy to assume that dung is always plentiful, but elsewhere in the world there is a highly competitive scramble for what is, in essence, an erratic and ephemeral food source. By carving off balls or nuggets of dung and hiding them out of sight, the adult beetles are ensuring the survival of their relatively immobile offspring. In non-British studies of *Copris lunaris* (Klemperer, 1982a,b), a female may lay only 10 eggs in her lifetime – perhaps five in a nest of five dung balls in each season of her short two-year life. Such is competition from interlopers and thieves that she will stay some time and guard the brood. The grubs devour the food stores from the inside, each having (hopefully) been supplied with enough to see it through to adulthood.

Carrion is another fairly particular type of organic decay, and attracts a more particular group of beetles. The dark, armoured, highly active onisciform larvae of various Silphidae are easily disturbed from small animal carcasses beaten over a plastic sheet. Less easily found are the paler, more maggoty larvae of *Nicrophorus*, the burying beetles. These distinctive black or black-and-orange beetles get their name from their habit of interring small animal corpses. Usually, a male and female work together to undermine the soil from beneath the body, piling it up at the sides until the excavated hole consumes the carrion and the spoil covers it over. Having excluded competitors, they mate and lay eggs, and then start to feed the hatchling larvae with decaying flesh that they have chewed and regurgitated. Eventually, the larvae are big enough to feed themselves, by which time the carrion is highly putrescent and almost semi-liquid in the beetles' underground bunker.

Later in the decay timeline of a corpse, the dry sinew and bone stage is almost abandoned except for hide beetles – *Anthrenus* and *Dermestes* (Dermestidae). The tufted, hairy eruciform/elateriform larvae are not so much feeding in organic decay as organic mummification, and are particularly adapted to eating the dry keratin of skin, fur and feather, which is about all that remains. This ability to survive on dry food has enabled hide beetles to invade the dry sanctum of the human home, and *Dermestes* are most often known as larder beetles. 'Lard' is the clue giveaway syllable here, because unlike the pantry, where bread (Latin *panis*) products were stored, the larder was for dried and cured meats like bacon and ham, on which larder beetle larvae could develop just as well as on shrivelled fox carcasses. *Anthrenus*, meanwhile, uses its hair- and fur-digesting abilities to attack the similar animal fibres found in wool and silk carpets; almost universally known as carpet beetles nowadays, the destructive larva, an animated knot of

FIG 68. *Nicrophorus vespilloides*, one of the burying or sexton beetles. This specimen was not photographed on carrion, but artfully posed on the remains of a lamb chop from dinner. It's called artistic licence.

fawn bristle and beige fluff, is sometimes rather ironically given the friendly sounding name woolly bear. This is one group of beetles that is despised, even by coleopterists; not being content with the Axminster underfoot, larvae of *Anthrenus verbasci* are quite able to complete their development devouring dead insects, and this beetle is now that scourge of insect collections and other museum displays – the museum beetle.[11]

A behavioural link, of a sort, occurs in *Ctesias serra* (Fig. 69). A very close relative of *Anthrenus*, *Ctesias* is a carrion beetle on a more diminutive scale, sneaking mouthfuls of the dead insects caught by *Amaurobius* spiders, which make their untidy trip-wire webs under the loose bark of old trees. Spiderwebs are likely to be the natural reservoir of several dermestids known mostly from domestic houses or stored products (Hinton, 1943). Even more bristly than woolly bears, the *Ctesias* larvae are able to avoid being eaten by their arachnid hosts. Dermestid larval hairs are, anyway, tipped by detachable spear-shaped heads, which break off

FIG 69. Scavenging on dead insect remains in the messy spiderwebs under loose tree bark, the larva of *Ctesias serra* presents an unpalatable mouthful to any spider.

---

11 Although the seemingly more appropriately named *Anthrenus museorum* can also be a pest in museum collections, it breeds mainly in the wild in Britain, occurring in bird, mammal, bee and wasp nests, where it eats moulted fur and feathers, and dead young.

FIG 70. A close relative of *Anthrenus* and *Ctesias*, *Trinodes hirtus* also feeds on dead insect remains in spiderwebs under old bark, but remains highly restricted to ancient woods in central England. Richmond Park, 26 May 1985.

in the mouths of any would-be attackers, but *Ctesias* needs an extra tactic, living, as it does, at the very dining table of its hosts. By rapidly vibrating its long tail tufts to an invisible blur, *Ctesias* seems to confuse the spiders, which are, after all, hunting by touch and feel on the silk strands of their webs in the darkness. This lovely creature, a veritable animated boot brush, is easy to find under the bark of large old trees, including the street limes near my house, and simple to rear on a diet of dead insects found on windowsills around the house.

GOING ON THE ATTACK — PREDATORY LARVAE

Predation has evolved in almost every insect order, and beetles' biting jaws (in both adults and larvae) have made carnivory a simple transition in the Coleoptera. There is a tendency for predatory larvae to have longer, more slender and more curved scimitar-like jaws, sometimes with a strong internal tooth, the retinaculum, near the middle of the cutting edge. There is a strong advantage in being a carnivore – animal bodies are high in protein, the lack of which is a limiting factor to satisfactory growth in an organism. There appears to be no single evolutionary switch to hunting, and predatory species occur in many different beetle families.

Perhaps the best-known predatory group are the carabid ground beetles, which have large-eyed, large-headed, large-jawed, long-legged active forms as both adults and typically campodeiform larvae. As with most predators, the answer to the question 'What do they eat?' is in reality 'Whatever they can get', so many predatory beetles and their larvae are generalists, snapping at any small invertebrate that inadvertently cuts across their path. There are, however, some specialist techniques.

Tiger beetles, *Cicindela* species, are well named for their sharp, multi-toothed mandibles and ferocious attacking style; the adults dart rapidly about on very long legs, snapping at low-flying insects by taking off like miniature jump jets. Their larvae are more sedentary, but equally aggressive, using a sit-and-wait gin-trap of unequalled ferocity. On the plastic sheeting, the oddly skewed larva looks broken and disjointed, with its head bent forwards at right angles and its jaws twisted backwards and upwards, but if it had been left in its vertical, chimney-like burrow in the sandy soil, its structure would make perfect sense. Lodged in its shaft-like tunnel, using paired bristly, warty growths on the back of its fifth abdominal segment to wedge it securely in place, the flat, flip-top head is now flush with the soil surface and presents the jaws opened to left and right, ready to snap closed on anything foolish enough to walk over it. The victim, often an ant or ground bug, is dragged underground and finished off, before the trap is reset to wait for another meal. Tropical tiger beetles, subfamily Collyrinae, use a similar gin-trap method, but dig vertical burrows in tree trunks and logs.

The active, typically campodeiform first-instar larvae of *Lebia* ground beetles need only their mobile hunting form until they find their much larger prey item, a plump chrysomelid leaf beetle larva or pupa. Once their attack begins, they moult to a soft eruciform or vermiform grub and take to feeding externally on the remains of their host; they need just the one. Similar to oil beetle hypermetamorphosis (see p. 74), their behaviour is less active because they have switched to become ectoparasitoids rather than predators. The widespread, but scarce bombardier beetle *Brachinus crepitans* is thought to have a similar life cycle, probably attacking pupae of other ground beetles in the genus *Amara*.

Staphylinid rove beetle and dytiscid water beetle larvae, also typically campodeiform, are significant predators in the undergrowth, both terrestrial and aquatic. On a personal note, a maggot of one of the diving beetles, probably a *Cybister* or *Megadytes*, is the only insect larva ever to have drawn my blood, when I incautiously picked one from the dip-net in a small stony stream in Costa Rica many years ago. It was huge, but even the more modest British *Dytiscus* larvae are capable of giving a bit of a nip.

Other predators are less obvious in their appearance. The larva of the common glow-worm (*Lampyris noctiluca*) looks to be a slow, cumbersome fusiform/onisciform creature, its small head, presumably with small jaws, tucked under the first thoracic segment, and yet it is a significant predator of slugs and snails. Admittedly, these are hardly able to outrun it, but their large size and copious sticky mucus might seem to make them formidable protagonists in a fight. *Lampyris* larvae do not need to struggle with their prey, however; they simply have to bite, injecting a digesting toxin that kills the snail, thus preventing it from smearing its attacker with defensive goo. It is often reported that *Lampyris* larvae detect and follow the slime trails of their intended prey. Snails are also attacked by the delightfully tufted eruciform larva of *Drilus flavescens*, but it is so rarely found that its full life history was not accurately reported until Crawshay (1903) related his various breeding experiments. Feeding is difficult to observe because the larva enters the shell of the snail; indeed, videos posted online show the larva wedged tight into the shell of the common garden snail (*Cornu aspersum*) as the prey victim itself tries to retreat. Unlike many predator–prey interactions, the snail is much larger than the *Drilus* larva. The sharp, pointed processes on the abdominal segments, each covered with a raucous brush of stout spines (burnt sienna in colour, according to Mr Crawshay), may serve to prevent the larva from becoming mired in the huge mollusc's thick defensive slime. Despite the seemingly easy availability of large protein-rich prey, *Drilus* spends two to three years as a larva before metamorphosis.

Equally unlikely looking as predators are the soft, velvety larvae of soldier beetles (Cantharidae) living in rotten wood mould and at the soil–root-thatch boundary. Here, they eat the tiny invertebrates – springtails, bristletails, insect larvae, nematode worms and other small fry – that probably form the basic diet of most generalist predators like carabids and rove beetles living in the leaf litter. That these typically eruciform, caterpillar-like cantharid larvae are dangerous predators ought not to come as a surprise, because they are closely related to, and look superficially like, organisms that are very familiar as predators – ladybird larvae.

FIG 71. The velvety larva of a cantharid, a common predatory denizen of the leaf-litter and soil layer, or under loose fungoid bark.

FIG 72. Larva of the seven-spot ladybird (*Coccinella septempunctata*), perhaps the most familiar predatory beetle larva, at least to some gardeners.

Although sometimes overlooked by gardeners, who know and celebrate the adult beetles only too well, immature ladybirds are some of the most easily found British beetle larvae. They shuffle about in the open, along with the adults, crawling on leaves or tree trunks, and can be observed wreaking destruction through the aphid colonies they inhabit. The first meal of the first instar is the eggshell from which it has just emerged, along with any other eggs and embryos that have not developed properly, or are just late in hatching. Actually, any other insect eggs nearby are also fair game. The second meal is the hardest, though, as the tiny larvule often clambers aboard the aphid, puncturing its victim's body with stout but pointed jaws and sucking out the internal fluids. As it grows (through four larval instars in Coccinellidae), the predator:prey size ratio decreases, until the final instar is a veritable monster in the greenfly midst. Thanks to the pretty patterns, familiar forms and 'helpful' horticultural assistance of ladybird larvae, we know a little about their appetites.

In the medium-sized *Hippodamia variegata*, during the 17 or so days it spends as a larva, roughly 100 aphids are eaten (Farhadi *et al.*, 2011). In the much larger (and decidedly more voracious) harlequin ladybird (*Harmonia axyridis*), up to 370 aphids are consumed (the modal number being 23.3 per day) during approximately 10 larval days before pupation (Koch, 2003). The Asian *Harmonia axyridis* has been deliberately released all around the world as a biocontrol agent against alien aphid pests that home-grown ladybirds do not tackle. However, in North America (where it has been regularly released since 1916) it was accused of being too vigorous – outcompeting native ladybirds to the point of their local extinction, and even eating their larvae. When it arrived in England, seemingly under its own steam, in 2003, similar expectations led to worrying tabloid headlines and genuine fears among British entomologists. *Halmonia axyridis* fourth instars are huge (15–18 mm long), about twice as large as those of the seven-spot ladybird (*Coccinella septempunctata*), and distinctively black and orange. They have regularly been recorded eating other ladybird larvae, each other, hoverfly and lacewing larvae (themselves also aphid predators), moth eggs and the occasional unlucky moth caterpillar. But predators do not get things all their own way. Initial ladybird larval mortalities (at least in the native two-, seven- and 14-spots) varied from 20 to 80 per cent (Majerus, 1994), mostly depending on whether or not first-instar larvae could find an initial aphid meal during the first 35 hours after hatching.

FIG 73. What does a predator eat? Anything it can get – including its own kind. Cannibalism is rife in nature, as exemplified by this two-spot ladybird (*Adalia bipunctata*) larva eating a conspecific pupa. Deptford, 1 June 1998.

Some ladybird species have overcome the problem of prey location by adaption of their larvae to feed on moulds and mildews, rather than on aphids. Their ecologies are still interlinked, though, since the mildews on which they feed are possibly an evolutionary result of aphid feeding (Giorgi *et al.*, 2009). Since aphids drink plant sap, which is notoriously low in protein, they must process vast quantities of it; however, it is high in sugar and water content, so these are passed almost unaltered in their copious honey-coloured liquid excrement 'honeydew'. The sugary liquid, spilled onto the leaves where they feed, is soon colonised by mould spores, on which the mycophagous ladybird larvae feed. Two British species, the 24-spot ladybird (*Subcoccinella vigintiquattuorpunctata*) and bryony ladybird (*Henosepilachna argus*), have evolved full herbivory, grazing the leaf surfaces of false oat-grass (*Arrhenatherum elatius*) and white bryony (*Bryonia dioica*), respectively, and many ostensibly aphid predators are regularly observed supplementing their diets with mildew, honeydew or plant material.

Not all predatory beetle larvae are quite so obviously proactive among their prey, and consequently their feeding ecologies have not been well understood. Numerous 'dung' beetles (*Cercyon* and *Sphaeridium*) are probably feeding on the bacterial soup as adults, but their plump, wrinkled, eruciform rather than campodeiform, grey/brown larvae are predators, eating fly and other dung beetle larvae in the pat. These beetles, in the large family Hydrophilidae, have close relatives that live in water, with likewise scavenging adults but predatory larvae in the muddy edges of ponds and lakes. As noted earlier, the great silver water beetle (*Hydrophilus piceus*) has a huge larva, dirty grey and wrinkled, like a severed, mummified finger. Although the larva and adult are fully aquatic, the final instar must heave its way out of the water to find a nook on dry land in which to pupate, and sometimes alarms those who see it writhing menacingly across waterside paths.

Carnivory is still evolving, and there are occasional predatory species in what are otherwise fully phytophagous groups. In Britain, the small mottled weevils *Anthribus nebulosus* and *A. fasciatus* are predators of scale insects, while the rest of the family Anthribidae feed on fungi, dead wood or plants. Although most click beetles (Elateridae) feed in fungoid wood, many are not above taking a small creature that crosses their burrow. The common *Dalopius marginatus* seems to feed mostly on soil-dwelling invertebrates, and the rare *Elater ferrugineus* probably feeds on wood-mould insect larvae, including maybe those of the lesser stag beetle (*Dorcus parallelipipedus*), since its known European prey, the scarab beetle *Osmoderma eremita*, does not occur in Britain. In Brazil, a female of the small weevil *Ludovix fasciatus* seeks out *Cornops* grasshopper eggs laid in water hyacinth (*Eichhornia crassipes*) stems. She then sucks out some of the contents of the eggs for her own nutrition, and lays eggs so that her grubs can finish the job (Zwolfer & Bennett, 1969).

A FINAL MISCELLANY

Standard beetle textbooks like to discuss neat certainties – straightforward food choices, standard habitat requirements, fixed niche types and regular life cycles. But as any biologist knows, in nature anything is possible. It is normal to find burying beetles breeding in carrion, but there is always the odd time that you find a pair under a particularly meaty-ripe fox scat, and you have to wonder whether they ever realised their confusion, or indeed whether they were ever successful in burying it and raising their young. I suspect not, but such aberrations in behaviour are part of the flexibility that gives enough variation for species to conquer new niches, spread to new geographic zones and switch to enticing new food-choice menus. Adult beetles, mobile, active and motivated to disperse, can end up almost anywhere, so it is unusual occurrences of their larvae that sometimes give fascinating insights into how their life cycles have evolved, or how they might adapt in the future.

I remember being surprised when I found many thousands of larvae (and adults) of the tiny mould beetle *Enicmus brevicornis* (Latriidae) under dead

FIG 74. Beetles don't get everything their way. This rove beetle has been killed by an entomophagous fungus.

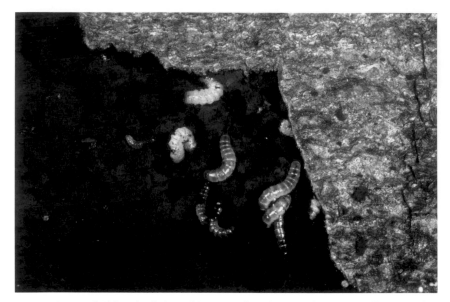

FIG 75. Larvae of *Diplocoelus fagi*, *Synchita separanda* and *Enicmus brevicornis* wallowing in the dense sooty spore mass of *Cryptostroma corticale* under dead sycamore (*Acer pseudoplatanus*) bark. One Tree Hill, Honor Oak, 19 September 1995.

sycamore (*Acer pseudoplatanus*) bark in a suburban London cemetery (Jones, 1993), when it was always claimed to be a species of ancient broadleaved woodlands, associated with rotten birches and beech (*Fagus sylvatica*). It was thriving in powdery black spore masses of *Cryptostroma corticale*, the fungus that causes sooty bark disease in maples. Since the 1990s, there has been an upsurge of sycamore deaths from this disease, exacerbated by water stress during hot, dry summers, and the beetle is now widespread on what is presumably a new foodstuff, after it was able to switch from some related, but much rarer, fungal infection in other broadleaved trees. Several other formerly scarce beech-specific beetles have also made the jump to sycamore victims of sooty bark disease, including *Diplocoelus fagi* (Biphyllidae), *Synchita separanda* (Colydiidae), *Salpingus planirostris* and *Salpingus ruficoillis* (both Salpingidae) (Fig. 75).

Some work has been done exploring whether changes in larval food intake can influence subsequent adult egg-laying sites, and therefore the nutritional choices of future offspring. If this did happen, it might explain how foodplant switches could occur. Evidence is patchy, but it seems that most adults go about their usual instinctive food selection, irrespective of what they were given as larvae. Nevertheless, foodplant switches do occur. Famously, the North American leaf beetle *Leptinotarsa decemlineata* munched away on buffalo-bur (*Solanum rostratum*), a spiny yellow-flowered nightshade, in its native Rocky Mountains for untold millennia, before making the jump to cultivated potato (*S. tuberosum*) plants in about 1840. The Colorado beetle has never looked back.

Beetles living in houses have made an ecological leap sometime in the last 30,000 to 50,000 years, since they must have been occupying some other habitat until early humans started to make permanent shelters. The cellar beetle (*Blaps mucronata*, Tenebrionidae), is a widespread scavenger, but despite its catholic food choices it is declining in Britain as cellars are no longer used for long-term food storage or have been converted to additional inhabitable rooms. I'm pleased that three of the houses I've lived in had this marvellous clunky animal in them; I suspect they were scraping out a living under the floors, where the occasional food morsel dropped through the cracks in the boards. Even here, fitted carpets, lino, vinyl and parquet finishes are shutting them off from sustenance. *Blaps* may not be native to Britain, but originally a Mediterranean cave species adapted to human food stores and then introduced here with trade sometime since the Roman occupation. It may well be on the way out now, as perhaps is its specialist predator *Sphodrus leucophthalmus* (Carabidae), which was last seen in 1979, in Cornwall, and may already be extinct in Britain. Both adults and typically campodeiform larvae of *Sphodrus* attack the elateriform *Blaps* larvae, but such specialism, either to a particular prey item or in a very specific niche, has meant that a tipping point arrived, where although the odd *Blaps* colony survives, it is too small or too far from the next to make *Blaps* predation a viable option for the doomed *Sphodrus*.

These species made an exploratory transition into the novel habitat of human dwelling places many thousands of years ago, and new ones continue to arrive. A small, narrow brown New Zealand weevil, *Euophryum confine*, is now the most widespread domestic 'woodworm' in Britain. Its larvae, which probably arrived by way of sodden wooden barrels and casks in the late nineteenth century, need dampness and the beginnings of fungal rot in the skirting boards, but they are now thriving in urban homes as well as Antipodean forests.

Human intervention has undoubtedly helped some species find new niches. The ridiculously named odd beetle (*Thylodrias contractus*), a Central Asian dermestid with 'odd' wingless females, might just replace the museum beetle *Anthrenus verbasci* as a pest in insect collections, and has also been found in wool and silk clothes and furnishings in houses. How it got from Outer Mongolia to the Natural History Museum in London or the Smithsonian in Washington is anybody's guess, but its tufty onisciform larvae are on every museum curator's watch list – even though they look cute, and can curl round into a ball.

Meanwhile, coleopterists should be prepared to find other odd beetles in odd places. Larvae of the cigarette beetle (*Lasioderma serricorne*) also eat cigars, it's just that the holes they leave in the white paper cigarette casings are more obvious. And if they cannot get tobacco, they will eat neat chilli powder, cayenne pepper or paprika in the kitchen spice drawer. This last instance is a good example of the extreme dryness that some beetle larvae can tolerate, the moisture content of the substrate being close to the 25 per cent minimum at which the beetle can survive (Howe, 1957); they can also survive in a partial vacuum (Finkelman *et al.*, 2006). The 250 females of the camphor shot-borer (*Cnestus mutilatus*, Curculionidae, subfamily Scolytinae) burrowing into plastic jerry cans full of two-stroke fuel in a Louisiana garage in 2011 were undoubtedly confused. They were supposedly attracted by volatile petrochemical aromas, which may have been reminiscent of the alcoholic scents they expected to find emanating from gently fermenting logs (Carlton & Bayless, 2011). This was a dead-end move for them, since those that successfully penetrated the 2.5 mm polypropylene were immediately poisoned by the petrol mix. On the other hand, a thriving Arizona colony of *Tribolium confusum* (Tenebrionidae) found breeding in several tonnes of confiscated cannabis at the local drug-enforcement office were simply adapting to the plentiful supply of mouldering seeds in the slowly decomposing bales. It turned out that the smugglers were rather slapdash when it came to weatherproofing their stash before it was shipped (Smith & Olson, 1982).

When I moved into my current house in south-east London in 1999, the many thousands of *Dermestes* larvae (and quite a few adults) I found under the rotting kitchen carpet were also taking advantage of a certain lapse in expected care and hygiene. Judging by the dubious stains from spilled slops on the once pink but now discoloured beige shagpile, they were probably eating the impregnated carpet itself rather than spilled food. A rich haul of four *Dermestes* species (*D. lardarius*, *D. maculatus*, *D. haemorrhoidalis* and *D. peruvianus*) and a rare bethylid parasitoid wasp (*Laelius microneurus*) showed this to have been a long-lived and mature colony. I still feel guilty about eradicating it.

CHAPTER 4

Special Features – Adult Quirks and Oddities

I F ONE WORD IS USED TO sum up the Coleoptera, that word must be diversity. The astonishing numerical variety of beetles is complemented by their sometimes perplexing life histories, barely credible behaviours, pretty colours and patterns, strange forms, odd shapes, bizarre ornaments and other structural peculiarities. Some are visible only down the microscope, while others have to be observed in life. Some, read of in books, seem hardly plausible, almost the stuff of science fiction. This chapter takes a look at a few of the oddball characters in the Coleoptera – structural, physiological and behavioural fancies that make beetles such an endearing and fanciful bunch.

FIG 76. *Scaphidium quadrimaculatum* shows the least elytral abbreviation in any British staphylinid.

WING-CASES – THE LONG AND THE SHORT OF IT

One of the most easily noted oddities is the shortening of beetle wing-cases. Although the standard norm may be for tight-fitting sheaths smoothly covering the abdomen and folded flight wings – the elytra, otherwise so distinctive to the Coleoptera – there are plenty of examples with shortened stubs. The rove beetles (Staphylinidae) are fine exemplars of this, with the majority of species (more than 1,100 in the British Isles), distinguished by having abbreviated elytra and a large portion of the abdomen exposed as the flexible hind body. The extent of elytral contraction can range from *Scaphidium quadrimaculatum* (subfamily Scaphidiinae), where little more than the tail-end two segments are revealed (Fig. 76), to *Micralymma marinum* (subfamily Omalinae), where the short, rounded elytra appear little more than epaulettes for the shoulders.[12]

An immediate advantage of this is to achieve even greater flexibility in the body, allowing staphs to push down under logs and stones, loose bark or rock crevices, into the grass thatch or soil, in dung, compost and other decaying organic matter, or indeed any tight corner where full-body-length elytra might be an impediment to movement. Most are still able to fly perfectly well, with their membranous flight wings just being tucked up a bit more concertina-tight underneath. That staphylinids fly well is obvious to any cycling coleopterist, who soon discovers that the 'fly' in your eye (or your mouth) is much more likely to be a tiny staph than any dipteran. In a nice example of convergent evolution, earwigs also have greatly abbreviated wing-sheaths and an exposed hind body. They, too, can insinuate themselves into tight crannies, and they, too, can fly.

Elsewhere in the Coleoptera, shortened wing-cases can sometimes confuse, with specimens of *Necrodes* and *Nicrophorus* (Silphidae), *Carpophilus* (Nitidulidae), *Ptinella* (Ptiliidae), *Malthinus* (Cantharidae) and *Meloe* (Meloidae) sometimes leading the novice coleopterist down the wrong identification pathway. It's difficult to know whether shortening of the elytra, or lengthening of the abdomen, is responsible for some of these cases.

Three rare British longhorns (Cerambycidae), *Molorchus minor*, *Glaphyra umbellatarum* and *Nathrius brevipennis*, stand out in this usually narrow, tapered or cylindrical-bodied family. Like many small species in this group, their larvae develop in dead plant stems, but the adults are frequent flower visitors and are remarkably wasp-like in their busy flight from bloom to bloom, an illusion enhanced by their short wing-cases.

---

12 Members of the subfamily Scydmaeninae are now considered to be minute staphylinids, but are outside this discussion since they have the elytra conventionally roundly and closely fitting to the abdomen, showing no abbreviation. They have previously been given separate family status.

SEXUAL DIMORPHISM

Wing-case reduction is taken to its extreme in the unusual females of *Drilus flavescens* (Drilidae) and the common glow-worm (*Lampyris noctiluca*, Lampyridae), which lack elytra completely. The male beetles look conventionally beetle-like, but so odd are these females that they are apt not to be recognised as beetles at all. As well as lacking elytra, they also lack any flight wings, and although *Lampyris* females have a vaguely sclerotised pronotal shield, the other thoracic segments are poorly developed and appear little different from the soft abdominal segments following on behind. When I found the bizarre bloated bag of eggs that is the female *Drilus*, I recognised her as such only because of the diminutive male attached precariously at the tail end, in the act of coitus (Fig. 77).

While these beetles are sometimes described as neotenic (retaining immature features into adulthood),[13] this is not quite appropriate. Both *Drilus* and *Lampyris* (unlike *Micromalthus*, discussed in the previous chapter) go through a true pupal stage, and although the resulting adult females are certainly larviform in their soft, undifferentiated bodies, they both have truly adult features such as appropriately segmented antennae and tarsi, and fully functional reproductive

FIG 77. This huge, bloated, wingless, larviform female of *Drilus flavescens* looks nothing like a beetle, and is recognisable as such only because of the coupling male. Downe, Kent, 18 June 2003.

13 Neoteny is more appropriate in the discussion of vertebrates like the axolotl, which is neotenous by virtue of its retained larval gills into adulthood.

tracts beyond simply possessing ovaries. Worldwide, several other glow-
worms also have wingless females, and in *Phosphaenus hemipterus* (very rare in
Britain) males are also wingless but have vestigial elytral stubs. As noted in
Chapter 3, *Drilus* and *Lampyris* are both snail predators as larvae (*Phosphaenus*
too, presumably), but why this should have led to the evolution of larviform
females is still open to conjecture. Neither species is very abundant, but both are
frequent where they occur. Nevertheless, males often seem to outnumber females
by 100 to one (my off-the-cuff statistic), probably because the males can at least
fly about and are swept from the herbage, but the wingless females remain
secretive and elusive.

Other, though less extreme, forms of sexual dimorphism are common
in the Coleoptera. As noted in Chapter 2, tarsal pads of male Dytiscidae and
Chrysomelidae aid mating, the fat hind legs of *Oedemera* flower beetles may be
used in courtship, and female weevil snouts are longer because they are the ones
drilling the holes for their eggs. Some dimorphisms seem to be visual cues to
mating, thus the broad black sutural band down the back of the elytra of the
common flower-visiting longhorn *Strangalia melanura* is much broader and more
pronounced, with a distinct apical spot, in the female. Care needs to be taken,
though. It was long assumed that all-black specimens of the bark beetle predator
Tillus elongatus were males, while the females had the thorax red, until Twinn
(2009) noticed some of the supposed red-thorax females had male antennae;
genitalia dissection confirmed them as males, and a check through museum
collections soon showed that the colour bar was not absolute.

FIG 78. Mating
14-spot ladybirds
(*Propylea quattuordecim-
punctata*), but although
the male is a paler
yellow than the
bright female, this
is an intraspecific
rather than a sexual
dimorphism.

In the click beetle *Athous campyloides*, the pronotum of the male is narrower and more parallel-sided than that in the conventionally rounded female, a sexual difference greater than between some species in this genus. In its near relative, the pretty *Denticollis linearis*, males are slim and parallel-sided, with the elytra a pale orange or straw colour, while the females are much broader and flatter, and the dilated elytra are usually black with a pale margin. They look completely different, and in early beetle tomes they were described as different species, the female as *D. mesomelas* and the male as *D. livens*.

Although the subtleties of these sexual differences are often lost on us, some distinctions are clear. The final abdominal segments contain the organs of sexual reproduction, so the undersides of male and female beetles often show differences. These may be ever so slight variations in colour pattern and surface sculpture (as shown in ladybirds – see Roy *et al.*, 2013), or marked structural forms useful in identification (e.g. the projections and clefts in male *Anaspis* – see Levey, 2009). The bizarre copulatory apparatus of male *Malthodes* (Cantharidae) are clearly visible in side view, and have long been used in identification guides and monographs. They comprise various extensions and bifurcations to the last dorsal tergites, and similar tails and ribbons on the seventh ventral sternite. *Malthodes* mate end to end, and it has been suggested that the strange formations at the end of the male abdomens are an important adaptation to this, the prongs and hooks being used as receptor and clasper for the female abdomen. Oddly, the female terminalia are described as being 'monotonously similar' (Fender, 1951), so notions of key-and-lock evolution do not seem to be supported here.

Perhaps the best-known example of sexual dimorphism in British beetles occurs in the iconic stag beetle (*Lucanus cervus*). Only the males of this dramatic and distinctive animal carry the large curved mandibles, and their multi-tined, antler-like shape is echoed in colloquial names across the world: *cerf volante* in French and *cervo volante* in Italian (both translating as 'flying stag'), *Hirschkäfer* in German (meaning 'stag beetle'), *kuwagatamushi* in Japanese (translating as 'stag worm') and so on. The males use their jaws to wrestle with each other, clasping an opponent around the thorax in an attempt to throw them off the rotten log or stump and thereby claim the decaying territory and attendant females. Male stags, generally larger than females, nevertheless show huge variation in body size: ignoring the mandibles for a moment, body lengths of 28–70 mm are quoted (Jessop, 1986). The larger the beetle, the larger the jaws, but this is not a linear relationship and closely conforms to a dimorphism within the males, such that there are large-horned 'majors' and smaller-horned 'minors', as occurs in the soldier and worker castes of some ant species (Clark, 1977).

It seems that developing very large jaws at metamorphosis into the adult stage is only an advantage if they do not detract from a large body. Smaller beetles (nutritionally deprived as larvae perhaps) benefit from forgoing super-massive jaw size in favour of redirecting some of that pupal body mass to give a slightly bigger adult body, with perhaps the hidden advantages of a more robust immune system, greater thoracic and leg musculature, or increased fecundity (Knell *et al.*, 2004). There is usually some pupal threshold size, above which males have plenty of latitude to grow giant mandibles and below which they had best reserve body mass to generate body tissue instead. There is evidence that emergence time can influence this. In an Italian study (Harderson *et al.*, 2011), the largest beetles were found in May, with a pronotal width of 1.74 cm marking the point at which giant mandibles were more likely to appear. By August, fewer males showed large mouthparts, only those with pronotal width over 1.9 cm displaying this tendency.

FIG 79. The magnificent and unmistakable form of the male stag beetle (*Lucanus cervus*). Alarmed by the approach of the camera, it has taken up the rearing, threatening pose it adopts in the arena.

Despite their menacing size, the jaws of the stag beetle are rather feeble and their pronged tips are unable to give more than a playful nip – the musculature in the head, broad though it is, just cannot deliver the leverage. Having said this, when I came across a spectacularly large example in Downham Woodland Walk, south-east London, in 1999, I was impressed by the massive sharp spike halfway down the inside of each jaw. In the interests of science, I stuck my right index finger between them and the beetle duly obliged by biting down hard, skewering me good and proper. The stabbing pain was exquisite, and I got an alarming amount of blood all over my pocket notebook. Sadly, I have never seen, first hand, a stag beetle battle, but I am reliably informed that when gripped in one protagonist's jaws, the audible creaking of the victim's body armour is quite sickening. Contrary to the popular notion of wrestling as some sort of casual romp, there is probably quite some considerable morbidity (possibly also mortality) during these violent contests.[14]

A strange dimorphism is well known in some bark beetles (Curculionidae, subfamily Scolytinae), but is seldom remarked upon. In genera like *Orthotomicus*, *Xylocleptes* and *Pityogenes*, the end of the male elytra are suddenly squashed inwards to create a near-vertical bowl-shaped depression – the elytral declivity. Female elytra are smoothly rounded or, at most, flatly abbreviated. The male declivity is often armed around the edge or sides with spines, hooks or short

FIG 80. Female bark beetle *Hylesinus crenatus* apparently using her heavy elytra to block her tunnel entrance to prevent access by the chalcid parasitoid *Entedon ergias*. Ivinghoe, Buckinghamshire, 25 July 1997.

14 Females sometimes suffer their own undignified mortality (Fig. 81), and are more often found crushed underfoot on paths than the males. There have been suggestions that this is because passers-by do not recognise them as the worthy conservation flagship species (they lack the eponymous jaws), and so they are mistaken for vulgar cockroaches (Miquel, 2005).

FIG 81. A sad end to a noble beast; a female stag beetle (*Lucanus cervus*) seemingly crushed underfoot on a path in Catford.

processes, and since these are species-specific, they make good characters in identification guides. Quite why this male-only elytral adornment should occur is not clear, although there are reports of the males using the declivity as a type of shovel to bulldoze away chewed sawdust in their burrows, or perhaps as a recognition signal in the narrow tunnels since male–female reverse nudging has also been reported. Equally plausible, since the spines might be defensive or aggressive, is the possibility that they are used in the tunnels in mate-guarding against marauding late suitors, or to repel predators and parasitoids (Kirkendall *et al.*, 1997; Vega & Hofstetter, 2015).

WHAT IS THE POINT OF HORNS?

We might not know much of what is happing at the tail end, but there is a little more information available for beetle head and thoracic horns. These are particularly obvious in dung beetles and much work has been done on exotic *Onthophagus*. Most British species are rather subdued in comparison to their exotic relatives, nevertheless I've spent many a happy hour poring over the microscope trying to work out whether the male horns on the *O. similis* I've found under dog (and human) dung in London brownfield sites are different enough to make them the very rare *O. fracticornis* instead. Sadly, I have yet to be rewarded for my efforts. The appropriately named *O. taurus* (with two long, sweeping bull-like horns on its head) is reputed to have once occurred in the New Forest and Exmouth, but the records are a bit suspicious; now that would be a find and a half. Again, it is usually the males that are furnished with these ornaments, and they come in all shapes and sizes. This has set taxonomists off in making *Onthophagus* one of the most hyperdiverse genera in the world, with an estimated 2,500 species. A good guess is that there are something like 10 standard horn shapes (ranging from bumps and ridges to spikes, spines, prongs, antlers, spears and spatulas) arising from 25 different zones on the head and thorax (Simmons & Ridsdill-Smith, 2011b). These are no mere show-off embellishments; they appear to be genuine weapons.

Two male *Onthophagus* fail to engage each other on the laboratory desk, but in the real world they tend to meet in a tunnel dug under dung. Here, instead of meeting politely face to face like mirror images, they can shuffle around and over, so that one is upside down compared to the other. Now they can lock horns, quite literally. It seems the species-specific form of the horn braces against similar species-specific nodes, dimples and ridges on the front of the head and thorax, and they wedge firmly in place. Much like the spike-and-bar arrangement of a hand-held bottle-opener bracing against the tight cap of a reluctant beer bottle, the beetles now press up to each other and engage is terrific push-and-shove battles in the confines of the burrow. These skirmishes can last 75 minutes, before the victor – usually the larger, more massive and more muscular of the two – eventually shows who's boss and the smaller, weaker beetle sneaks away (Knell, 2011).

Just as with stag beetles, *Onthophagus* horn length and body size are related, but not fixed. Again, the largest males have the longest horns, but intermediate beetles, which might be expected to have intermediate horns, are just as likely to have short horns or lack them altogether. The reallocation of pupal body mass into long prong or not is not straightforward, but the costs are not insignificant. Longer head horns are directly linked to comparably smaller eyes (Harvey & Godfray, 2001), smaller wings and less powerful wing musculature (Moczec & Nijhout, 2004), and even smaller testes (Simmons & Emlen, 2006). Hornless male *Onthophagus* also have the advantage of stealth. Although they do not engage in macho posturing or physical clashes underground, their hornless (feminine?) appearance means that while the large horned 'major' males are fighting each other, the more diminutive 'minor' males can sneak into the brood tunnels and mate with the females.

Similar horns occur on the heads of the male *Copris lunaris* (a beautiful but very rare dung beetle, probably extinct in Britain) and *Odonteus armiger* (also rare, and more likely a subterranean fungus-feeder). The more widespread minotaur beetle (*Typhaeus typhoeus*) is aptly named and shows equally impressive horns (Fig. 82). The male has three prominent spikes on the front of the pronotum, a shorter one projecting over the head and a longer chisel-shaped spine on either

FIG 82. The unmistakable form of a male minotaur beetle (*Typhaeus typhoeus*), showing the long, protruding thoracic horns. I was very pleased to dig this one up from under deer dung in Knole Park, Sevenoaks, 26 February 2017.

side. The female has two or three vague tubercles or thorns in their place. It seems likely that the males compete for ownership of burrows, dung stores or females, or all three, and again, some males are better endowed than others. Males and females occur under cow and horse dung, but will also bury rabbit crottels; they are active during the winter months, from October onwards, and because they are attracted to lights, they often appear when moth traps are tentatively put out by eager lepidopterists in March. Thus, even moth enthusiasts get a chance to wonder at spectacular horned beetles.

SIZE MATTERS

The commonest form of obvious sexual dimorphism is in overall size. Leaving aside the minority groups like stags and dung beetles, where males are usually heavier and larger, there are plenty of good examples where females are generally bigger than males. Very often this has to do with the simple biological certitude that females carry large eggs, while males produce microscopic sperm. Oil beetle (*Meloe*) females are huge, distended egg-carrying transport sacks. One *M. proscarabaeus* was recorded laying a total of nearly 40,000 eggs, with each clutch of 8,000–10,000 eggs taking up 40 per cent of her body mass (Lückmann & Assmann, 2005). Oil beetles are very scarce (I've only ever seen two alive in my life), but similar distended females occur in leaf beetles, particularly the widespread metallic green *Gastrophysa viridula* (Fig. 83), and the sombre black *Galeruca tanaceti*. They lay much larger, but far fewer eggs (approximately 1,000 and 1,200, respectively), but after egg-laying their abdomens are noticeably less inflated.

FIG 83. Mating pair of *Gastrophysa viridula* showing the grossly distended female abdomen packed full of eggs. (Penny Metal)

Size variability is sometimes great in individual beetle species, so much so that, as in the *Denticollis* click beetles noted above, males and females can look very different. In the common hawthorn weevil *Neocoenorrhinus aequatus* (Rhynchitidae), males at 2.5–3.0 mm long look dwarfed by the females at 3.5–4.5 mm. And when I found a tiny (4 mm) hazel leaf-roller (*Apoderus coryli*), it took some time to convince myself that it was not an exciting new species, just a stunted male specimen at the bottom end of the usual 6–8 mm range. Throughout the weevils (Curculionidae etc.), leaf beetles (Chrysomelidae), longhorns (Cerambycidae) and ladybirds (Coccinellidae), it is almost always the females that are larger, even though there may be plenty of overlap.

Towards the end of July, hogweed flowers (*Heracleum sphondylium*) are frequented by large numbers of the very common red soldier beetle (*Rhagonycha fulva*, Cantharidae), mostly in mating pairs. The female, underneath, is usually slightly larger than the male atop her, but the most striking difference is the larger and obviously much more bulging eyes of the male (Fig. 84). In many flies (and the domestic honeybee), males have larger eyes than females, a trait associated with their hovering, territoriality or mate location; quite how eyesight in *Rhagonycha* might be sex-specific is not recorded.

FIG 84. A mating pair of the red soldier beetle (*Rhagonycha fulva*), showing the male's bulging eyes. Wayside flowers drip with thousands of these beetles from July onwards. (Penny Metal)

BEETLE ALCHEMISTS

As I may have mentioned, I hardly ever find oil beetles (Meloidae), yet these are large, obvious, striking insects, and to my increasing frustration, they are regularly found by general naturalists simply out for a walk in the countryside. I regularly see photos of them in cupped hands, usually captioned with the plea: 'What's this then?' Magnanimously, I reply with an identification, and also with a caution: 'Did they bleed on you? Be careful – the liquid they ooze is a skin irritant.' As far as I know, none of my correspondents has ever suffered ill effects. This is quite lucky, because oil beetles get their name from the toxic oil they use as a chemical defence.

The active ingredient is cantharidin, $C_{10}H_{12}O_4$, an odourless (to humans at least), colourless sesquiterpenoid, barely soluble in water but easily soluble in organic solvents and oils, including those that give the beetles their name. Cantharidin's highly technical descriptor is 2,6-dimethyl-4,10-dioxatricyclo-$[5.2.1.0^{2,6}]$-decane-3,5-dione, a monoterpene comprising two isoprene units, but it has long been known by its alchemical title Spanish fly, after the common name of a pretty metallic green oil beetle from which the drug has been derived since antiquity. Spanish fly (*Lytta vesicatoria*) is a very rare vagrant in Britain, but well known in southern Europe and around the Mediterranean. The beetle's toxin was known to the ancient Greeks and Romans, and has a chequered history as a poison, a blistering agent (when such things were in vogue) and an aphrodisiac (Moed *et al.*, 2001). Pliny the Elder, Dioscorides and Galen all make mention of insects we now recognise as *Lytta* or close relatives, and give recipes for manufacturing potions to treat sores, alopecia and warts, or to remove brand marks. Taking this substance internally was already known to be dangerous, but it was prescribed as a diuretic, to treat dropsy (oedema) and jaundice, to promote menstruation or to cause abortion of a dead foetus (Beavis, 1988). Tacitus, writing in the first century CE, reports that the Empress Livia used Spanish fly to elicit sexual activity in members of the royal household that she could later use to blackmail them. In 1772, the Marquis de Sade was accused of poisoning prostitutes with a draught containing Spanish fly; nobody died that time, but he was found guilty and sentenced to death in his absence, after which he fled to Italy to evade justice. If swallowed, some cantharidin is taken into the blood, is filtered out by the kidneys and causes a tingling sensation in the urinary tract; in men, it can be priapic, causing a prolonged, painful erection.

Cantharidin is highly toxic, and despite the supposed viagran attractions, the LD_{50} in humans (the dose that will kill half the test subjects) is 0.03–0.5 mg/kg, in the same range as strychnine. Its main danger comes from its complete inhibition of protein phosphatase, an important enzyme in the metabolism of glycogen. Even if a dose does not kill the victim, it causes great tissue damage, pain, respiratory difficulties, convulsions, gastroenteritis, confusion and long-term health effects. Today, cantharidin is sometimes used as an external application on skin complaints, to treat dermatitis or to remove warts, and its cell-destroying qualities soon become obvious as the skin reddens, dies, blisters and peels (the Meloidae are often also called blister beetles). Although they are unseen, the effects of taking the compound internally are the same, on the gastrointestinal tract, inside blood vessels and in the urethra. In a thorough review of insects used as aphrodisiacs (primarily Spanish fly), Prischmann & Sheppard (2002) finish with a caveat emptor warning – don't try this at home, folks.

Thankfully, British meloids, if handled respectfully, do not seem to exude enough cantharidin to cause skin problems; my request to the British beetles online newsgroup elicited few reports of irritant effects. Most people admitted to being thick-skinned when it came to handling oil beetles; perhaps it would have been a different story if they had tried to eat them. Caution should still be exercised. Johannes Lückmann confirmed that he had blisters on his fingers when he handled the beetles regularly, and sores around his lips when he accidentally touched his mouth. He illustrates some examples of skin irritation in his book on oil beetles (Lückmann & Niehuis, 2009). Although we cannot be sure of the exact species used in ancient times, ground-up oil beetles mixed in a draught of wine were given to the condemned at their execution – a dreadful, painful way to die. In North America, horse and cow deaths are sometimes attributed to a dead oil beetle (probably an *Epicauta* species) accidentally ingested when the animals are given hay in which the beetle has been trapped. If this happens, the meat from the dead animal cannot be consumed by humans or pets, since it retains its toxicity. Sadly, there appear to be few serious reports on oil beetle distastefulness to potential predators. Great bustards (*Otis tarda*) are known to eat oil beetles, and the birds may have some resistance to cantharidin, utilising its toxicity to self-medicate against intestinal worms and bacterial infections (Bravo *et al.*, 2014).

If *Lytta vesicatoria* is handled roughly it will exude droplets of its cantharidin-laden haemolymph (which is pale blue-green) through special one-way pores at its knees. This reflex bleeding is a rapid and effective way of getting the toxin into the mouth of any predator foolish enough to attack the beetle, and similar defences can be seen in a number of more common British beetles. Ladybirds are the easiest to annoy, and will obligingly squeeze their orange-yellow blood through similar knee pores at the junction of femur and tibia. As noted on p. 55, their toxins seem to range in potency according to the advertised warning colours of their bodies. Likewise, the bloody-nosed beetle (*Timarcha tenebricosa*) drenches the fingers of anyone who picks it up with red haemolymph, but this time through pores in its mouthparts.

Another famous concoction is pederin, often reported as being the most complicated beetle toxin known – it's a vesicant toxic amide with two tetrahydropyran rings, $C_{25}H_{45}NO_9$, with a chemical name extending to 80 syllables that I won't bother to repeat here. Along with the similar pederone and pseudopederin, it is secreted by the brightly coloured orange-and-black rove beetles *Paederus*. There is no doubt that the adult beetles are strikingly (i.e. warningly) coloured, but they do not reflex-bleed; the beetles need to be crushed to release the poisons. In the tropics, where bigger and more virulent *Paederus* species occur, the skin and eye lesions caused by the toxin are sometimes known as pederosis (although this term is also sometimes used, rather confusingly, for paedophilia). Coleopterist Tony Drane once made the mistake of using the wet-finger technique (see p. 390) when a rogue *Paederus* escaped a tube and took off over the study desk one evening; he forgot about it, but must have later touched his lip, which 'blew up beautifully and tingled and numbed for two or three days, gradually subsiding'. I think we should take heed from his cavalier behaviour. Many other rove beetles contain noxious chemicals, perhaps an evolutionary response to their exposed vulnerable hind bodies; phylogenetic studies suggest that these and their glandular defence systems have evolved separately and independently (Dettner, 1993).

Anybody who has ever uncovered a devil's coach-horse (*Ocypus olens*) from beneath a rock will remember the menacing way it curls up its tail and exudes a foul-smelling droplet from its anus. Defensive defecation is an easy start, but the staphs are known for having a number of different gland set-ups in different genera, producing different chemicals. *Eusphalerum* produces (Z)-2-hexenal and 3-methylbutanoic acid, which repels predatory insects and spiders; many Aleocharinae produce toxic quinones; Oxytelinae use toluquinone; Staphylininae produce iridodials; Schistogenini use methyl- and ethyl-esters of octanoic, decanoic and dodecanoic acids; *Zyras* make hydrocarbons; and Aleocharinae produce aldehydes – defensively repugnant all. The distinctive, pretty, flat *Deleaster dichrous* has, uniquely, two pairs of glands on its abdomen, and when the quinones from the ninth tergite gland mix with the iridodials from the tenth, they polymerise on contact with air to produce a thick, sticky glue that coats and clogs small invertebrate predators. A useful review of beetle chemosynthesis is given by Dettner (1987).

The numerous and distinctive goggle-eyed staphs in the genus *Stenus* produce a veritable cocktail of chemicals, and although one of these (N-ethyl-3-

(2-methylbutyl)-piperidin) is the mildly toxic stenusine, it is primarily used as a spreading agent on the surface of the water where these rove beetles often occur. Like many marshland insects, they can walk on the meniscus if they fall onto the water surface, and in dire need eject onto the surface layer the contents of their abdominal glands, which spread out rapidly, jet-propelling the insect forwards and away from danger.

Stenusine also appears to deter invertebrate predators and may additionally be a defence against microorganisms. This multiplicity of functions is quite typical of insect chemicals and possibly part of the explanation why they have evolved. An original function of quinones is in the sclerotisation tanning of beetle cuticle. Tenebrionids use quinones in a water phase (seen as a primitive characteristic), whereas other more evolutionarily advanced species are able to synthesise organic solvents to improve their delivery or toxicity, and which in themselves are also toxic or irritant. Elsewhere, water beetles of various families secrete defensive chemicals that are also fungicides, prevent microorganisms attaching to their cuticle surface, are toxic to fish and other vertebrate predators, assist in the wettability of their bodies after venturing onto dry land, or are alarm pheromones. When they scurry out from under a corpse or an animal dropping, the stout rove beetles (*Aleochara*) curl up their tails and emit repellents, but in small quantities these same chemicals are a male mating stimulant, acting together with female pheromones as an aphrodisiac. The musk beetle (*Aromia moschata*), a scarce longhorn (dramatically metallic green, much like *Lytta vesicatoria*), gets its name from the delicate scent it gives off when disturbed. The chemicals, produced by paired glands on the metasternum, may have originated as mate-finding pheromones, but they are also powerfully distasteful, containing monoterpenes; one of these, rose oxide, apparently reminiscent of lychees and Gewürztraminer wine, is an important fragrance in perfumery. Indeed, cantharidin may have originated as a sex pheromone; it seems only male oil beetles synthesise the chemical, but they transfer it to the female as a nuptial gift when mating (Sierra *et al.*, 1976). It may have retained its original purpose in ant beetles (Anthicidae), which actively seek out oil beetles (and cardinal beetles, Pyrochroidae, which also produce it); there are plenty of reports of anthicids riding aboard the much larger beetles, which they seem to treat as cantharidin tankers, licking the surface of the body or feeding on dead meloids (Tylden, 1865).

Although many offensive chemical weapons are synthesised by insects themselves, sequestering defensive poisons from foodplants is a good energy-saving alternative. Masters of this are the leaf beetles, Chrysomelidae, so warningly obvious for their contrasting colours, bright patterns or striking metallic sheens.

FIG 85. The nationally scarce (notable) cardinal beetle *Pyrochroa coccinea*, demonstrating a stark warning with its garish colours: do not eat.

Able to feed on strong-smelling plants rich in potent volatile substances that have evolved to deter herbivory, many leaf beetles accumulate the poisons for their own defence. The bloody-nosed beetle stores anthraquinones from its bedstraw foodplants, and several willow-feeding *Phyllodecta* species create salicylaldehyde from salicin in the leaves. Incidentally, salicin is used to produced aspirin and the salicylic acid used medically to remove seborrhoeic warts and verrucas.

An insect's dissuasive powers of foul taste or outright poison are pretty straightforward; emphasised by alarm displays, aposematic colours and stridulation (see p. 135), the lesson to the foolhardy attacker is soon learned. This may be backed up with warning scents – allomones. Although there are only four British net-winged beetles (Lycidae), this group is well studied worldwide for their bright colours (usually orange or red, and black), their repellent taste, and the fact that they are mimicked by other beetles taking advantage of the unpleasant chemistry lessons suffered by would-be predators. The Australian genus *Metriorrhynchus* gives off highly volatile warning alkylpyrazines (which away from the Coleoptera smell of roasted peanuts, freshly baked bread, chocolate and popcorn), and these are enough to dissuade an attacker from even having a sample bite. Beetles from several other unrelated families copy *Metrorrhynchus* colour patterns and reinforce this warning by secreting almost identical alkylpyrazines (Moore & Brown, 1989).

There is one curious example though, where a lethal compound produced by beetles is not toxic to vertebrate or invertebrate predators. One of the most deadly insect poisons is diamphotoxin, produced by the South African leaf beetle genera *Diamphidia* and *Polyclada*. The toxin is incompletely known, but appears to be a single-chain polypeptide with a molecular weight of around 60,000 (de la Harpe *et al.*, 1983). Oddly, it appears not to be toxic if ingested (the polypeptide is simply digested by proteases in the stomach), but is dangerously haemolytic if injected into a mammalian blood system. The protein is also found in the beetles' obligatory parasitoids, members of the ground beetle genus *Lebistina*, which feed on the larvae and/or pupae. Quite how the toxicity of diamphotoxin was first noted is a mystery, but this chemical lore was passed down by the !Kung Bushmen of the Kalahari, who excavate the larvae and cocoons of leaf beetles and carabids from the soil around the foodplants, to use their crushed bodies to make arrow poison. As a safety measure against accidental self-injury, the pointed head of the arrow is left unanointed, but the crushed beetle remains are smeared into the sinew bindings just behind, where the head is attached to the arrow shaft. The poison enters the hunter's quarry when it is punctured deep by an arrow; it may take a day or two, but one poison arrow will bring down and eventually kill a half-tonne giraffe.

The continued development of novel compounds, and novel mixtures, allows new combinations of substances to create new chemical reactions and better defences in an escalating war against predators. These give rise to the deadly poisons discussed here, and also sometimes unexpected yet marvellous consequences.

A LIGHT IN THE DARK

A truly alchemical offshoot of noxious predator-repelling chemicals is the famous light production in glow-worms and fire-flies – which, of course, are not worms, or flies, but beetles. Bioluminescence, the production of cold, heat-free light by organisms, is rare but widespread across a number of groups, including springtails, fungus gnats, millipedes, krill, squid, jellyfish, bony fish, fungi and bacteria. However, it is best known in beetles. Light is produced by the oxidation of luciferin (a small heterocyclic molecule, $C_{11}H_8N_2O_3S_2$, with a molecular weight of about 280) by the enzyme luciferase (a complex polypeptide enzyme with a molecular weight of about 63,000). These are both generic terms, and although the chemical structure of glow-worm/fire-fly luciferin is more or less constant across different species in the family Lampyridae, each organism has its own slightly different version of luciferase, a feature utilised in constructing evolutionary phylogenies. The glowing organs occur on the underside of abdominal segments 5, 6 or 7, or some combination of them.

Despite its name, the common glow-worm (*Lampyris noctilua*, Lampyridae) is actually relatively uncommon, but it does occur through most of Britain. Its pale green glow is well enough known to have brought the beetles and their larvae to the attention of the country folk who called them grass-stars, ground fairies, will-o'-the-wisps and a whole slew of other common names. All stages of the beetle contain lucibufagins (distasteful steroids) and are unpleasant to would-be predators. The larvae also have eversible pale glands down the side of the body, which they expand if attacked by ants (Tyler & Trice, 2001). It is generally accepted that the glowing is a nocturnal warning, where strongly contrasting colour patterns would be invisible in the dark. There is also evidence that luminescence evolved in larvae first, and only later did beetles retain the ability into adulthood (Branham & Wenzel, 2001, 2003). In adults, the glow additionally serves males when finding a mate in the darkness – one of those wingless larviform females, trundling through the night grass. Since there is only one widespread British species, no confusion arises and the males are simply attracted to the continuous glow. The very rare 'other' glow-worm, *Phosphaenus hemipterus*, is diurnal, but the female glows in response to being molested, perhaps an echo of the original warning display of glow-worm evolution.

In North America, the fire-flies (also Lampyridae) take this communication to a new level by flashing their brighter yellow-green lights on and off to a predetermined code rhythm that is species-specific. The light is under the beetle's physiological control, but Pliny the Elder thought that the beetles achieved flashing by opening and closing their wing-cases. Lloyd (1983) contemplates light communication across various insect groups. Codes vary from abrupt 40-ms flashes to continuous glows, with one to 11 flashes of varying lengths in each code pattern. A male initiates by signalling a flash starter message (= 'where are you') and the female responds with a corresponding answer (= 'I'm here'). As they approach each other, the signalling may become more intense; males may try to interfere with each other's signals by adding extra flashes as a rival is communicating – a kind of fire-fly jamming ruse. Eventually, though, they find each other. The beetles can be tricked into landing near an entomologist who is armed with a small torch and who knows the right measure of female on–off timings. Lloyd (1983) is inspiringly enthusiastic about this low-tech approach to fire-fly study. The females can get confused, too. On the Costa Rican shore of the Pacific Ocean, I watched in wonder as lightning from a distant thunderstorm off at the horizon was enthusiastically answered by massed female fire-flies in the trees above the beach, all answering in unison, but all fated to unrequited

FIG 86. Successful coupling of Florida *Photinus* fire-flies.

trysts with the uncaring clouds over the water. The beetles can also be tricked by more sinister fraudsters. Females of *Photuris* mimic the flash responses of various *Photinus* females; a male *Photinus* homing in on a duplicitous *Photuris* thinking it is a receptive female is met by a hungry predator instead of a mate.

One of the largest light-producing beetles, the click beetle *Pyrophorus noctilucus*, occurs around the Caribbean and Central and South America; and although light communication in this species is not fully understood, it was long claimed to produce the brightest light of any insect (Langley & Very, 1890). These claims (later verified) had the unexpected consequence that *Pyrophorus* was particularly sought after, not by the world's entomologists, but by its nutritionists. My informant is Dr Joan Stephen, who in 1954 was working at the London School of Hygiene and Tropical Medicine, and who helped set up the Tropical Metabolism Research Unit in Jamaica, where *Pyrophorus* was affectionately known as a 'winkie'. This was a time when the chemical understanding of nutrition (or malnutrition in this case) and metabolism was being unravelled. A field station in an impoverished and undernourished Commonwealth country was one of many ways to find out how poor living, poor diet and metabolic imbalance could be a leading cause of death, especially in children. Coincidentally, during the 1930s and 1940s biochemists had finally worked out the basis of photosynthesis, as light photons hitting the chlorophyll in green leaves released electrons and protons (hydrogen ions) through a cascade of reactions that finally promoted adenosine diphosphate (ADP) to the energetically more available adenosine

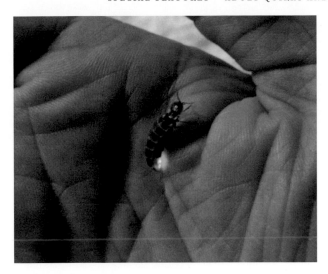

FIG 87. Female common glow-worm (*Lampyris noctiluca*) or 'grass-star', found in the grass at a Dordogne holiday gîte, 11 August 2012.

triphosphate (ATP). ATP exists through all living organisms as the molecular currency of energy – energy used in every metabolic reaction – and a keystone to understanding how metabolism, and in its turn nutrition, works. In today's high-tech labs, it's easy to forget that, in the 1950s, chemical assays relied on boiling up substances in test tubes with various reagents, and often manually checking subtle colour changes against printed colour charts. This was fine for roughly measuring the carbohydrate content of a pain au chocolat, or how much alcohol is in a bottle of Guinness, but ATP and ADP are highly reactive molecules that occur in truly minute amounts. An assay to measure their activities required a much higher level of analysis. And it was available thanks to glow-worms.

The cold[15] light given off by these beautiful beetles, without heat and without combustion, is the exact opposite of photosynthesis. During a special chemical conversion, that energy-storing molecule ATP gives up its electron to produce ADP, and as the reverse cascade of chemical reactions continues, it produces back those photons, to shine out into the night sky to try to attract a fire-fly or glow-worm mate. Back in the 1950s, that small light output in the test tube could be carefully measured using recently developed photometry devices, which gave a fixed numerical value to spectacularly minuscule chemical reactions. The light-producing chemicals of the fire-fly, luciferin and its enzyme converter luciferase,

---

15 The usual quoted statistic is that the heat from fire-fly or glow-worm light is 1/80,000 that from a candle of the same brightness, and that compared to incandescent bulbs, which give out 5 per cent of the energy as light (95 per cent as heat), bioluminescence is nearly 100 per cent efficient.

offered a unique and super-sensitive assay for the newly discovered ATP/ADP energy reaction. A similar assay to detect ATP as an indication of extraterrestrial life was considered by NASA, should a mission to Mars ever launch (Dudley *et al.*, 1971).

In a low-tech twist, though, luciferin could not then be synthesised, but needed to be harvested from adult beetles. Back in Jamaica, the light-producing *Pyrophorus* beetles were now in great demand, laboriously collected in the Jamaican hillside forests. However, rather than the scientists doing the collecting themselves, Joan Stephen was in charge of marshalling the local children; they collected the winkies and she paid them with pennies, she, of course, being known as the 'winky woman with the pennies'.

GOING WITH A BANG

Another splendid innovation arising from repellent chemical synthesis occurs in the bombardier beetles (Carabidae, subfamily Brachininae), so named for the explosive cannon-like report clearly heard when they blast off, and clearly seen in the smoky vapour that is released. Paired glands inside the abdomen secrete hydrogen peroxide (H_2O_2) and hydroquinones (benzene-1,4-diol and various derivatives), which are stored in a tough, muscular vesicle just inside the anus. Either of these would be suitably repulsive if exuded onto an attacker, but the beetle is a supreme pyrotechnist.

When it feels threatened, the bombardier beetle contracts the storage reservoir and the contents are squeezed through a one-way valve into a thick-walled reaction chamber, often referred to as a crucible. Oxidative enzymes lining the crucible cause the hydrogen peroxide to decay rapidly to give water and oxygen, which combines with the hydroquinones to produce p-benzoquinone, a powerful vertebrate and invertebrate deterrent. The chemical reaction is strongly exothermic, producing enough heat to vaporise a significant portion of the reactants, which are discharged at up to 100 °C and under pressure (Aneshansley *et al.*, 1969). The toughened crucible contains the pressurised explosion, which closes the inlet valve and is instead released through the anal outlet valve. As the pressure is released, the outlet valve closes and the inlet valve can reopen to allow another reaction to take place. This repetitive control of chemical reactions creates a series of about 70 pulses in one-tenth of a second.

Most bombardier beetle enemies are invertebrates, with ants suggested as the major aggressors. High-speed video photography of a constrained beetle shows that if individual legs are tweaked by fine tweezers, it will fire in the direction of each leg attack. Similarly, if it is approached by ants it will spray directly at each ant head rather than rely on a general explosive repulsion. The extreme apex of the beetle's abdomen is highly flexible, and can be directed to left, right, behind or underneath the beetle's body to target its attacker. It can also bounce the spray off a pair of reflectors, part of the armature of the abdominal apex, to direct it over its elytra and douse any attacker above its pronotum (Eisner & Aneshansley, 1999).

When I took a bombardier beetle along to the natural history club at south London's Ivydale School, I pontificated about cannons, bombardment, chemical explosives and the useful deterrent ability of being able to secrete stable neutral chemicals from abdominal glands, mix them inside a reinforced internal crucible reaction chamber, and then squirt a directable jet of boiling-hot quinone gas from your anus into the face of your would-be attacker. There were a few sceptical looks, and maybe two seconds before one 10-year-old quipped, 'What, it farts?' I had their attention now; getting them to quieten down to listen took a bit of time. Of course, the beetle gets its name from the audible pop of its defensive mechanism. But as it is only 7–8 mm long, even its loudest output would need the complete silence of the group of 20 giggling schoolchildren in order to be heard. I waited while the beetle shuffled about in its tissue-lined petri dish, and the general hubbub subsided to calm expectation. Using some fine watchmaker's tweezers, I gently tweaked its back leg and there came the report. The looks on the assembled faces were priceless. Not a pop, not a bang, not a fizz, but a tiny, clear, slightly musical 'toot'. Perfect.

HOP, SKIP AND JUMP

The conventional escape mechanism of just running away from an enemy is not always fast enough. Likewise, it takes a concerted moment or two to flick open elytra and flex the flight wings, let alone get them beating enough to take off. Sometimes the equivalent of sudden side-stepping is the best avoidance technique. Hopping, skipping or jumping has evolved several times in insects, notably in fleas, springtails, grasshoppers and leafhoppers, so it is no surprise to find beetles at it too. Perhaps the best known are the flea beetles, partly because they are very common, but also because many are well-known horticultural or agricultural pests. As noted earlier (p. 45), the thick hind femur contains an energy-storing scroll-shaped coil, the metafemoral spring, where it articulates with the hind tibia. The coil tenses under a build-up of gradual muscular power, but can be released suddenly. The jerk snaps the femora back, pushing down through the tibia and into the tarsus gripping the leaf or stem surface, and the entire beetle is propelled away at breakneck speed.

Comparisons with human athletes make good reading, but are flawed because the organisms differ in size by several orders of magnitude, and they have wholly different physiologies and body structures. Nevertheless, I will offer the following: a flea beetle achieves a take-off velocity of 3 m/s (10 kilometres per hour), accelerating at more than 270 g, and can easily jump 100 times its own body length (Brackenbury & Wang, 1995). The jump of a flea beetle is magically eye-deceiving; one moment it is there, the next it is gone. The sudden ping of an emergency escape does not always favour the insect though. A traditional method of controlling *Phyllotreta nemorum* flea beetles on turnips, cabbages and kale was to push a light wheeled framework of heavily tarred boards along the edge of the crop; the startled beetles jump and are caught on the sticky tarred surfaces (Anon., 1895). Launch is not always arbitrary or random, and the trajectory may be intended, aided by flight wings (which open 50 ms before take-off), towards a target landing spot.

An almost identical adaptation occurs in the Scirtidae (two British *Scirtes* species), the flea weevils (Curculionidae, subfamily Curculioninae, tribe Rhamphini) and several other exotic beetle groups, to similar effect (Furth & Suzuki, 1992). In each of these, the enlarged hind femora also contain a scroll spring mechanism, allowing the beetles to jump violently away.

FIG 88. The deep keeled body and large hind legs of *Mordellochroa abdominalis* make it look elegant and pert when the beetle is perched on a flower feeding, but any disturbance sends it dropping to the ground, where its jerking movements and smooth lines soon hide it in the dense herb layer.

The hind legs of some flower beetles (Mordellidae) are enlarged, with stout muscular femora, and with broad tibiae and tarsi lengthened and armed with neat rows of tiny pegs to give the beetle purchase when, with rapid skipping motion, it jerks out of danger. The adults often feed on flowers, but should a shadow fall over them, they quickly hop off and disappear deep into the herbage. Mordellids are slightly flattened laterally, with deep boat-shaped body forms and keel-shaped undersides, so they naturally lie on their sides. In the sweep net, this characteristic escape response sends the beetle scuttling around in circles, but on the flat surface of a beating tray their erratic skipping is difficult to follow and they are quick to escape. A similar skipping occurs in *Orchesia* (Melandryidae); this genus is flattened dorsoventrally rather than laterally, but the jerking movements are equally effective when they are discovered under loose fungoid bark, or beaten onto a plastic sheet from a decaying bracket fungus.

The archetypal skipping beetles have to be the clicks (Elateridae), so named for the noise they make when jumping. Typically, a click beetle that has fallen on its back[16] on a flat surface will jackknife its body so suddenly that the momentum flips it into the air, allowing it to right itself on landing. The flexion is achieved by bracing a spine on the underside of the thorax (on the prosternum) against the edge of a pit-like cavity just behind (on the mesosternum). Muscular tension

FIG 89. The sleek, bullet-shaped form of *Ochesia undulata* feeding under fungoid bark. If disturbed, it skips itself away using its strong back legs.

16 Many beetles similarly upended onto their backs can right themselves simply by flicking their elytra open for a moment. Inevitably, this puts them right way up, and off they scurry.

creates stored energy, and the sudden release of the peg into the pit at this hinge jolts the beetle's centre of gravity upwards suddenly enough to take the entire insect spinning into the air. Typical quoted figures are six to 10 somersaults and a similar number of rolls, taking the insect a distance 25 times its body length at an acceleration of 380 g (Evans, 1972, 1973).

There seems to be little control over where the click beetle is going, but as an escape response it is superb (Ribak & Weihs, 2011). The sudden lurch is often enough to fling it out of danger, and if it is picked up in the beak of a bird the powerful twitching and loud clicking is quite startling. My first click beetle was probably the common *Athous haemorrhoidalis* (Fig. 90), but to an eight-year-old, feeling its remorseless pinging between finger and thumb, and hearing that clearly audible click, it was a wondrous revelation. I remember running to show my mum and dad, convinced that I had discovered a fascinating phenomenon – which indeed I had.

FIG 90. The distinctive, elegant elongate form of the click beetle *Athous haemorrhoidalis*. The junction of pronotum and elytra forms a natural hinge, a bit like a baguette that has been tightly squashed between finger and thumb.

Skipping and jumping beetles are fascinating to watch, and marvels of evolutionary engineering, but many can evade their enemies, including humans, by the simple expedient of just relinquishing their hold and dropping to the ground, where they will be lost in the herbage. This is sometimes combined with thanatosis – feigning death – when the beetle draws its legs in and remains motionless for some minutes. The lack of movement makes it much more difficult to locate once it has fallen into the jumble of dead leaves, twigs, lichen, soil particles or whatever forms the underlying substrate of its habitat. The small ground-dwelling weevils *Trachyphloeus* and *Cathormiocerus*, often encrusted with bits of sand and soil, are all but invisible when searched for under the cliff-top vegetation they favour.

The large black seaside ground beetle *Broscus cephalotes* exhibits a strange angular thanatosis if handled roughly; it sticks its legs out, rigidly, at strange angles, and can be balanced on its hind legs or propped up against a hard surface like a toy. This catatonic rigidity, also called catalepsy or tetanic immobility, may be making use of two survival ploys. With its clawed legs jutting out at unreasonable angles, the beetle may be less easy to swallow, or it may lull its attacker into a false sense of security when, thinking the beetle dead, it puts its victim down for a moment or loosens its grip, allowing *Broscus* to revive miraculously and run off. Similar ideas have been suggested for why possums play possum.

MAKING A NOISE ABOUT IT

That first click beetle made a lasting impression, and the sense of wonder has never left me. I made a similar personal discovery when, as a small boy, I first picked up a dumbledor, probably *Geotrupes spiniger* (Geotrupidae), while walking with my father over the chalk downs at Bishopstone near Seaford. Before it flew away, and while I was holding it tight in my clenched fist, I could feel it vibrating in my fingers; I could hear it too, making a low, annoyed squeak before I released it. This memory was recently reawakened in me when I saw a short video on the Internet of *Copris lunaris* (Jersey, not Britain, sadly) scrabbling about in someone's hand, chirruping like a pet mouse.

Although the video commentary claimed the beetle was singing, and that dung was fun, there was a much more prosaic and serious reason for its noise – it was trying to persuade its handler to drop it. While the sound was audible, it was the vibrations the beetle hoped would do the persuading, and although it felt danger from being in a friendly human hand, imagine its predicament in the beak of a hungry bird or the mouth of a curious badger. It may not have long before it is crushed and swallowed – perhaps a second or two. Most predators 'test' their prey on first gripping it, just to make sure it is not poisonous or venomous. If the beetle can startle its enemy into relaxing its jaw grip for even a moment, it may have a chance to escape. A grating, vibrating squeak in the mouth is not a pleasant thought – it's a bit like chewing a clockwork toy, and anyone who shrinks at the sound of chalk squeaking on a blackboard, or a serrated cheese knife vibrating on the edge of a china plate, knows the sensation intimately.

There are at least 14 different sound-producing organs in the Coleoptera (Wessel, 2006). They all share a common mechanism in that a plectrum (usually a raised chitinous strip) is rubbed over a series of parallel grooves and ridges, the pars stridens. This is somewhat analogous to rubbing a knife-edge down the fine teeth of a comb, and is similar to the structures used by other sound-producing insects like crickets and bush-crickets. Sometimes the distinction between plectrum and comb is blurred because both sound-producing surfaces comprise multiple ridges of varying sizes that rub across each other.

Geotrupes has a plectrum on the underside of the metasternum, against which it rubs the pars stridens, located on the corresponding adjacent surface of the hind coxa. *Copris* has a plectrum near the end of abdominal segment (tergite) 6, against which it rubs the pars stridens on the underside of the elytral apex. Other sound-producing combinations occur between the pronotum and mesonotum (longhorns, Cerambycidae), abdomen and femora (chafers, Cetoniinae), and head and pronotum (Endomychidae).

Stridulation is remarkably common in the tough-shelled Coleoptera (in 30 beetle families, according to Wessel, 2006), although humans may be able to hear or feel only the larger species. It's difficult to find accurate measurements, but perhaps the loudest British species is *Hygrobia hermanni* (Hygrobiidae), a medium-sized, rather stout water beetle, which like *Copris*, rubs the ridged underside of the elytra against raised pegs on the abdominal tergites. Although called, rather feebly I feel, 'squeak beetle' in at least one modern identification guide, *H. hermanni* is far better known as the screech beetle, for its loud, unearthly noise.

There is growing evidence that stridulation is not just an emergency startle response, but somehow encodes communication between individuals. Quite what the noises mean is still to be determined, but there is growing evidence to suggest that adult burying beetles, *Nicrophorus* species, communicate with their larvae in the buried carrion, and that the larvae communicate back by a similar stridulatory structure on the mandibles and maxillae of the head capsule (Niemitz, 1972).

Sound communication in beetles has been known for centuries, but was not necessarily recognised as such until recently. The deathwatch beetle (*Xestobium rufovillosum*), which we left on p. 94 gnawing its meagre way through dry timber beams, needs some way to find a mate in the dark labyrinthine voids it and its many generations of ancestors have excavated deep in the large supporting joists of old buildings; this it does by a Morse code, banging its head hard down onto the wood to create the familiar eerie taps in the night. The male usually initiates the drumming. The strike of the forehead against the wood is not a single bang, but a series of about nine to 13 percussions in the space of 1.0–1.2 seconds. There is then a pause of a few seconds before another tap. It's all very sinister and clock-like. The sound can be mimicked by dropping a pencil, tip first, onto a hard surface from a height of about 2 cm. Females answer back in the same way, and using each other's sound signature in the darkness to locate one another, the males and females continue head-banging until they eventually meet up.

Light, poison, sound and some strange body structures are just a few of the better-studied beetle oddities, but for every one that is known or has been

FIG 91. At rest, the orange ladybird (*Halyzia sedecimguttata*) withdraws its head beneath a transparent rim around the front of the pronotum. Is it keeping an eye out for danger? This small winter huddle, orange tears around the eye socket of a carved stone angel memorial in Nunhead Cemetery, is surely a metaphor for something.

recorded, there are hundreds still waiting to be observed or understood. What is the significance of the large pointed teeth found under the femora of many weevils, when other species in the very same genera have none? What does the ant beetle *Notoxus monocerus* (Anthicidae) use its large, vaguely gnarled thoracic horn for? What is the function of the raised, shining rectangular mirrors along the elytral suture of *Elaphrus* ground beetles? What does the orange ladybird (*Halyzia sedecimguttata*) need to see through the transparent visor-like rim along the front of its pronotum (Fig. 91)? What nutrition or hormones does a female *Hypebaeus* (Malachiidae) get when she's nibbling the strange, twisted earlobe-like growths at the tip of a male's elytra? If I had the time, I could devote years of research to each of these. Meanwhile, further notes on strange specialist structures and bizarre behaviours associated with particular families or genera are given in the taxonomic analysis of British beetle families in Chapter 7.

CHAPTER 5

Beetle Flight

BEETLES FLY

Because of their hard elytra, and the fact that they are most often seen running around on the ground or crawling over herbage, it is easy to forget that most beetles fly, and that this is part of their secret to world domination. But beetles do fly, and they fly very well. In studies of aerial plankton, where suction or net traps are balloon-hoisted many hundreds of metres into the open air, beetles make up a small yet significant sector of the haul – 5 per cent over land in Chapman *et al.* (2004) and 2 per cent over open ocean in Bowden & Johnson (1976). Mostly these are small fry, but Chapman *et al.* reported several ground beetles 4–7 mm long, and a specimen of the large (15–18 mm) water beetle *Colymbetes fuscus* weighing 160 mg. Small (2 mm) pollen beetles (Nitidulidae) and staphylinids made up the majority of beetles in that particular study in Bedfordshire, but even these small, light species are not just being blown about passively like dust – they are flying, wings open, in the true sense of the word, and if they close their wings, they fall earthward.

Some of the best-documented aerial clouds of insects are of beetles – ladybirds – because they are so easily recognisable. There are regular reports of light-aircraft pilots flying through swarms of ladybirds at 500–1,000 m, and on landing having to clean out hundreds of them from the canopy and air intakes. In the record-breaking hot summer of 1976, cars parked in the winding streets of the Brighton Lanes were bedecked with hundreds of thousands of seven-spot ladybirds (*Coccinella septempunctata*). As I looked up through the tall buildings to the narrow strips of bright blue sky above, I could see endless pinpoint black specks as this cloud of flying beetles stretched up and away out of sight. This phenomenon, matched elsewhere around the country that summer, made national news headlines, as did the crimson tide of dead ladybirds washed up over the next days on the beaches around southern England. A conservative estimate (Majerus, 1994) was of 650 km of strandline comprising 23.6 billion beetles. Becoming airborne in search of new food sources, they had migrated from the exhausted hinterland until they met the impassable barrier of the sea, where they congregated until their strength eventually failed (Muggleton, 1977).

Twenty-six years later, in 2002, I found a similar strandline wash-up on the Normandy coast. This time, it was mostly winged males (and a few queens) of the black pavement ant (*Lasius niger*), but among my estimated 25 million or so dead insect bodies per linear kilometre of beach, I found 69 species of beetle, ranging from tiny *Atomaria* (about 1.5 mm long) to several medium-sized (5–7 mm), chunky weevils like *Pissodes pini*, *Curculio glandium* and the wasp beetle (*Clytus arietis*) (Jones, 2013). They were swept up in the massive offshore air currents of the hot, humid June day, before being caught in a dramatic thunderstorm that evening and dashed into the sea.

Similar reports punctuate the literature, from hapless bark beetles landing at an altitude of 3,000 m on the Oregon and Washington snowfields, high above the treeline and many miles from the nearest trees (Furniss & Furniss, 1972), to a specimen of the dung beetle *Onthophagus taurus* that was dissected from the stomach of a herring caught in the vastness of the Great Australian Bight (Berry, 1993).

Most beetle flight is not random or handed to blind fate; it is calculated and coordinated. I live in south London, and I feel immensely privileged to see stag beetles (*Lucanus cervus*) flying over my garden every June. These huge insects look and sound a bit like model aeroplanes when they drift past, about 5 m up in the air. Their flight is not fast – in fact, it's rather sedate and dignified – but it is, on the most part, controlled and determined. They know where they are going. Although there was that time when I saw a giant male, flying at attic height, take a wrong turn to crash into the roof-top and tumble a few rolls down the slates into the gutter; perhaps that was just a misjudged landing. While a beetle may not have a mental map of where it is going, it does set off purposefully, with the aim of finding food, or a mate, or a new habitat, and it uses the cues of the physical environment to guide it. As a boy, living near the coast in Newhaven, I heard a similar loud, clockwork buzz from the great silver water beetle (*Hydrophilus piceus*) as it flew over the grazing meadows and water-filled dykes of the River Ouse floodplain thereabouts. The beetles flew at dusk, and were sometimes difficult to see in the gloaming, but their sound was quite distinctive – a whirring rattle as they flew past, then a moment of silence as they stopped and folded their wings in mid-air, then a loud plop as they dropped into the water of a ditch.

GETTING AIRBORNE

The first thing a beetle needs to do when launching itself into space is to unlatch and flip open the elytra. A two-headed condyle (the ball of a ball-and-socket joint) at the base of each elytron articulates with the lateral part of the mesonotum. Hard, and more or less inflexible, the elytra do not flap, and apart from perhaps acting like baffles or a static aerofoil, they do not contribute to the flight power. Mostly they are simply held away from the body as the beetle flies; in stroboscopic video analysis, they look tantalisingly as if they may act as rudders, but the twists and torsions seen in the photographic frames may simply be an artefact of air pressure and vortices generated by the membranous flight wings behind. In some cases, the elytra may swing at very low amplitude, at a similar frequency to the hind wings (about 50 Hz), but whether they are acting as counterbalances, aerofoils or guiding vanes is still not obvious. Conversely, the hemelytra of hemipteron bugs are reckoned to be functional aerofoils, flapping together with the hind wings to produce propulsion and lift. In some beetles, notably the rose chafer (*Cetonia aurata*) and its near relatives (subfamily Cetoniinae), the elytra are closed back down again during flight anyway, but the hind wings are still able to flap because a curved notch along the outer edge of the wing-cases accommodates their movement, and the beetles can still steer well enough.

Insect flight is not like that of an aeroplane but more akin to hovering, so the stiff, outstretched elytra should not be considered analogous to the wings of a 747. In all other insect groups, flight is powered, controlled and steered using the flexible aerofoils of the flight wings. Flies have only one pair of wings, yet they are the supreme insect fliers; indeed, they are named after the very act of flying. In Lepidoptera and Hymenoptera, the most advanced flying groups with four wings, the front and rear wings are coupled together by rows of hooks (the frenulum and hamuli in each order, respectively) to form a single aerofoil too. Another way to look at this is to consider that beetles fly despite their elytra, not because of them. A tremendous overview of beetle wing structure is given by Sun & Bhushan (2012).

The elytra move by direct muscular control only when they are flexed opened or closed shut (Frantsevich *et al.*, 2005). In some species, the elytra hinge fully

FIG 92. Beetles fly readily. The elytra are flicked open and the membranous flight wings are unfurled. Seen here are a soldier beetle, *Cantharis rustica* (left), and a cardinal beetle, *Pyrochroa coccinea* (below). (Penny Metal)

open, up, out and well forwards of the abdomen; in burying beetles, *Nicrophorus* species, the elytra are twisted backwards and held upside down, nearly meeting, over the back of the abdomen. Even in rose chafers, where the elytra are shut during active flight, they must flex open (sideways and upwards) about five degrees to allow the hind wings to be extended.

Leaving aside elytra for a moment, beetle flight wings owe their tremendous success to their combination of light weight and strength. Although appearing flimsy and membranous, the wings are reinforced by narrow ribs, traditionally called veins. Elasticity, especially at points of folding or where extra flexibility is required, is provided by resilin, a tough but elastic protein (Haas *et al.*, 2000). This construction of a diaphanous membrane, elastic joints and semi-rigid struts gives a wing the braced flexibility of, say, an expensive umbrella.

Another nice analogy I've borrowed from the excellent book by Brackenbury (1992), and one that is useful when looking at powered movement later, is the comparison of the wing-bearing third thoracic segment with four sides of a box section, made up of a top plate (the metanotum, along the back of the beetle), a base plate (the metasternum, on its underside) and two side plates (the left and right metathoracic pleurites). The wing itself is a double shelf-like outgrowth from the top and side plates, where they meet to form the top edge of the box. This is exactly how the wing is produced during beetle metamorphosis, so that it comprises a double membrane, one from each plate.

The evolutionary origin of this double membrane occurred long before beetles emerged as a separate insect lineage; indeed, it seems that wings evolved just the once during insect evolution, and that all modern insects are derived from a winged ancestor. The process by which insect wings originally appeared is a speculative venture that is beyond the scope of this book. Various possibilities have been suggested, from retention of broad, flat aquatic larval gill-flaps, extension of thoracic flanges into broad articulated lobes, flattening of pre-existing appendages attached to the base of the leg plates, or some combination of these (Brodsky, 1994). What is certain is that, unlike bats, birds and pterosaurs, the evolution of insect flight structures involved no loss of walking legs.

The next philosophical hurdle to overcome is how or why a patently non-flying precursor insect should bother evolving apparently non-functional flaps that might or might not at some time millions of generations in the future evolve into wings. Did the nascent flaps have a function? Again, speculation is the mother of all good storytelling. Possibilities include thermal regulation, pattern or shape camouflage, leg or spiracle protection, or sexual display. The behaviour associated with the leap from non-flying winglet to powered wing is a bit easier to guess at, and suggestions include: passive floating or gliding in the air, paragliding after falling or jumping, an adjunct to fast running, and surface sailing whereby aquatic insects skim across the water.

Whatever the origin, insects were the first creatures to fly, and this act was as revolutionary as the first organism leaving the primordial murk of the Cambrian swamp around 500 mya. Wings offered insects the Earth, and the sky, and they took it. Beetles continue that quest.

WING STRUTS AND WING SUPPORT

Reaching out into the wing, between the membranes, are the reinforcing veins, which are connected to the beetle's body cavity to supply blood and nerves. The wing membrane itself is tough and resilient, but the basal attachment, where it articulates with the thorax (the pleural wing process – usually a knob, ridge or lobe on the side) remains soft, flexible and rubbery, to allow constant flapping without brittle fraying or breaking.

Historically, each group of insects has its own set of vein names or numbers, and a separate terminology for cells, folds, coloured patches and thoracic attachments. Various detailed studies of beetle flight wings are available (Forbes, 1924, 1926; Hammond, 1979; Kukalová-Peck & Lawrence, 1993), and by comparison between insect orders, it is possible to identify how beetle wing veins relate to those in other orders through the unified naming system first proposed by Comstock & Needham (1898, 1899) and later revised by Kukalová-Peck (1978, 1983).

The basic insect wing appears to have eight veins – precosta, costa, subcosta, radius, media, cubitus, anal and jugal – although in almost all extant insect groups the first two are fused to create the solid, heavy strut along the leading edge of the wing, and the final jugal vein is seldom apparent. In beetles (and also to some extent in flies, bees, wasps and ants), these linear veins are especially grouped towards the middle half and front edge of the wing, and become thickened towards the base. Here they attach, via a series of hard seed-like nodules (the basal sclerites), to individual muscles in the thorax. By individual adjustment, at the articulation with the pleural wing process, these muscles alter the angle of attack of the wing blade during flight; using the umbrella metaphor again, it's a bit like individual spokes being twisted, tightened or relaxed to alter the curvature and shape of the fabric canopy. At its simplest, the membrane of the wing is twisted one way to catch the air during the powerful downstroke, then adjusted to allow some deformation during the recovery upstroke.

Flapping is more, though, than an elegant dance with billowing membranes and twisting wing bases; it has to deliver huge power, rapidly, to generate enough lift and thrust to get the beetle going. This is the job of the powerful internal thoracic musculature. Two sets of muscles create the flap energy: longitudinal muscles running from front to rear of the segment, and dorsoventral muscles running from top plate to bottom plate. Unlike the situation in birds or bats, insect wing-flapping it not about direct muscular pull on the wing base, but has more to do with elastic twanging of the wing attachment at the thorax. The box of the box-section analogy introduced earlier does not have four straight sides; instead, they are all gently bowed. As the dorsoventral muscles contract, effectively flattening the thoracic segment slightly, the wing-attachment base along the top edge of the box is pulled down, levering the wings up in the recovery stroke. When the dorsoventral muscles relax, the longitudinal muscles contract and the reverse happens, allowing the top plate to spring back to its starting position and become slightly hunched, causing the wing base to spring up and the wing blade to flap its thrust-producing downstroke.

Meanwhile, the side pleural plates are bowed outwards during each upstroke or downstroke of the wing. Inside the thorax, at the point of greatest stress (corresponding to the pleural wing process on the outside), a tough but flexible flying buttress of chitin, linking side plate to base plate internally, reinforces the snap power of the thoracic segment.

The stretching is least in the wings-fully-up or wings-fully-down position, but in the middle of each sweeping wing movement there is a point of greatest elastic tension, when the buckled pleurites are at maximum distortion. As the wing movement passes this critical moment, the elastic energy stored in the pleurites and internal buttress is suddenly released, jerking back to give extra energy to the up- or down-flap of the wing.

This constant cycle of back-and-forth snapping as the wing flaps up and down saves energy and also allows more rapid wing movement. The wing-beat becomes a bouncing oscillation in the thorax, almost automatic, and much faster than nerve impulses could possibly direct. Instead, the oscillation self-triggers, so that internal stimuli cause the opposing longitudinal and dorsoventral muscles to contract and relax at startling rates. In laboratory experiments on flies, wing-beat frequencies approaching 2,200 cycles per second can be achieved, but in living beetles a flap rate of 30–200 beats per second is more normal – the common cockchafer (*Melolontha melolontha*) is reported at 220 beats per second and seven-spot ladybird (*Coccinella septempunctata*) at 75–91 beats per second. These are similar to those in hemipteran bugs, which fly at about 20–80 beats per second.

The very smallest beetles, the ptiliids, have slender paddle-shaped wings, usually fringed all round the edges with long hairs – hence their common name, feather-winged beetles. The aerodynamics of flapping is very poorly understood here. The resemblance of this wing type to water beetle legs (which also have hair fringes) led to the suggestion (Horridge, 1956) that ptiliids row through the air using wing drag instead of lift, but Wootton (1992), reviewing beetle flight, pointed out that there is no evidence for this.

MASTERS OF ORIGAMI

Beetle hind wings are much larger than the elytra, and have to be folded, sometimes in a complex origami fashion, to fit beneath their protective shells. In other insect groups, the evolutionarily simpler fold is the corrugated concertina-type, like a hand-held Japanese paper fan. Even in non-folding insect wings (flies, bees, butterflies, etc.), there is a distinct corrugation, as the longitudinal wing veins are alternately concave or convex. This is very distinct in the pleated appearance of dragonfly wings; the Odonata do not fold their wings at all, but instead hold them flat, and such architecturally regular semi-pleating is noticeable even in fossil specimens. Full pleating is also the basic folding mechanism seen in the quadrant-shaped rear wings of mantids and grasshoppers, which collapse their flight membranes beneath their long leathery forewing sheaths when at rest.

To fit their long flight wings neatly beneath their shorter elytra, beetles also use lateral cross-folds. This has several consequences in terms of wing morphology and use. If the wing is to be cross-folded, the linear spoke-like veins that give it rigidity in flight need to be broken or hinged at the point of the fold. These joints, appearing as breaks next to thickened shoulders in the veins, give beetle hind wings their distinctive appearance, making them look very different from the elegant tracery of vein-enclosed cells familiar to anyone studying flies and bees. The major cross-fold is usually marked by the marginal joint, a break in the main rib along the costal fore-edge of the wing. In many insects this is the place where a darkened swelling, the pterostigma (often just stigma), is situated. According to Wootton (1992), the pterostigma in dragonflies is a blood-filled sinus that acts as an inertial counterweight, opposing the tendency of the wing to pitch in the wind. In many bees, wasps and flies, the stigma is a distinct wing mark where the veins thicken, helpful in identification, and possibly having some similar balancing role. It would seem logical that the aerodynamic flexion properties of the wing beyond this weighted node are different from the well-strutted basal two-thirds, and that the evolution of a folding point here was inevitable. The point beyond the fold is still reinforced by veins, but the struts are now disjunct, braced by elastic connections rather than a solid connective scaffold. There is a lovely illustration in the early paper by Forbes (1924) showing how this works using two fingers held apart, but connected by two twisted elastic bands, and a matchstick braced in their spiral twists to make a third, half-digit, extending further beyond the fingertips (Fig. 93).

A transverse fold in the wing membrane usually works in conjunction with a longitudinal fold and obliquely angled counterfolds, often described in origami

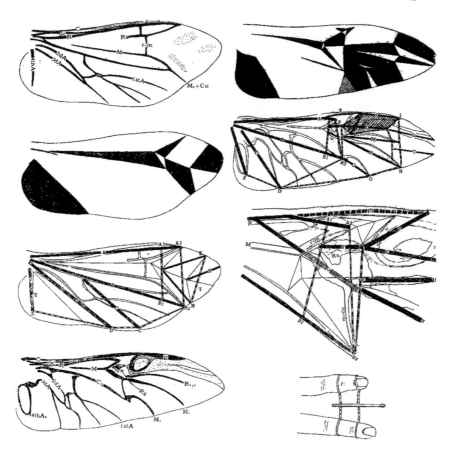

FIG 93. This illustration from Forbes (1924) is still useful when trying to understand how beetles fold their wings. The black-and-white origami diagrams can be re-created as paper models, and the picture showing twisted rubber bands and matchstick is a masterful analogy.

as inside-reverse folds. An early worker in wing folds, William Forbes delightfully describes his interest (1924), from his initial 'mistake' of pinning beetle specimens through the scutellum and setting the wings pinned out flat much like butterfly and moth enthusiasts. When he started to make wing maps, he utilised the still-held convention of identifying the straight wing folds, and colouring in the wing fields black or white depending on whether these regions retained their original horizontal bodily alignment, or are reversed upside down by the folding. His wing maps (Forbes, 1926), and later ones, look just like origami models, and indeed such models are easy to make from the published diagrams.

It was long ago recognised that the membrane of the beetle hind wing contains no musculature, so when it folds it must do so automatically, by some sort of natural spring mechanism. It was also realised that unfolding (flattening out) of the wing was by direct muscular movement, a splaying out of the vein ribs (mainly radius and cubitus) where they attach to the thorax via the basal sclerites, and that relaxation of these allowed the wing to spring back to its normal folded, resting position. A simple home experiment uses a piece of stiff paper and one of those inside-reverse origami folds. Four creases – three concave and one convex – meet at the origin, or in paper-folding parlance, three valley folds and one mountain fold meet at the knot. This creates four flat panels hinging around the origin. The paper (wing) naturally folds in on itself at rest, but a gentle pulling apart of the base edge, stretching the effector panels, levers open the two distant panels, which flip into position to give the wing its full length. In life, beetle wings usually comprise several series of folds – valley and mountain – and several knots, but they all interconnect through the leverage applied by the longitudinal veins (Haas & Wootton, 1996). This splaying/levering action also explains why there are no significant cross-veins in beetle wings. In many other insect orders, cross-veins between the longitudinal veins add strength and rigidity to the wing blade, but such rigidity would prevent the flexible splaying needed to unfurl a beetle wing.

The flexing of the flight membrane takes place in two movements, although they can blur into each other or be carried out independently. The 'promotion' of the wing is the movement forwards and outwards into the flight position, then the unfolding of the wing flattens the membrane ready for flight. Using our umbrella analogy again, these movements correspond to pointing it, furled, up into the air, then pushing the mechanism to raise the fabric canopy by splaying the ribs into full inflated and locked position. In a relatively large African chafer, *Pachnoda marginata*, Haas & Beutel (2001) measured the time to promotion as 80–260 ms, with flexion of the wing membrane into a flat blade, as the effector panels flip open the wing-tip panels, taking a further 80–300 ms. There was wide variability in both actions, with left and right wings working independently. Refolding, or placing the folded wing back under the elytra, is termed remotion.

Folding mechanisms are relatively uniform within beetle families, and have been used to confirm or adjust beetle classifications. In the Adephaga (predatory ground and water beetles), a more powerful spring-fold, nearer the base of the wing, allows more extensive folding, and perhaps a 50–60 per cent reduction in wing length as it is tucked away (Hammond, 1979). A more distant spring-fold in the Polyphaga (most other beetle families, see Chapter 7) creates a less convoluted wing origami. The *Cantharis* type, named after the common soldier beetle genus, is the simplest example, saving only about 15–20 per cent on the wing length, while an intermediate configuration gives about 40 per cent reduction in some click beetles and leaf beetles. A progression of folding complexities is also echoed in the decreasing number of cross-veins in the wing. As noted earlier, cross-veins act like braces in many other insect groups, but in beetles they have been largely lost, broken or reduced over evolutionary time. Sure enough, the primitive Archostemata usually have four cross-veins in the radial sector, but in the evolutionarily 'advanced' Polyphaga, there are only two.

Not all wing folding is spring-loaded, a fact easily observed when watching beetles landing from flight. In some groups, the wings flip out of sight in a trice, but in others the tips remain sticking out behind the elytra for a few moments. The beetle appears to be shuffling its hind end about, a bit like me when I try to struggle into my shirt after emerging slightly damp from the shower. The beetle quite clearly flexes its rear abdominal segments in and out, usually three to 10 times, and the wings, by stages, disappear under the elytra. Dissection shows that the upper surface of some abdominal segments (e.g. segments 5 and 6 in ladybirds) have small oval patches of microscopic spines (spicules), 150–300 per linear millimetre, pointing inwards and forwards. These act like gripping patches to push the wings back into their tightly folded resting positions. High-power magnification of wing membranes shows that insect wings are also often covered with microscopic hairs, called microtrichia. In beetles that use abdominal spicule patches, the microtrichia on the wings are longer and thicker, and directed in different alignments across the wing to mesh with the spicules where they make contact during the shuffling; this is sometimes visible as the slightly darkened sub-cubital cloud called the fleck or medial fleck on beetle wing diagrams. In species relying solely on automatic spring-folds, the microtrichia appear to be randomly or uniformly aligned (Hammond, 1979), and the darkened flecks are

FIG 94. Despite their short, stubby elytra, staphylinid rove beetles still manage to furl their wings completely when at rest. This *Tachyporus* takes to the air after unfolding its membranous flight wings, which are about four times as long as the wing-cases.

absent. In rove beetles, the flexible curvature of the hind body is used to help lever in the long wing-tips. Small staphylinids fly very well, and their wings are often three or four times as long as the abbreviated elytra. Almost inevitably, when they land, the wings are held flat together over the hind body, but with a careful turning up of the tail, and with left and right twists, they easily succeed in shuffling the membranes back into their convoluted, folded position.

It has been suggested that other series of spines along the hind edges of some abdominal tergites may clean and groom the wings as they are pushed back into their storage folds underneath the elytra. It seems a shame that the term 'toiletry devices' has not been more widely adopted for these structures (Hammond, 1985). There are precious few published descriptions of wing-grooming in the Coleoptera, but it is possible that early reports of beetles using their legs to help pack away their wings were really wing-cleaning episodes using the often bristly tarsi or edges of the larger leg segments.

WHAT IS THE POINT OF FLYING?

Beetles seem to get on with life very well by scrabbling about in the herbage or running over the ground. But long and active as their legs may be, running cannot compete with flight when it comes to moving further afield. Movement might just be to the next field, or it might be a lot further – indeed, insect flights range from a few feet to thousands of kilometres. Students of insect flight break down this scale, slightly arbitrarily it has to be said, into three overlapping zones, which they term trivial, swarming and migratory flight (Brodsky, 1994).

Under this scheme, a single beetle, moving around its own private locale, seeking food, shelter and a mate, or avoiding danger, is trivially engaged – although it may be of life-or-death importance to the individual. This might be the short hovering hop from one flower to the next for *Oedemera nobilis* in search of nectar or pollen; their slow, ponderous motion is heavy, bobbing, casual and obviously not urgent. Green tiger beetles (*Cicindela campestris*), on the other hand, make exciting trivial flights, running fast over bare ground and then taking to the air for a few metres, before landing and dashing off into the undergrowth. They are notoriously difficult to follow, which shows how good they are at evading enemies, including curious coleopterists. They also fly to snatch at their prey, often a fly or another beetle, itself flying past low to the ground.

Although the ladybird hordes of 1976 (and other years) were often referred to as 'swarms' (see above), this is not entirely correct. These were just large numbers of ladybirds that all happened to be doing the same thing at the same time as a result of individual action on the part of each specimen. Swarms are more than just large numbers of insects in a cloud. Technically, only honey bees swarm, when a queen and a large cohort of workers gather together in a great flying multitude and a pheromonally controlled huddling mass, to fly off to found a new colony. Without being too pedantic, though, swarming here really means a concerted gathering, a coming together to enact some primal biological purpose – usually mate location. Well-known insect examples occur in flies, where groups of males, often called leks, congregate and wait for females; similar male gatherings occur in social wasps and bumblebees.

Few examples of swarming are recorded for beetles. I'm tempted to suggest that male stag beetles (*Lucanus cervus*), flying in and gathering together to fight each other might, at a push, be classified as swarming. There are good video reports of up to 10 males fighting on the brick walls of houses. The choice of a brick wall seems more to do with the fact that other males are there, rather than (as might happen on a log) females being present or it looking like a suitable egg-laying site. On 27 June 1990, I watched a cloud of beetles (well, maybe 20–30 of them) flying tantalisingly high and out of reach around the dead branches of a stag-horn oak on Ashtead Common, Surrey; they were all males of the scarce *Lymexylon navale* (Fig. 95). Now, whether they were deliberately gathering en masse

FIG 95. Peter Hodge uses a long-handled net to try to capture flying *Lymexylon navale* on Ashtead Common, while Ian Menzies patiently looks on.

to mate, or were simply each attracted on this lovely warm, sunny day to female activity in the dead wood, is debatable. This species is renowned for having highly complex, branched palpi, which have chemosensory function instead of the antennae; the males could have been attracted to a female pheromone, to dead-timber scents or to each other.

Back in the Brighton Lanes of 1976, what those ladybirds were really doing was migrating. While this was not perhaps in the way that swifts and cuckoos migrate together many thousands of miles to predetermined winter or summer destinations, honed by millions of years of instinct, it was at least a deliberate attempt to make a long-distance move, either to find new resources and exploit new territories, or to escape the hardships of famine back home. One of the best-known ladybird migrations is in the North American *Hippodamia convergens*, which, as its name suggests, converges in great huddles to overwinter, usually in rocky hollows and small caves. Although these ladybirds have only a yearly life cycle, they are attracted to the same sheltered overwintering sites as their parents, grandparents and long-distant ancestors by a safety pheromone, which lingers from one year to the next, and which is reinforced by new migrants each autumn. In one such congregation, an estimated 30–40 million beetles were gathered

FIG 96. Although British ladybirds do not normally congregate in deep, multitudinous hordes, the 16-spot ladybird (*Tytthaspis sedecimpunctata*) can regularly be found in small huddles of 10–20 specimens. Here, one specimen shows 'blushing', adopting a winter reddish tinge to its normally pale mustard-yellow coloration.

together. In spring, the ladybirds emerge and fly down into the valleys a few kilometres away to find new aphid colonies.

For most insects, wings enable them to disperse, away from the small plot on which they spent their larval life; to move away from predator and parasite pressures, which soon build up in a static population; to escape overcrowding and limited food supply in what may be a small, transient or ephemeral microhabitat; and to avoid the dangers of inbreeding by finding genetically more diverse mates and increasing the depth and variety of the gene pool. That beetles fly off to new territories is clear to anyone who has dug a garden pond. Within days (sometimes within hours), dytiscids are suddenly swimming around in the new water. They did not know your pond was there, but since stillwater pools are transient in the natural world, filling with silt or plants, or shifting during oxbow and meander readjustment, their instincts tell them to fly off to see what they can find. Likewise, dung beetles fly in to examine a pat within moments of it being deposited (Jones, 2017); the chances are that there is precious little left of the dropping in which they fed up as grubs. And it is not long before a dead fox is being investigated by a wide variety of carrion scavengers. The distinction between a trivial search and a migratory exploration is a fine one, but to a small beetle any movement beyond a few metres is a journey into the dangerous unknown.

Hopefully, using antennal chemosensors and vision, a flying beetle has at least some idea of what it has set off in flight trying to reach – be this a pheromone-emitting female, a volatile-laden manure heap, fungus-fermenting dead wood or a brightly coloured flower. The rare jewel beetle *Melanophila acuminata* is dicing with death when it sets off on a mission – to find fire. The beetle is pyrophilous – it detects and flies towards large-scale forest fires and lays its eggs in still-smouldering pine logs. The adults eat insects killed in the flames and the larvae take advantage of being the newest colonists in recently dead timber; the fact that dead pine trees do not exude dangerous sticky defensive sap is an added bonus. *Melanophila* detects the infrared radiation given off during combustion, using 50–100 globular sensilla in each of two small pit organs next to the coxal cavities of the mesothoracic legs, which are exposed when the beetle is in flight. Heat-induced expansion of the fluid in these spheres is detected by the beetle on the subnanometre scale (Schmitz *et al.*, 2007), potentially allowing it to detect a forest fire from 100 kilometres away. That's quite an ambitious migratory plan.

Flying is not without its dangers. Taking to the air introduces several hazards that already hard-pressed beetles just do not need. Away from the shelter and confusion of herbage, flight in the open makes targeting by a predator, or a coleopterist, a lot easier. While flying with their wing-cases open (the vast

FIG 97. Flying is a dangerous business, with predators and traps waiting everywhere. The summer chafer (*Amphimallon solstitiale*) is well known for its aerial jaunts, but even a large and powerful beetle like this can fall victim to the ever-present spiderweb. The spider's venomous jaws can more easily find a chink in the armour when the hard elytra are flexed out during flight. (Penny Metal)

majority of family groups), beetles no longer have elytral protection against predators, water loss or heat loss, making flight a dangerous and exhausting activity. As shown from aerial trapping, small beetles are likely to be wafted far up and away, even out of sight of land, and although some will make useful landfall in suitable new territories, the majority will perish. So while having wings to get about is sometimes an advantage, it can also be a serious nuisance.

Flightlessness occurs sporadically across many beetle groups, but the low numbers of truly flightless beetles among the flying majority suggests that this is a marginal adaptation. It is also potentially an evolutionary dead end, since re-evolving wings at some later geological date would seem highly unlikely. Large, lumbering beetles, which might find it energetically difficult to get airborne anyway, often lack wings, hence the oil beetles (*Meloe*), large *Carabus* ground beetles and the bloody-nosed beetle (*Timarcha tenebricosa*) all lack wings, and in the last of these the elytra are fused together down the suture.

Insects living on small islands often lack wings, a feature well known to island biogeographers the world over, and typified by such wonders as the Lord Howe Island stick insect (*Dryococelus australis*, colloquially known as the 'land lobster'), the giant wingless weta orthopterans of New Zealand, and the multitude of flightless flies, moths, crickets, beetles and wasps of Hawai'i. At first this seems counterintuitive, because the ancestors of these creatures most likely arrived on the islands in the first place as part of the aerial plankton. The reasoning goes that although their antecedents may have been normally winged, there would have been an increasing evolutionary advantage to any offspring that had shorter, less efficient and less powerful wings, because they would have been more sedentary, less adventurous and so less likely to be blown out to sea to their deaths. A similar effect occurs in mountain-top insect species, too. Over time, such advantages would culminate in complete atrophy, with wings reduced to stubs, or missing completely, and a purely earthbound existence.

Island archipelagos offer a bounteous supply of wingless or flightless insect groups, evolved from some winged ancestor, but now radiated into a diverse fauna of closely related yet distinctly different genera and species. Wollaston (1854) was one of the first to notice this when he reported that nearly 50 per cent of the beetles of Madeira were flightless. In Britain, Lundy, being both small and relatively remote, is home to the endemic Lundy cabbage (*Coincya wrightii*), foodplant for a short-winged, flightless form of the widespread flea beetle *Psylliodes napi*. Although the short-winged form also occurs on the mainland, it is always mingled with the normal long-winged forms; only on Lundy has it evolved a permanently grounded population (Compton *et al.*, 2007). Lundy cabbage is also host to the closely related *P. luridipennis*. This beetle is winged, and flies, but is found nowhere else in Britain; perhaps it just doesn't fly well enough to get over the Bristol Channel.[17]

Reports that some apparently winged beetles do not fly should really be taken with a pinch of salt. Many reports stated that the deathwatch beetle (*Xestobium rufovillosum*) did not fly, even though it is fully winged. Received wisdom had it that colonies gnawing away in the construction timbers of ancient buildings never seemed to fly from one area to the next; they certainly never flew in through the window. It was suggested that the beetles merely continued to exist in their birth site, or perhaps crawled a short distance to the next beam. Those that fell to the ground (providing opportunities for coleopterists to find adults in old houses) also never seemed to fly up to the woodwork again. But I have seen

---

17 A pale-legged form (*pallipes*) of the weevil *Ceutorhynchus contractus* also feeds on the Lundy cabbage, and is possibly a species in the making. It, too, flies, but perhaps poorly.

them fly with my own eyes. On a bright, sunny 21 May 1980, while I was working as an engineer's assistant on the Cuilfail Road Tunnel at Lewes, one landed on the collar of the foreman's white shirt as he and I were discussing theodolites and benchmarks. He was bemused, but not surprised, when I popped the beetle into a tube in my pocket. And in April 2009, the holiday house in which I was staying near Chale on the Isle of Wight was often a cacophony of taps and rattles from the beetles in the beams (see p. 137), and respite was achieved only by vigorously attacking the woodwork with a shoe to shut them up. Several times during the evenings I caught sight of the beetles landing on bare woodwork from a short-haul flight. Thankfully, Deathwatch flight patterns are now fully recorded in the literature (Belmain *et al.*, 1999).

Wing dimorphism, where some specimens are short-winged and others long-winged, occurs in many beetle species – at least 200 British species according to Hammond (1985). Sometimes, as with *Psylliodes napi*, the wingless beetles are regarded as a local morph, under some sort of genetic or nutritional control in a particular population. At other times they are regarded as different enough to warrant separate specific status; for example, the small, pale, common ground beetle *Bradycellus verbasci* is always winged, but the extremely similar *B. sharpi* is usually wingless.

There has been some work to examine whether long-winged forms of a particular species are more likely to occur where dispersal is advantageous (in unstable, ephemeral habitat patches) and short-winged specimens where movement is unnecessary (where there is plenty of permanent, continuous habitat). The picture is highly complicated. Wing dimorphism in heteropteran bugs is well known, with macropters (long wings), brachypters (short wings) and micropters (small wings) often occurring together at the same time, and with no clear habitat, seasonal, nutritional or geographical pattern apparent. These forms are obvious because in the Hemiptera the front wing-sheaths (the hemelytra) vary in length along with the hind wings. Wing dimorphism in beetles is often hidden by the uniform lengths of the elytra across all forms, but can be reflected in loss of musculature and sometimes in the elytral shape.

A neat concomitance occurs in the small, unspotted ladybird *Rhyzobius litura*, which frequently has short-winged forms, comprising 24–30 per cent of individuals in British and European populations. In brachypterous specimens, the abdominal spicule patches that usually help tease back the wings into their folded positions under the elytra are reduced in size and spicularity. Oddly, though, the spicule patches of brachypters of the very closely related *R. chrysomeloides* are exactly the same as those of long-winged specimens (Hammond, 1985).

THE LIGHT AT THE END OF THE TUNNEL?

My first dung beetle was the common *Acrossus* (*Aphodius*) *rufipes*, found not in a dung pat but on the windowsill of my parents' Newhaven house. My father had alerted me to the noise, a soft 'tick, tick, tick' against the glass, one late-summer evening. The beetle had flown up from the grazing meadows of the River Ouse floodplain a few hundred metres off, but had been sidetracked from its nocturnal wanderings to explore, instinctively and insistently, the bright light shining out from our lounge. Many insects are attracted to artificial lights – notably moths – so much so, in fact, that trapping them using bright mercury-vapour or ultraviolet bulbs is now a standard sampling technique all over the world, from beach shore to jungle mountain top. The attraction has long been known – the Greek Nicander writing in the second century BCE describes moths darting around lamps in the evening, and the Roman author Columella (4–70 CE) describes a lighted moth trap used to protect beehives from wax moths (Beavis, 1988).

Beetles, too, are attracted to lights. When, in 1771, Carl Linnaeus saw an engraving by the French naturalist Louis-Jean-Marie Daubenton of a massive longhorn in Count Buffon's famous *Histoire Naturelle* encyclopedia (Leclerc, 1749–89), he named it *Titanus giganteus*, even though he had not seen a specimen (Hogue, 1993). He'd never be allowed to get away with that nowadays. The reason he'd never seen the beetle was that this was, for centuries, one of the rarest insects known. A specimen would turn up dead every so often, either washed up on the shores of the Rio Negro near Manaus in Brazil, or cut from the belly of a fish hauled out of the river. The first living beetles were not seen until 1958, when they were attracted to the electric street lights recently installed in towns and villages in the area.

The fact that insects are attracted to lights is exploited by electrocutor pest traps used in commercial premises, where an electrified mesh cage surrounding a bright light source zaps any flying insect that touches it. These are mainly used to control flies and cockroaches in food stores and restaurants, but no doubt they kill larder beetles and woodworm too. Occasionally, this has odd consequences, and there is a report of larvae of museum beetles (*Anthrenus*) feeding on dead flies in the tray at the bottom of such a trap (Jones, 2015).

Quite what the beetles (and moths) are thinking when they crash into the light is open to a great deal of speculation. At one time it was thought that lights interfered with long-distance navigation during the migratory flights of moths (many of which do cross huge distances in regular seasonal movements). The calculation was that a moth using the moon as a navigational beacon would fly by keeping a constant angle between its intended flight path and the lunar light. But as an artificial light, whether electric street lamp or glowing Grecian lantern, is nearer to hand, the moth would quickly pass it by, then seek to readjust its heading to keep the light at a constant angle to its eyes. In effect, this would soon bring the moth spiralling in towards the light source. The trouble is that moths don't really appear to migrate using the moon, and nor do beetles. Another suggestion was that some light wavelengths (a lit candle, for instance) mimic the frequencies found in luminescent moth sex pheromones (Callahan, 1977), so perhaps the moths are looking for a mate. The trouble is that male and female moths both come to light. And the same is true for beetles.

The most sensible current theory (in my limited opinion) is that bright lights do not necessarily attract the insects, but they do confuse them. The assumption is that the strong light overloads the chemically primed sensors in the compound eye's rhabdom arrays. The light is so strong that the signalling ability of the light receivers and their associated neurons in the eye becomes exhausted. With no 'light' messages being sent to the brain, the insect interprets the very centre of the source as being dark, so moves towards it expecting to find a hidden niche in which to secrete itself. It's a bit like a rabbit being dazzled in the headlights of a car.

In truth, we are still some way off fully understanding why insects come to lights, but it is obvious that they do. I occasionally put out a bright mercury-vapour light over a white sheet in the garden. My cats chase the Jersey tiger moths (*Euplagia quadripunctaria*), but thankfully ignore the small beetles that I find much more interesting. The carrion beetle *Trox scaber* is a regular, breeding, I suspect, in discarded meat and bone kitchen waste in compost bins rather than under dry, sinewy animal carcasses. I doubt it migrates, but I would expect it to be attracted to small, dark places, which might be the nesting holes of owls and other birds in old tree trunks, in which it also regularly breeds.

WHEN TO FLY?

Obviously, it is only night-flying beetles (and moths) that are attracted to lights – when it is dark. However, there is a wide range of times when beetles are on the wing. Flower visitors are active during bright sunshine, but even dung beetles and dead-wood-feeders fly during the day, usually taking to the wing at some critical temperature. With just a little practice, flying beetles can be picked out against the sky and easily distinguished from the bees, wasps and flies that are also active. Beetles have a much more direct flight – straighter, more even, less swerving and less flighty. The air is obviously their second home, not their first, and they are less manoeuvrable here than their more aeronautic distant cousins.

Having said this, there are several astonishingly clever wasp mimics that easily fool the untrained eye. The wasp beetle (*Clytus arietis*) and wasp longhorn (*Rutpela maculata*) are brightly patterned black and yellow on pronotum and elytra, and perhaps gain some measure of protection from predators because they are so warningly coloured. But when they fly, usually in the middle of the day, they have the same hovering, bobbing flight as social wasps hawking about over blossoms and among the herbage. When they land on a branch or log, they have a similar jerking movement as wasps, which land on wood to chew pulp for their paper nest cartons. The deception is remarkable. Top tricksters, though, must be the bee chafers (*Trichius*). When I decided to examine a few bumblebees in a south London wood in 1996, I was mightily surprised when one of the orange-and-black carders I snatched at as it flew around a bramble (*Rubus fruticosus*) flower transformed in the net and turned out to be the rare vagrant chafer *Trichius zonatus* (now *T. gallicus*).

The importance of beetles in pollination has not been fully quantified since it generally occurs in non-specialist flowers where many different insect groups are active. However, there are a few known cases where orchids are pollinated solely by beetles (e.g. Steiner, 1998), and Bernhardt (2000) listed 34 plant families where at least one species is primarily pollinated by beetles. There was some recent scientific excitement when a specimen of an extinct oedemerid, *Darwinylus marcosi*, caked in pollen grains, was described from amber dating back to the time of gymnosperm dominance 125–90 mya, before the evolution of the flowering angiosperms that these beetles pollinate today (Peris *et al.*, 2017).

Dung beetles fly well during the day, and grazing meadows can be abuzz with them as they zigzag in on fresh droppings. Because of their ecological (and financial) significance in clearing away animal droppings, and the ease of studying them at baited traps, there is a wealth of research on their flight activity. Different species fly at different times of the day, thus avoiding inter-specific competition (Koskela, 1979; Feer & Pincebourd, 2005). In large African *Kheper* species at least, early-morning beetles can raise their body temperatures by rapidly vibrating their powerful thoracic flight muscles. The wings do not expand, nor are they flapped, but the metabolic energy released as heat by this shivering warms the beetles to the threshold of about 34 °C where flight is possible. This allows them to take advantage of the morning bowel evacuations of the local elephants ahead of their cooler competitors.

Many beetles (dung and otherwise) fly well into the late evening. Plenty of the old monographs (and new guidebooks) refer to evening sweeping as a technique for finding some of the more unusual species. Crepuscular flight may avoid some of the daytime bird predators, which are just settling down for the night come evening, and also bats, which have not quite roused from their roosts. There is also, in southern Britain at least, a typical change in the weather as the sun sets of a summer evening; winds drop and there is a period of still calmness before night begins to cool the air. I have only come across the pretty woodworm *Hedobia imperialis* by catching it on the wing just before dusk, and the tiny, probably wood-boring 'dung' beetle *Saprosites natalensis* (found in something like five or six British localities) flies through my garden on June evenings when I contemplate the world with a late cup of tea. It's tempting to speculate whether the scarce species taking to the air at this time of day are rarely seen because they live secretive lives and fly only during this brief thermal/behavioural/ecological window, when most coleopterists are having a good dinner.

CHAPTER 6

Beetle Habitats and Natural History

BEETLES GET EVERYWHERE

One of the great delights of insects, and the Coleoptera in particular, is that
no matter where you are, you can always find them. Every entomologist has
his or her favourite tale of finding a strange insect in an unexpected place. On
12 October 1975, the top of Beachy Head was bathed in autumn sunshine, but
as it was pitching up a howling gale we[18] retired to the sheltered, rock-pooled
foreshore, where the sun reflecting off the white chalk cliffs soon had us in our
shirtsleeves. On a whim, we started to pick at the rocky outcrops well below the
high-water mark with stout trowels and penknives, only to discover the small,
pale, wingless carabid *Aepus robinii* in the silt-filled cracks. This is one of only
a handful of insects that can tolerate this harsh habitat, covered over by the
sea twice a day. The only time I have ever found the saw-toothed grain beetle
(*Oryzaephilus surinamensis*) was in a sugar bowl in London's Hotel Russell on
7 September 1981, when I was attending a book fair there. My specimen still
appears to have a couple of sugar grains adhering to its body. And on a cold
30 November 1984, I heaved a sodden log from Crummock Water, up in the Lake
District, to find it riddled with Britain's own rhinoceros beetle, our smallest stag,
the handsome *Sinodendron cylindricum*.

---

18 'We' in this case included my father, the botanist and naturalist Alfred Jones (1929–2014),
'eccentric' entomologist and bookseller Laurie Christie (1930–2001), and 17-year-old schoolboy me.

These are my personal contributions to the standard entomological rhetoric that insects dominate the land, from the seashore to the mountain top. These far-flung insects are nearly always beetles. During the depths of winter you can still find beetles (this is the best time to go grass-tussocking; see p. 391), and the mantra of the hardened entomologist, 'See a log, turn it over', is ever rewarded no matter the weather, and no matter the season. When I moved to north London after a semi-itinerant post-university year, the first insects I found, on 7 December 1980, were several specimens of the small blue-black ground beetle *Ocys quinquestriatus*, under flowerpots, rather than logs, on the flat roof of my Willesden apartment. And when I was recently teaching some entomological basics to London Wildlife Trust trainees, on what was surely the wettest summer solstice in living memory, we were still able to find insects in the deluge, including the handsome lesser stag beetle (*Dorcus parallelipipedus*) and a single specimen of the tiny spiny weevil *Acalles misellus*, a 'dead-hedge' specialist when such boundary constructions were in vogue. My credo, that you can always find beetles and that coleopterists never get a day off, was intact.

The abundance, diversity and tenacity of beetles make them ideal subjects for ecological study, and they can reveal much about a specific habitat or a particular locality. Some beetles have very narrow habitat requirements or very precise geographical ranges; these tend to be the rather more uncommon species, which attract attention because they are not everyday, and are themselves associated with rare plants, unusual geology, precise climate or some other limiting factor. We can often make reasoned judgements about why they occur in some places and not others. On the other hand, common or garden species can occur everywhere and often occur anywhere; these are the chaff among which, to mix metaphors horribly, we can search for pearls.

CONCEPTS OF RARITY

It's always exciting to make a new find; to discover a species you've not seen before. One of my first published beetle articles was a short note (just half a dozen lines) in the *Entomologist's Monthly Magazine*, bragging that I had found *Hydnobius perrisi* (now known as *Sogda suturalis*), around the flooded gravel pits at Rye Harbour on 2 October 1977, new to me, and possibly the seventh British locality (Jones, 1979). Such notes are the currency of the field entomologist, reporting unusual discoveries to far-flung colleagues, and they make fascinating reading down the decades – even centuries later. Charles Darwin famously reported that his greatest pleasure, as a young man, was to see some of his own modest beetle records quoted by top coleopterist of the day James Francis Stephens in the supplementary appendix of the monumental *Mandibulata* volumes (Stephens, 1829). Before the advent of uploadable database record-keeping, these short published notes ensured that the observation got into the public domain, to be trawled and logged next time a monograph covered the group. They continue to be an important source of ecological and distributional information for journal readers and recording-scheme organisers.

After a while, even to the untrained eye, it is quite obvious that some insects occur all over the place while others are very restricted. Two basic ecological concepts are apt here: range and distribution. These concepts are similar, but not quite the same. The range of a species is the area in which it might occur if everything went its way. There are often geographic or climate limits to a beetle's range, thus the red-and-black click beetle *Ampedus elongantulus* and the small brownish downland chafer *Omaloplia ruricola* (Fig. 98) share the same range through southern England to about the Severn–Humber line. These are both rather scarce species, but the distribution of the *Ampedus* is limited to acid or sandy soils, and that of *Omaloplia* to base chalk and limestone. They are almost mutually exclusive. Meanwhile, the small ground beetle *Syntomus obscuroguttatus* has a similar range, but broadly overlaps both soil types.

FIG 98. The nationally scarce (notable) downland chafer *Omaloplia ruricola*. This is the all-dark colour form, lacking the pale brown elytra, found at Mt Caburn, Lewes, 19 June 1993.

The usual convention was (and still is) to describe uncommon species as either very local or rare. Effectively, 'very local' means found in only a few places, usually obvious or well-known localities, but a visitor to those sites might be able to uncover a specimen if they look hard and long enough. 'Rare' just means that specimens turned up seemingly at random, and that no amount of diligent searching could necessarily guarantee success on the hunt. These were often subjective measures, prone to misunderstanding, personal interpretation and local variance. 'One of our rarest Coleoptera' was a particular favourite phrase in the old books and monographs, regurgitated each time the writer reported a choice discovery, and exactly how Fowler (1889) described *Hydnobius perrisi*, much to my teenage delight. With the advent of computerised databases, burgeoning national recording schemes and a growing awareness of ongoing environmental monitoring, the task was set, during the 1980s, to try to standardise rarity, and give a real numerical measure to an individual species' abundance or scarcity in the landscape.

The first British insect Red Data Book (RDB) (Shirt, 1987) included 546 beetle species – about 14 per cent of the 3,900 species in the British beetle fauna of the day. This total was second only to the Diptera (827 species, 13.7 per cent of British fauna), although the book did not include many of the poorly understood and barely recorded parasitic Hymenoptera. Criteria for allocation of the various RDB statuses were necessarily varied and complex, but they were generally based on an individual assessment of how precipitously a species appeared to be declining, or perhaps just about hanging on, and this was hung on the framework of the numbers of hectads (10-km squares) on the Ordnance Survey national grid from which a species had been recorded. The landmass of Britain (England, Scotland and Wales) comprises 2,823 hectads; Northern Ireland and Eire (together comprising 1,007 hectads), the Isle of Man (parts of 14 hectads) and the Channel Islands (small portions of 15 hectads) were rather sidelined to start with, but recording and data-storage efforts have since sought to rectify this. On the whole, RDB statuses hinged around whether a beetle occurred in one, or up to 15, of the hectads.

Later came the Coleoptera species reviews (Hyman & Parsons, 1992, 1994), which built on the RDB categories, but which also included the next rarity echelon down, of species deemed to be 'nationally scarce' (also termed 'notable' – in other words, worthy of a short note in an entomological journal according to my reckoning). Again, notable statuses were mostly based on the numbers of hectads from which a species was known – nominally recorded from up to 100 hectads since 1970. My *Sogda/Hydnobius* species was thought to be very restricted (it was known from only three recent localities), but the Leiodidae are small,

The one and only... the enigma of the unique specimen

We should not lose sight of the fact that we are extremely lucky in Britain, having as we do a long tradition of 'amateur' naturalists, busy finding, describing and recording our biodiversity. We have just about the best-studied fauna on the planet. Elsewhere in the world, especially in the virtually unknown tropics, things are very different. In a sample stocktake of published articles on beetles from *Zoological Record* (Stork & Hine, unpublished, quoted by Stork, 1993), 45 per cent of species were described from just a single locality, and 13 per cent were from single unique type specimens. That's very nearly every eighth beetle species in the world. They were seen once, and may never have been found again.

When I was recently asked what was the world's rarest insect, I could have chosen any number of species, each based on one of those unique type specimens, but thought the neotropical water beetle *Megadytes ducalis* offered a useful historical lesson. Described in 1882 by distinguished coleopterist David Sharp (1840–1922), at nearly 50 mm long, 30 mm wide and 15 mm deep it is still the largest water beetle ever found. In the usual fashion of the day, Sharp described it in both Latin and English, gave its locality (Brazil) and gave the name of the collection (Saunders) in which it resided. In the concise 200-word entry for this species, he also states: 'I have seen only a single individual of this species'.

That single individual, a male, now housed in the Natural History Museum, London, remains the only specimen ever found. The data label attached to the specimen offers the same meagre locality information as Sharp's monograph. Brazil is the fifth-largest country in the world, with an area of 8.5 million square kilometres, and the largest system of freshwater rivers on Earth. The story goes that the specimen was found in the bottom of a dugout canoe. How is it possible to go back and rediscover it now? All we can do is wait and see whether any more turn up. It may already be too late. In 1994, barely a hundred years on from its original description, *Megadytes ducalis* was declared extinct.

poorly studied and seldom recorded, so although the beetle probably qualified for some sort of RDB status, nobody was really quite sure which one; consequently, it was dumped into the caveat RDB-K category – rare, but 'insufficiently known'.

Presently, rarity statuses are being reassessed using criteria laid out by the International Union for Conservation of Nature (2013). These are similar to the previous RDB statuses, and there is some overlap. Other rarity statuses are reflected in several National Environmental Research Council (NERC) lists for England, Scotland, Wales and Northern Ireland, of species of 'principal importance for the purpose of conserving biodiversity', and species that have their own priority biodiversity action plans (BAPs) to identify and halt factors in their decline. These more formal-sounding statuses are regulated for use by government, conservation bodies and planning departments, and have legal or political weight. For the field entomologist, RDB and notable statuses usually suffice. My *Sogda* now labours under the status 'red list, GB, pre-1994, insufficiently known'. It's still very rare as far as I'm concerned, and I've never found it again – I'm not sure many people have. Technically, the IUCN lists (being international) and BAP groupings do not include mere nationally scarce species, but the ongoing reviews of various family groups continue to reassess each taxon on a species-by-species basis (e.g. Hubble, 2014; Telfer, 2016). If this all seems very complicated, it can be simplified down to the notion that statuses reflect existing known distributions, from records identifiable in the literature and on databases, and on the way certain groups of organisms have reacted to land-use changes during the last however many years. In other words, are they rare, and are they becoming rarer?

One of the side effects of hunting out rare species, and disproportionately recording them (and publishing short notes on their discovery), is that every time a rare beetle is found it is, by necessity, less rare than before. This way, statuses can be downgraded – something that has happened widely with each new review of a family. Some species previously reckoned to be very uncommon are now shown to have been overlooked, and are actually much more widespread. Previously, the flea beetle *Longitarsus dorsalis* was considered to be nationally scarce on the grounds that it was known from only 24 hectads, but by the time of the reassessment review (Hubble, 2014), this had increased to 115 recorded squares. Often, all it takes is for someone to go hunting in likely looking but seldom-visited sites, or to go armed with a bit more knowledge of the beetle's biology. *Longitarsus dorsalis* is a brownfield specialist (see below) and started to appear quite regularly when abandoned, derelict sites were visited as part of legally prescribed environmental impact assessments, as they became scheduled for redevelopment. Some species are genuinely spreading, increasing in numbers and geographical range. The small subcortical colydiid *Cicones undatus* (Fig. 99) was given RDB-1 (endangered) status when it was first discovered in Windsor Great Park in 1984, but 30 years later (Alexander *et al.*, 2014), hindsight concluded it to be a recent introduction to Britain, associated with sooty bark disease in sycamores (*Acer pseudoplatanus*) (see p. 106); its modern immigrant status therefore precluded any long-standing native conservation status. We are now approaching the point where the regular assessment of beetle records might be useful in monitoring climate change, or mapping the spread of invasive species. More on this in Chapter 10.

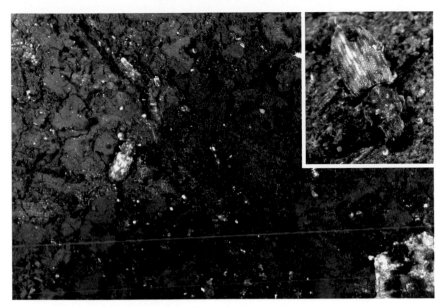

FIG 99. *Cicones undatus*, feeding in the black sooty spores of *Cryptostroma corticale*, the fungus that causes sooty bark disease in sycamores (*Acer pseudplatanus*).

Conversely, checking whether a particular species is being recorded less and less is a more difficult exercise. On 26 August 1981, I was skulking about on Rye Golf Course near Camber Sands, picking at deeply lichen-impacted stones around the edges of the rough, when I unearthed specimens of the small black colydiid *Orthocerus clavicornis*. Described simply as 'local' in books of the time, it was only with the centralised database storage of records that the realisation eventually dawned – the beetle is actually very rare, and has declined dramatically. It was listed as nationally notable (B) by Hyman & Parsons (1992), but this has since been upgraded to nationally rare (RDB-3) by Alexander *et al.* (2014). So, records continue to be monitored, and statuses assessed and reassessed, and eventually consensus arises – some rarities are downgraded, while others are promoted.

Today, environmental surveys and published reports of interesting finds are usually qualified by reference to the rarity statuses published in the various formal reviews. Nevertheless, there is always room for subjective interpretation of scarcity. Britain is not a uniform landmass, and it is very easy for a person in one county to find an uncommon species often, while a colleague in the neighbouring province has never seen it. This is more than me finding the pretty red-and-black lycid (net-winged beetle) *Platycis minutus* in the broadleaved woods of southern

England, while my Aberdonian colleagues find only the closely related *Dictyoptera aurora* in the Caledonian forests of the Cairngorms. Such geographic disparities are only to be expected. No, this is Peter Hodge finding the widespread but very local *Chrysolina brunsvicensis* on St John's-wort fairly regularly in Sussex, Kent and Surrey, but me never having seen it once. I'm not bitter, but I am curious to know why it's never turned up in my neck of the woods. When it does, I hope I can recognise the subtle ecological significance of the locality.

A WALK IN THE WOODS

Lowland Britain is chequered all over with small woodlands. Until about 1940, the start of intense mechanised farming and a major turning point in rural economies, these woods would have been managed by coppicing, regular thinning or some other forestry regimen, to harvest poles, brushwood and building timber by the local populace. Many of these oddments now look rather scrappy, but appearances can be deceptive. Although they contain few large ancient trees, they nevertheless represent truly ancient vestiges of the landscape, managed and maintained in the same way for centuries, if not millennia, and it is the beetles that have helped reveal and confirm this.

Beetles that breed in dead wood, fungus, fungoid rot or fermenting decay form part of this distinctive saproxylic wood-rot community. Some of Britain's rarest beetles occur only in large, internationally important ancient woodland sites, like the flightless weevil *Dryophthorus corticalis* (Fig. 100) and the mottled click beetle *Lacon querceus*, until recently both known only from Windsor Forest, and the strange malachite beetle *Hypebaeus flavipes*, known from only a few trees in Moccas Park, Herefordshire. These beetles don't necessarily need huge, ancient or decrepit trees; what they need is the long-term (over several centuries) continuity in the supply of fallen or standing dead timber, and fungal wood decay in all stages. The longer such a habitat exists, the more complex, diverse and species-rich it becomes. If, on the other hand, a woodland is clear-felled, it will take decades to build up the range of dead and fallen timber, with its varied decay fungi, and it may take centuries for the specialist saproxylic beetle fauna to recolonise the area to anything like the previous diversity. By the same token, a hedgerow, though it contains only scrubby growth, may yet be ancient. Such boundary hedges may have been in place for many hundreds of years, but do not necessarily contain any large centuries-old trees; this is because the trees within them are cut through every so often, and woven back into the layering of the hedgerow. On the other hand, pasture woodlands, where animals were

FIG 100. One of Britain's most restricted insects, the Red Data Book endangered weevil *Dryophthorus corticalis* occurs under loose fungoid bark, often in company with the brown tree ant *Lasius brunneus*. It is known from only a few old trees in Windsor, Richmond and Langley Park, Gloucestershire. Windsor Forest, 28 April 1985.

traditionally grazed under large old trees, often look more like fancy ornamental parks, but they contain much fallen timber, and dead standing limbs still attached to the sometimes gnarly old pollard trees punctuating the picturesque oil-painting landscape. These three disjunct habitats are the homeland of Britain's important dead-wood beetle fauna.

Harding & Rose (1986) emphasised this faunal association when they carefully grouped 196 beetle species that (i) were known only from documented, truly ancient woods and pasture woodlands; (ii) were known mainly from ancient woodlands, but also occurred in other well-wooded areas; and (iii) occurred widely in wooded habitats, but were collectively associated with ancient woods. These indicator species soon received considerable attention from coleopterists, and following their publication others attempted to compare and contrast different ancient woodland sites. This they did by constructing numerical scores and indices based upon the saproxylic beetle species lists. For example, Alexander (1988) and Harding & Alexander (1994) suggested simple scores for the indicator species, those most linked to ancient woodland scoring higher than those only collectively associated with it, thus: three points for each species in group i, two points for group ii, and one point for group iii. They then produced a table of tallied saproxylic quality scores for well-known ancient woodland sites in Britain. Large and well-studied ancient woods scored very highly – Windsor Forest scored 233, the New Forest 183 and Moccas Park 129, these being Britain's top three sites. However, most ancient woods scored much lower (Harewood Forest 22, Nettlecombe Park 23, Bookham Common 20, Ashtead Common 33, etc.), and a value of 20 or higher appeared to identify the most important sites in a national league table series.

Further interest in saproxylic beetles has produced more work. Fowles *et al.* (1999) gave a much longer list of 596 beetle species, and included common saproxylic beetles as well as the scarcer indicators. They suggested a scoring system based upon national scarcity, those RDB and notable statuses appearing in the literature, thus:

Score	Status
1	common
2	local
4	very local/uncertain
8	nationally scarce (notable B)
16	nationally scarce (notable A)/RDB-K (insufficiently known)
24	RDB-I (indeterminate)/RDB-3 (nationally rare)
32	RDB-1 (endangered)/RDB-2 (vulnerable)/RDB appendix (extinct)

They tested the reliability of these scores against 126 extensive species lists from various well-known woodland localities across Britain, and satisfied themselves, and other colleagues, that the scheme had value in genuinely comparing and contrasting woods in terms of the quality of their dead-wood beetle faunas. A minimum threshold of 40 qualifying species was thought sufficient to ensure that unreliable (short and therefore unrepresentative) lists could be excluded. To overcome the bias of recording effort, especially at well-known and oft-visited sites, they suggested the calculation of a saproxylic quality index (SQI), as well as a total saproxylic quality score. This was achieved by dividing the total score by the number of qualifying species and multiplying by 100. A hypothetical site where only 'common' species occurred (each scoring one point) would produce an index of 100 – this acts like the baseline against which other sites can be compared.

Again, not surprisingly, Britain's top sites had huge scores. Windsor Forest, for example, produced a list of 365 qualifying species, a total score of 3,092 and an SQI of 847.1. An SQI approaching 600 seemed to denote a site of international importance (seven British sites) and an index greater than 500 could denote a site of national importance (a further eight British sites listed).

Seventeen years on, and species lists can now be uploaded to an SQI calculation website (www.khepri.uk) and, as of 15 August 2017, the rankings of 203 woodland sites were available online. Those top sites still head the list, with the New Forest just edging in front to first place with an SQI of 857.1 and Windsor being knocked to the number two slot at 850, although it has a slightly longer

FIG 101. The nationally scarce (notable) fungus-feeding darkling beetle *Prionychus ater*. Downham Woodland Walk, Bromley, 26 June 2002.

overall species list. Ancient parklands dominate this table, with famous seats like Bredon Hill in Worcestershire (849.3) and Langley Park in Buckinghamshire (777.8) jostling for podium positions ahead of famous deer parks like Richmond (709.4) and Bushy (707.5), and royal forests like Hatfield (686.2) and Epping (599.6). But what is most enlightening is that seemingly lowly pieces of woodland can still show that they are part of the ancient natural landscape, as demonstrated by the rare dead-wood beetles they continue to support.

My local manor, Sydenham Hill and Dulwich Woods (432.1), is a respectable 74th, and that was without claiming all the tantalising 'Forest Hill' historical records that punctuate the old beetle monographs (Jones, 2002). Most surprising was Downham Woodland Walk (354.3), a narrow wooded trackway, remnants of ancient hedges and field-edge windbreaks, zigzagging through a dense 1930s housing estate in urban south-east London (Jones, 2003), which is currently at number 125 thanks to such treasures as *Abdera quadrifasciata* (Melandryidae), *Euglenes oculatus* (Aderidae) and *Prionychus ater* (Tenebrionidae; Fig. 101).

ELSEWHERE IN THE WILD

Using scores to rate sites is now commonplace in environmental surveys, and as with ancient woodlands, it works on the basis that scarce and unusual beetles are scarce and unusual because they live in specialist habitats that are themselves scarce and unusual, and usually threatened by human destruction or degradation. The saproxylic beetle fauna is a good model, and relatively robust. A slightly adjusted species list (highlighted in Fowles *et al.*, 1999) has been suggested for Caledonian pine forests of Scotland, which vary significantly from the

FIG 102. The handsome chocolate-coloured longhorn *Arhopalus rusticus* is a rare native of Caledonian pine forests but probably a recent immigrant to southern England, where it occurs widely in planted conifers.

broadleaved woods of lowland Britain, but retain a similarly complex beetle fauna because of their ancient continuity in the landscape. Beetles like the longhorns *Arhopalus rusticus* (Fig. 102) and *Asemum striatum* are genuinely associated with ancient Scottish conifer woods, but elsewhere in Britain they simply occur in modern pine plantations. Similar scoring schemes have been used for water beetle communities (Foster & Eyre, 1992) and as part of a wider analysis of freshwater quality (Howard, 2002).

In fact, any habitat can be scored using the species recorded. Drake *et al.* (2007) suggest a variety of score systems similar to those used in assessing ancient woodlands, with a sliding scale ranging from one point for common species up to 32 for an RDB-1 endangered insect. Previously, Ball (1986), who first developed the concept of nationally scarce British species, suggested 100 points for an RDB species, 50 or 40 points for notables A and B, and 20 points for a local species. Any system is useful, as long as comparisons are made using like-for-like scoring and indexing.

Beetles lend themselves to ecological surveys: with something approaching 4,120 British species (Duff, 2017), of which nearly a half qualify for RDB or nationally scarce status, any half-decent site should be awash with rare and very local Coleoptera. And despite the earlier propaganda about beetles occurring everywhere, there are some key habitat types where interesting beetles excel. Below is a quick run through some of my personal favourites. A book could be written about each of these habitats; indeed, many volumes of the *New Naturalist* series are just that. It is impossible to be all-inclusive, so just a very few sample beetle species are given for each.

COASTAL COMMUNITIES

The coast offers two key ecological bonuses for beetles: warmth and environmental hostility. The two are connected. Coastal sites tend to be exposed, windy, sometimes salt-sprayed, and often crumbling or unstable; under such situations herbage is stressed, and unlikely to grow thick and rank, with scrub invasion limited, and with large patches of well-drained bare ground in full sun. The sea (especially where the North Atlantic Drift and Gulf Stream have influence) acts as a thermal buffer and coastal breezes are consequently relatively mild, keeping the land's edge mostly frost-free. Many beetles can tolerate hot, dry, sparsely vegetated conditions, and while other species find this too gruelling, they can thrive free of much of the usual competition.

There is a certain frisson of danger when chasing the pretty little tiger beetle *Cylindera germanica* on the precipitous undercliffs of the southern Isle of Wight, and there is always the distant possibility of unearthing other enticingly rare ground beetles, like *Drypta dentata*, *Chlaenius nitidulus* or *Philorhizus vectensis*. Down on the shingle foreshore, the yellow horned-poppy (*Glaucium flavum*), which can survive the constant salty gales, supports the mottled and almost invisible *Ethelcus verrucatus* around its roots; at least its pretty, white-lined congener *Mogulones geographicus* is a trifle easier to see under the basal leaf rosettes of viper's-bugloss (*Echium vulgare*). Almost any extreme salt-tolerant plant on the beach will have local beetles feeding on it: sea rocket (*Cakile maritima*) has the orange flea beetle *Psylliodes marcida* feeding on its leaves, while the seedpods support the scarce weevil *Ceutorhynchus cakilis*, extremely similar to the widespread *Ceutorhynchus typhae* (formerly *Ceutorhynchus floralis*), which feeds on a variety of more inland Brassicaceae. Further down the beach, piles of rotting seaweed on the high-water strandline might support large numbers of small beetles; largest of these are *Cafius* rove beetles, highly distinctive with their broad heads and mottled hind bodies; of the three British species, only *Cafius xantholoma* is widespread.

Hands-and-knees searching is a prerequisite for sand dunes, where the windswept marram grass offers little shelter to the Coleoptera or the coleopterist.

FIG 103. Coastal cliffs are prone to disturbance and subsidence, creating the constantly changing habitat that is the undercliff. Newhaven Fort, photographed in 2014; despite stabilisation works, erosion still occurs. This site is home to the scarce ground beetles *Polistichus connexus*, *Ophonus cordatus* and *Brachinus crepitans*.

Fingertip grubbing in the loose soil can produce the smoothly black ground beetle *Broscus cephalotes*, no doubt a predator; the smoothly black root-feeding weevil *Otiorhynchus atroapterus*; or the rugosely black tenebrionids *Opatrum sabulosum* and *Melanimon tibialis*, both probably scavengers. Footprints in the dune sand tend to act like pitfall traps, and the tiny scarab *Aegialia arenaria* can sometimes be found hopelessly clawing its way up through a never-ending trickle of tumbling grains.

Britain has a significant share of Europe's saltmarshes, and although these often appear barren, damp and uninviting, they are home to a fine array of unusual beetles. The purslanes, glassworts and samphires here are rarely tall enough for effective use of a sweep net, so this time hands and knees get muddy. The beautiful golden-violet weevil *Pseudaplemonus limonii* is found under sea lavender (*Limonium vulgare*), and the truly noble tortoise beetle *Cassida nobilis*, with its golden-streaked elytral suture, occurs on glassworts and oraches, although I've most often found it sheltering under driftwood. Logs and timbers

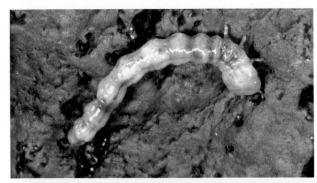

FIG 104. Larva and adult of the wharf borer (*Nacerdes melanura*). The larva was dug from rotten wooden fenders lining the quayside of Deptford Creek, a good 50 cm below the high-water mark.

washed up along the strandline shelter plenty of saltmarsh ground beetles like *Masoreus wetterhallii*, and *Pogonus* and *Dicheirotrichus* species, but are also riddled by the burrowing wharf weevil *Pselactus spadix* and the wharf borer (*Nacerdes melanura*, Oedemeridae; Fig. 104), both blamed for being notorious naval and mercantile pests in the days of wooden ships, and still doing damage to wooden groynes, pilings and fenders along British quaysides.

Just inland from the saltmarshes, the fields and meadows might look just like those of the deepest Home Counties, but the maritime habitat still exerts a powerful influence here. Ragwort and hawkweed flowers attract the brilliant bluish-green *Psilothrix viridicoerulea* and the dusky olive-green *Dolichosoma lineare* (both Dasytidae). These both have predatory larvae living in the root thatch and leaf-litter/soil layer, but what ecological factors limit them to a few seaside meadows and verges is still a mystery.

FRESH WATERS – AQUATIC AND RIPARIAN

Netting for aquatics is an entirely different pursuit from chasing terrestrial insects. It needs a much stronger net to suffer the indignities of dipping into muddy water and dragging through the dense pond herbage; and the fauna is different enough for some entomologists to specialise solely in aquatic invertebrates. The smooth lines of Dytiscidae and Haliplidae are immediately recognisable, but because they are the same shape, some of the larger genera (*Hydroporus*, *Agabus*, *Hygrotus*) can be awkward to identify; there are, though, plenty of prettily marked species, and seeing the dainty markings of *Brychius elevatus*, *Platambus maculatus* or *Graptodytes pictus* is always uplifting. Some water beetles are spectacularly picky in their habitat requirements, and theoretically very useful when monitoring environmental (particularly water) quality. The discrete, narrow, rusty-red *Hydroporus elongatulus*, which if not flightless is definitely flight-averse, has a strange disjunct relic distribution in Britain, occurring in fen pools in Scotland and pingo[19] ponds in East Anglia. The variegated black-and-white *Laccophilus poecilus* (formerly *L. variegatus*) was last found in a ditch on the Lewes Levels in 2002. This endangered beetle was regularly monitored since the creation of a dual-carriageway bypass through the marshes; fears for its survival are exacerbated by worries about pollution run-off from the road surface, and agricultural 'improvement' of the surrounding meadows.

Most water beetles have to be dipped for, using a pond net, but the whirligigs are active on the surface, and can be scooped out by the quick swish of a kitchen flour sieve, tea-strainer or (at a push) a conventional insect net. Most look the same in the field, glossy black, but the mostly north-western *Gyrinus minutus* and the southern and eastern *G. urinator* have a reddish underside. *Orectochilus villosus* is, as its name suggests, covered all over with short villose pubescence, giving it a dull grey appearance. Many books claim it is mostly nocturnal, but the only time I ever found it, dozens were skidding about the water surface in the bright midday sunshine.

To some extent, water beetling is a bit like thrashing about in the dark. It is only very rarely that you can spot a diving beetle swimming gently through the water and observe it directly; although many waterbodies are clean, they are cloudy with suspended silt, or obscured by light reflections. Despite being upside down under a strongly refracting surface, water beetles are restless and

---

19 Pingo comes from the Inuit word for a dome-shaped mound of earth over a lens of ice. When the last ice-age cover retreated from Britain 15,000–10,000 years ago, the ice mounds melted and drained, creating crater-shaped water-filled depressions in the Norfolk Breckland. Some (e.g. Foster, 2010) suggest these features should really be called palsa scars, a palsa being a frozen hummock of bog moss, smaller than a true pingo, but nevertheless a geological remnant of periglacial activity.

wary, and even a large *Colymbetes fuscus* idling in a flooded cart rut will disappear into the mud and roots if a shadow falls anywhere near. Most often, you have to trawl for them; this does have the advantage of throwing up some very unusual things. The riffle beetles (Elmidae) are secretive and concealed, usually clinging to the underside of stones and pebbles in flowing water – streams and rivers – rather than in ponds. Holding the net downstream and then kicking up the bed with booted feet often dislodges them in a cloud of detritus. Many elmids are northern and western species, preferring the fast-flowing waters of upland rills rather than the sluggish meanders of lowland rivers. *Oulimnius tuberculatus* and *Limnius volckmari* are about the commonest and most widespread British species, but some, like *Macronychus quadrituberculatus* and *Normandia nitens*, are very widely scattered. Getting in with boots and kicking up the herbage is, nevertheless, a useful technique anywhere, and as well as true water beetles expect to find things like the peri-aquatic weevils *Eubrychius velutus* and several *Bagous* species.

At the water's edge, thick reed beds and exposed banks all have their specialist faunas. Just standing and watching in the sunlight can be mesmerising, as dozens of small shiny ground beetles (*Bembidion*, *Elaphrus* and *Pterostichus* species) scuttle about. Burrowing species like *Dyschirius* (Carabidae), *Bledius* (Staphylinidae) and *Heterocerus* (Heteroceridae) can be tricked into showing themselves by splashing the banks with water, or by gently submerging marginal vegetation; they soon move, responding just as they would to one of the inundations that must frequently wreak havoc with these marginal habitats.

Reed beds and stands of bulrushes have their own special faunas, too. Flattened or narrow ground beetles like *Odacantha melanura* and *Demetrias imperialis* are able to insinuate themselves into the tight leaf sheaths. The startlingly red malachite beetle *Anthocomus rufus* and the related soldier beetle *Crudosilis ruficollis* are both confined to reed beds, and there are plenty of leaf beetles and weevils closely attached to waterside foodplants, like the attractively striped *Prasocuris phellandrii* on marsh-marigold (*Caltha palustris*), and the pale, dusky grey *Tapinotus sellatus* on yellow loosestrife (*Lysimachia vulgaris*). Two beetles that hold special pondside memories for me are the bronze-and-red *Chrysolina polita* and brilliant, metallic turquoise-green *Chrysolina herbacea* (formerly *Chrysolina menthastri,*); mental images of these leaf beetles are dragged up from the recesses of my mind whenever I detect the heavy scent of water mint (*Mentha aquatica*) from the crushed herbage underfoot. The former was one of the first beetles I ever found, at the pond in Plashett Wood near Lewes in the late 1960s, and the latter is memorable because of an unusual rich golden-brown variety my father and I found on water mint near Danehill, on 28 August 1971 and again on 3 August 1974, suggesting that it was a persistent local genetic form.

LOWLAND HEATHS (INCLUDING BRECKLAND)

Once derided for being agriculturally worthless (Shakespeare's 'blasted heath' resonated for centuries), the expanses of heather growing on low-nutrient sandy soils can be poor in terms of overall species numbers; they are, after all, more or less dominated by just a few endlessly repeating plant species. A frightening 86 per cent of Dorset heaths were lost between 1750 and 1980 (Stubbs, 1982), a statistic that can be extrapolated across any of the southern counties once so wealthy in this habitat. Today, lowland heaths across all of England have become fragmented, corrupted and isolated, so any specialist heathland insects (and there are plenty of them) have also declined drastically. Even the common heather weevil (*Strophosoma sus*) and heather beetle (*Lochmaea suturalis*) are actually rather local, found, as they are, only on *Erica* heaths and ling (*Calluna vulgaris*).

Today, heathy remnants can be hard work. With few plants other than heathers and ling, and even fewer blooms, most phytophagous beetles feed on birches, gorse (*Ulex europaeus*) and broom (*Cytisus scoparius*); yet despite these being common and widespread species, those growing on heaths can yield rare treasures. Several pot beetles are heathland specialists, and whereas *Cryptocephalus biguttatus* actually feeds on cross-leaved heath (*Erica tetralix*), so is more obviously confined to heathlands, the similar *Cryptocephalus bipunctatus* and *Cryptocephalus coryli* are found on heathland birches.

The scarce click beetle *Ampedus sanguinolentus* is one of many strikingly coloured black-and-red species that breed in dead and decaying birch stumps, but most adults are found by grubbing at the roots of heathers. Here, too, may be the startlingly metallic green ground beetle *Poecilus lepidus* or the inappropriately named wood tiger beetle (*Cicindela sylvatica*), which occurs only in those woods that have grown up on overgrown heaths. There are plenty of other scarce heathland ground beetles, including *Carabus nitens*, *Amara equestris*, *Harpalus anxius* and *Cymindis macularis*. These are all fairly generalist predators, so are not limited in their distributions by association with heathland plants or heathland prey; there is something about the microclimate of heathlands – hot, well drained (although sometimes boggy), sparsely vegetated and nutrient-poor – that favours these beetles, or allows them to thrive where others struggle.

The threatened hide beetle *Trox sabulosus*, a scavenger in carrion when it reaches the dry, sinewy skin-bone-fur-and-feathers stage, is known from only a handful of heathland localities. Its congener *T. scaber* finds animal corpses in the right state all over the place, and will make do with bits discarded in owl nests, or cooked remains of the Sunday roast chucked onto the compost heap. There is some, as yet unidentified, extra that *T. sabulosus* needs in a corpse.

FIG 105. The attractively mottled, domed shape of *Attactagenus plumbeus*, a weevil of light soils, especially sand dunes.

When I found several individuals on Ambersham Common in West Sussex on 23 August 1987, they were caught in the folds of some ancient mouldering hessian sacking; I still sometimes ponder on what the sack might once have contained to be so attractive.

Hot, dry heathlands do favour some plants other than heathers. The only place I've ever found the scarce weevil *Hypera dauci* is on the East Anglian heaths, the Norfolk Breckland, where it feeds at the roots of the common stork's-bill (*Erodium cicutarium*), a widespread plant, but one generally found in warm, dry sandy places. The very uncommon and delightfully chevroned weevil *Cleonis pigra* is also a Breckland (and coastal dune) speciality, although its larvae feed in the stems of several more or less ubiquitous thistle species.

CHALK DOWNS AND LIMESTONE HILLS

The precipitous decline of chalk downland has mirrored that of heaths, and for similar reasons. Modern agriculture fails to make much of a profit from the steep hillsides, where only costly, labour-intensive sheep- and cow-grazing regimes can be made to work. On the gentler slopes, the thin, nutrient-poor topsoil needs heavy input of synthetic fertilisers to support arable crops. Consequently, much of this once rich land has been abandoned to scrub invasion or lost to intensive arable farming under the harrow and plough. I grew up on the South Downs of

Sussex, where the rolling hills full of orchids, yellow-rattle (*Rhinanthus minor*), butterflies and bumblebees are now often seas of common hawthorn (*Crataegus monogyna*) and dogwood (*Cornus sanguinea*) thicket, or endless yellow fields of rape (*Brassica napa*). However, where chalk habitat does survive, it is rich with scarce beetles. The cute (I almost hesitate to use the word) chafer *Omaloplia ruricola* (Fig. 98) flies like a miniature bumblebee over the short, flowery turf; pale orange and blackish-brown colour varieties abound. The pale grey *Dascillus cervinus* is sometimes described as the orchid beetle, but its relationship with this group of plants is unclear beyond general observations that the larvae feed at roots in the upper soil layer, and that chalk turf is often rich in orchids.

Both bloody-nosed beetles feed on bedstraws and crosswort (*Cruciata laevipes*) in open situations, and although I used to find the large *Timarcha tenebricosa* quite regularly in my boyhood downland walks (it turned up in the garden once), it is the scarcer, more restricted lesser bloody-nosed *T. goettingensis* (I only ever saw it in Friston Forest), that is more closely attached to chalk soils. Other fairly strictly chalk-based leaf beetles include *Cryptocephalus bilineatus* on kidney vetch (*Anthyllis vulneraria*) and *Chrysolina sanguinolenta* on common toadflax (*Linaria vulgaris*). The flea beetle *Hermaeophaga mercurialis* feeds on dog's mercury (*Mercurialis perennis*), but even though this woodland plant occurs widely right across the whole of Britain, it seems to support the beetle readily only when it grows on calcareous soils.

Plenty of weevils feed solely on typical chalk grassland plants, from the diminutive *Squamapion atomarium* on thyme, to the giant, glossy black *Liparus coronatus* on cow parsley (*Anthriscus sylvestris*). I used to regard the pretty little black-and-red chevron-marked *Mecinus labilis* as being fairly common; it feeds on ribwort plantain (*Plantago lanceolata*), and I would regularly find it in the net as I wandered the South Downs as a boy, but I hardly ever see it now I have left the chalk. The same is true of *Hadroplontus litura* on creeping thistle (*Cirsium arvense*), *Miarus campanulae* on harebell (*Campanula rotundifolia*), *Baris picicornis* on wild mignonette (*Reseda lutea*), *Romualdius scaber* at plant roots and *Hypera arator* on campions; it's a red-letter day for me if I find any of these now, although they are not rare to other, more chalk-based coleopterists.

FARMLAND AND ARABLE

Modern intensive arable farmland is notoriously poor for wildlife, but this was not always so. The edges of ploughed fields were often a good place to look for ground beetles. The edge of the sod, where the plough has gouged its final cut through the soil, offers the perfect instance of an ecological edge effect. Fingertip grubbing at this edge is very productive, because although the same species probably occur well into the unploughed headland, and even up to the hedgerow, they become concentrated at the very extreme of what is to them a cliff top. This was where I always found *Synuchus vivalis* and *Stomis pumicatus*; again, not especially rare beetles, but ones I'm sure I see less of nowadays. On 3 June 1976, on a field edge near Alfriston, I turned over a broken plank of wood where the plough had crashed into a fallen fence post, and upwards of 100 of the pretty blue-and-pink *Anchomenus dorsalis* scuttled off. Because of the amenability of arable field edges to pitfall trapping studies, some attempts have been made to use carabid species richness in such locations to monitor things like organic versus conventional management (e.g. Holland & Luff, 2000; Döring *et al.*, 2003) and crop types in relation to aphid predation (Hance, 1990), or to see if unploughed headland buffer zones make a difference to biodiversity (e.g. Thomas, 1990; Irmler & Hoernes, 2003).

Obviously, any plant-feeders in arable fields are likely to be regarded with suspicion by farmers, but crop monocultures do not favour many species other than the ubiquitous pollen beetles, *Meligethes aeneus* and *M. viridescens* which sometimes cloud above fields of rape (*Brassica napus*) in their countless billions. In times past, the chunky carabids *Curtonotus aulicus* and *Zabrus tenebrioides* were regarded as minor nuisances in cornfields, because they feed on the heads of wheat. *Curtonotus aulicus* continues to be common and widespread, in rough meadows, verges and fallow land, feeding on almost any grass seeds, but by the 1970s *Zabrus* had all but vanished. When naturalist Roger Dumbrell found a dead and broken specimen among cliff-top debris near Brighton in the mid-1970s, he realised he'd probably never find another. He painstakingly reconstructed the beetle as best he could, resting the loose elytra on a knot of twisted tissue paper in place of the absent abdominal segments, and replacing the several missing legs and antennal segments with those from a large *Curtonotus* to make a fair reproduction of a presentable museum specimen. I know that some museum-based entomologists will shudder at this behaviour. The beetle appears to have enjoyed some resurgence of late (Oakley, 2004).

NON-ANCIENT WOODLANDS

Woods do not need to be ancient to be interesting, and plantations, overgrown hedges and abandoned copses all provide good beetle habitat. The curious and attractive (for a silphid) *Dendroxena quadrimaculata* lives in the foliage of

FIG 106. *Rhagium mordax* is a handsome longhorn found in woodlands throughout the British Isles. This one is fresh from the pupa, and although its upper body, legs and antennae look well coloured, its poorly sclerotised abdomen remains soft and pale.

broadleaved trees, mostly oaks, and attacks moth caterpillars, especially those of the green oak tortrix (*Tortrix viridana*). During mass outbreaks of this moth, the broad, dark carabid *Calosoma inquisitor* can also occur.

Modern orchards are sometimes little more than intensive monocultures, where dead limbs are ruthlessly pruned and dead trees uprooted to make way for new stock. If traditional, more relaxed management occurs, there is always a chance of seeing the noble chafer (*Gnorimus nobilis*), which breeds in the heart-rot of dead trunks and branches. It was always scarce, but is now reduced to the point where it is a NERC species of principal concern and has its own BAP. Being a bright metallic chafer (although narrower and more elegant than the rose chafer, *Cetonia aurata*), it is sometimes discovered by casual observers, and the occasional tweeted photo of the beetle or its frass-filled burrows is enough to alert coleopterists, such that viable breeding colonies have been located (e.g. Harvey *et al.*, 2014). The old apple-hub cities of Gloucester, Worcester and Hereford seem to be strongholds for this rare beetle, and also for the pretty weevil *Ixapion variegatum*, which feeds on the mistletoe (*Viscum album*) growing on the same old orchard trees. *Gnorimus* is undoubtedly a long-standing native species, but the sudden appearance of *Ixapion* (Foster *et al.*, 2001) and its subsequent discovery throughout the region might imply that it is a recent arrival, introduced from the Continent with the Christmas trade in decorative mistletoe sprigs.

Planted trees offer varying degrees of beetle richness. The supposedly scarce weevil *Dorytomus ictor* occurs on black poplars, most of which are planted cultivars in Britain. Rarely found by beating the leaves on which their larvae feed, the pale, mottled adult weevils can often be discovered sheltering under loose flakes of mildewy old bark around the bases of the trunks during the winter months. Sycamore and London plane (*Platanus* × *acerifolia*) trees are often derided as being ecologically worthless, but as noted on p. 106, the sooty bark disease now makes them more palatable to various beetles. Plane trees, the urban street tree of choice 150 years ago, have the advantage that their deciduous peeling bark flakes are easy to examine for sheltering beetles beneath. The common flattened ground beetles *Dromius quadrimaculatus* and *D. agilis* are regulars, and the cobweb beetle (*Ctesias serra*; see p. 98), once thought to be an ancient woodland indicator, occurs once the old trees have flakes large enough to shelter its *Amaurobius* spider hosts.

Overgrown hedges with large old hawthorn bushes are almost certainly home to the pink hawthorn jewel beetle (*Agrilus sinuatus*). So named for the distinctive sinuous burrows its larvae bore under the old tree bark (Fig. 107), this beetle's activity can be discovered by locating the D-shaped exit holes the adults chew when they emerge. All the old books claim that *A. sinuatus* is very rare,

FIG 107. The winding burrows of the ruby-coloured hawthorn jewel beetle (*Agrilus sinuatus*) under the bark of a dead common hawthorn limb.

FIG 108. The elusive, perhaps short-lived adult of the hawthorn jewel beetle (*Agrilus sinuatus*). (Penny Metal)

and certainly the adults are rarely found (Fig. 108). I suspect they are extremely short-lived as adults; I've only ever seen one, beaten from an old hawthorn hedge in Morden Cemetery on 15 July 1998. However, the widespread occurrence of burrows and exit holes shows that this is not an RDB-2 (vulnerable) beetle as was originally proposed, nor nationally scarce (the status to which is was later revised), and it is now regarded as merely 'local' (Alexander *et al.*, 2014). I've still only ever found the one though.

MOUNTAINS AND NORTHERN PEAT BOGS

Even more than southern lowland heaths and chalk downs, northern and western mountains and moors were regarded as agriculturally useless, but since they were mostly inaccessible to the overlords of industrial farming, they have survived better. Their inaccessibility does mean that they are often poorly studied, but when visited they are inevitably home to unusual and scarce insects found nowhere else. Perhaps the best known, though least often seen, is the rainbow leaf beetle (*Chrysolina cerealis*), which is restricted to a few stunted patches of wild thyme (*Thymus polytrichus*) growing in Snowdonia. This is one of only six[20] beetles protected in the UK by Schedule 5 of the Wildlife and Countryside Act 1981 (it is illegal to harm, kill or collect them). Since the foodplant is common and widespread across the British Isles, the beetle must have some other, as yet unidentified, habitat requirement.

There are plenty of beetles with distinct northern and western distributions, and these geographic spreads are often a direct result of the mountainous landscapes of Wales and Scotland. Many water beetles (including the elmids mentioned above) prefer hillside streams, not just because they are fast-flowing, but also because they are in less densely urban or industrialised areas, so high water quality is clearly a contributing factor. Typical mountain-dwelling water beetles include the whirligig *Gyrinus opacus* and the diminutive but well-named *Oreodytes alpinus*.

Hillsides are notoriously tree-free, but birches are typical fast-growing invaders (as is true of lowland heaths). The bee-chafer (*Trichius fasciatus*) breeds in mouldering rotten birch stumps in Wales and Scotland, and can be quite common where it occurs. Most adults are apparently found on thistle flowers.

Northern moors are often dominated by heather-covered hillsides or boggy mosses. On the drier hillsides, fingertip searching might discover the carabid *Miscodera arctica*, but the very rare *Bembidion humerale* is known only from Hatfield and Thorne Moors in Yorkshire. These sites are also the only known localities in Britain for the tiny rotund pill beetle *Curimopsis nigrita* (Byrrhidae). When it was first discovered here in 1977, it was felt necessary to insist on an English name, so that simple politicians and journalists would not be flummoxed by complicated Latin. I'm pompous enough to disapprove of this behaviour, but if a common name must be used then please let it be mire pill beetle and not the alternative, the ghastly, puerile, fatuous 'bog hog'.

Sadly, Britain's mountain ranges are severely impoverished when it comes to underground cave systems, or at least the insects living in them. Several species have been found in British caverns, but there is no true troglodyte here. The small ground beetle *Trechoblemus micros* has been found regularly in caves, but it is also widespread in damp grasslands, possibly in association with small mammal burrows. There are plenty of cave systems, and cave beetles, in Europe. These are mostly long-legged carabids and darkling beetles (Tenebrionidae). The bizarre, long-legged bulbous leiodid *Leptodirus hochenwartii* was the first truly cave-dwelling

---

20 For completeness, the others are *Graphoderus zonatus* (Dytiscidae), *Hypebaeus flavipes* (Melyridae), *Hydrochara caraboides* (Hydrophilidae), *Paracymus aeneus* (Hydrophilidae) and *Limoniscus violaceus* (Elateridae).

beetle to be described (in Schmidt, 1832), from caves in the European karst. It lays huge eggs that hatch into non-feeding larvae, these pupating quickly and emerging as adults that lack both eyes and wings.

BROWNFIELD SITES

In 1997, when I was asked to carry out an environmental survey of Deptford Creek in south-east London, I peered tentatively over the balustrade of the A200 Creek Road bridge and my heart sank at the bleak industrial dereliction all around. But I persevered, and I'm glad I did – because the scales fell from my eyes. Despite the unprepossessing state of fly-tipped heaps of rubbish, derelict buildings, bulldozed piles of rubble, overgrown walls and broken hardstanding, I managed to find 287 invertebrate species, including 63 beetles. And contrary to my expectations, they were not merely common or garden species, but some real oddities – several RDB and nationally scarce species among them. Throughout the various sparsely vegetated sites, yellow hawkbit flowers were dotted with small convex, shining black beetles – *Olibrus flavicornis* (Phalacridae). This RDB species had not been seen in Britain since it was found at Camber Sands in 1950, but was all over the place in derelict Deptford. I was not the first person to notice this London/ Thames Estuary resurgence for the beetle, and in fact it was quite widespread in the area. But this was not a simple sudden increase of a previously overlooked small beetle, or newly arrived immigration; nearly 20 years on, and *O. flavicornis* is still confined to the south Essex and north Kent sides of the estuary, on into Docklands London. This area is particularly pockmarked with endless derelict brownfield sites, where former maritime mercantile industries went into decline in the mid-twentieth century. It soon transpired that this brownfield land (deliberately named to contrast with the more attractive-sounding green fields of the countryside) held a secret promise of invertebrate biodiversity.

The key thing about brownfield sites is that, having lots of bulldozed rubble, crushed brick and broken concrete underfoot, the substrate is well drained and sparsely vegetated, with areas of bare ground that become very warm in direct sun. This favours warmth-loving or warmth-tolerating species – often those with a more Mediterranean distribution in Europe, and that in Britain are also found in warm, well-drained, sparsely vegetated sites like sandy heaths, chalk downs, sea cliff tops, dunes and dune-slacks. These, as discussed earlier, are vulnerable and declining habitats that support plenty of unusual beetles. Brownfield sites are the new reservoirs for many of these unusual beetles. A top statistic quoted by Gibson (1998) is that an estimated 12–14 per cent of Britain's nationally rare (RDB) and

FIG 109. The streaked bombardier beetle (*Brachinus sclopeta*), known from only a few brownfield sites on the east side of London and close to the Thames Estuary.

nationally scarce (notable) insects are recorded from brownfield sites – a higher proportion than is known from ancient woodland or chalk downland. Beetles, in particular, favour the harsh, dry, warm conditions so prevalent on brownfield sites.

Regulars on London brownfields include the bombardier beetle *Brachinus crepitans* (and, recently, the streaked bombardier, *Brachinus sclopeta*; Figs 109 & 110); the ground beetles *Panagaeus bipustulatus*, *Amara montivaga* and *Harpalus*

FIG 110. Brownfield site near Pudding Mill Lane Station, Stratford, where the streaked bombardier beetle (*Brachinus sclopeta*) was found in numbers on 21 July 2014.

anxius; weevils like *Otiorhynchus raucus* in the root thatch, *Taeniapion urticarium* on common nettles (*Urtica dioica*), and *Rhinocyllus conicus* and *Larinus planus* on thistles; the ladybird *Hippodamia variegata*; the leaf beetles *Chrysolina banksi* on black horehound (*Ballota nigra*) and *Chrysomela populi* on sallows; and the longhorn *Agapanthia villosoviridescens*.

Brownfield sites are often rich in annual plant species, since the soil (if there is any) is so dry and poor that perennials have a hard time surviving the regular droughts; this keeps the habitat in an early successional stage, preventing the rank plant growth or scrub invasion that now threatens sandy heaths and chalk downs. This often makes brownfields especially flowery. Wild mignonette flowers are usually dripping with the small anthribid *Bruchella rufipes* (scarce, but spreading), and I was more than pleased, recently, to discover the lovely bronze-and-red leaf beetle *Chrysolina marginata* by grubbing at the roots of the plant. Annual mercury (*Mercurialis annua*) is a plant typical of disturbed ground, and on brownfield sites in south-eastern England is host to the scarce *Kalcapion semivittatum*. Other typical brownfield beetles include the weevil *Pseudostyphlus pillumus* on mayweeds, *Mordellistena acuticollis* on wormwood (*Artemisia absinthium*) and the leaf beetle *C. hyperici* on St John's-wort – although not *C. brunsvicensis* apparently.

Disused quarries are technically brownfield sites, although they have often blended back into the landscape, having been flooded and greened up, and often used as leisure facilities. Such delights as the curious rotund ground beetle *Omophron limbatum*, burrowing *Bledius* rove beetles and *Heterocerus* mud beetles can often be dislodged from the edges of the pools by splashing water up onto the banks. Away from the water, typical sandpit plants like broom can be attacked for weevils like *Exapion fuscirostre*, or horsetails for the pretty *Grypus equeseti*. Maybe even the occasional *Sogda suturalis* will make an appearance.

Other brownfield designations are used for road verges, railway embankments and gardens, although these tend to be more naturalistic, at least in the green scenery sense. They are often useful wildlife reservoirs because they are sometimes less intensively managed than the surrounding agricultural land. The first place I ever found the scarce weevil *Exapion difficile* was on a drift of its foodplant, dyer's greenweed (*Genista tinctoria*), on a broad roadside verge near Uckfield, although the plant is widespread on the Ashdown Formation geology than dominates the central Sussex Weald thereabouts. Perhaps it was something to do with the sheltered sweep of the steep bank, and the fact that municipal mowers had trouble keeping the sward under control.

Interesting beetles turn up in gardens all the time, mostly because interested coleopterists tend to keep an eager eye out when they're sitting out on the patio with a cup of tea or weeding the veg patch. Britain's smallest (3–7 mm) longhorn, *Nathrius brevipennis*, landed on me when I was reading the newspaper at the end of my south London garden on 18 July 2010. Although the species has long been considered a sporadic vagrant, with singletons usually recorded, sometimes emerging from wickerwork, another crawled over my metal garden table on 14 July 2014, making East Dulwich a long-term breeding colony as far as I'm concerned. And I was always jealous of my friend Roger Dumbrell's good fortune to find (on several occasions) the large, brassy green and elegantly sculptured ground beetle *Carabus monilis* in his small garden in Wilmington, East Sussex, where the soft outline of the River Cuckmere floodplain meets the undulating ridges of the South Downs.

FIG 111. Female stag beetle (*Lucanus cervus*). Every year we get stag beetles flying about in our south London garden. Although I claim this one as my own, it was actually the cats that nosed it out in the long grass near the pond.

INVADING OUR PRIVATE SPACE

Beetles don't stop at the garden; there are plenty that have adapted to live almost exclusively in human habitations. And the moment they cross the threshold to invade our private domains, they risk immediate persecution for their temerity. These are not, though, creatures from the garden taking a wrong turn at the back door, but often cosmopolitan insects that have been transported around the world in foodstuffs and manufactured goods, and that nowadays occur almost only inside buildings (Jones, 2015). Buildings may be warm and well stocked with food, but in Britain at least it is their dryness that dictates which species can survive indoors and which cannot. Water loss is an important danger to small insects and one of the prime movers behind tight-fitting elytron evolution, so it is no surprise that those beetles that evolved in dry places succeed better indoors. Many indoor beetles are notorious food pests; those like the grain weevil (*Sitophilus granarius*), flour beetles (*Tribolium*), the biscuit beetle (*Stegobium paniceum*), the mealworm (*Tenebrio molitor*) and the cellar beetle (*Blaps mucronata*) eke out quiet livings on spilled scraps of food in dusty corners, but if left uncontrolled in food stores, they can become pests of biblical proportions. Many of these are originally from the biblical Middle East, North Africa or the Mediterranean, and first invaded 'from the wild' with the dawn of agriculture sometime around 15,000 years ago.

Today's modern houses, with air-conditioning, double-glazed windows, fitted carpets, fridges, Tupperware and Hoovers, are even more of a challenge to domestic wildlife, so coleopterists rarely come across these pest species unless they are in the pest-control industry. Unfortunately, with an eye on potential litigation, lax bakers and unhygienic restauranteurs are often unwilling to let

FIG 112. Something's been eating my food. A well-established colony of biscuit beetles (*Stegobium paniceum*), which has reduced the medium egg noodles to ticker tape and dust.

entomologists go scraping around in their storerooms looking for interesting infestations. Instead, letting it be known that a slightly eccentric entomologist with a sympathetic interest in domestic insects lives in the vicinity can sometimes bring the beetles to the coleopterist. So it was when I was asked by a parent at my children's school to identify the things crawling up her skirting boards and drowning in her cat's water bowl. They were the rice weevil (*Sitophilus oryzae*), emerging from some rice-like seed in the wild bird-food mix in a kitchen cupboard. I'd never seen it before.

Whole monographs have been written on the beetles infesting homes, warehouses, stores, shops and offices (a good place to start is Hinton, 1945). Some groups, like the Dermestidae (*Dermestes, Anthrenus, Attagenus, Trogoderma* and *Megatoma*) have wholeheartedly embraced this interloping lifestyle, but new contenders continue to arrive. The pretty spider beetle *Ptinus sexpunctatus* was usually regarded as scarce, a scavenger in solitary-bee nests, feeding on pollen stores and dead brood, and is perhaps sometimes associated with pine woods, but a recent request canvassing opinion about its nationally scarce status confirmed that most modern records were from inside houses. I sometimes used to find it in the bath, and I always imagined it to have fallen through the light-fitting hole from the loft, where old birds' nests could have provided much dusty nutrition in the form of spilled food, shed feathers and the odd dead hatchling. I recently came across several hundred of them in a mouldering carton of fish food in my garden shed. Along with mundane ingredients, which included cereals, vegetable-protein extracts and derivatives of vegetable origin, were fish and fish derivatives, yeasts, sugars, algae and, rather startlingly, molluscs and crustaceans. The food was obviously a good source of animal protein, which is presumably one of the key nutrients contributing to the beetle's success.

On the other hand, some previously common household beetles now have nowhere to live. Archaeological excavations show that the small (1.6–2.2 mm), blind *Aglenus brunneus* (Salpingidae) was once widespread, indeed common, in domestic houses (Kenward, 1975). It occurred in the crushed rush and straw stems (and other sweet-smelling herbs) strewn across the trodden earth floors that were once standard in low-status medieval homes. With the advent of raised wooden floors, where flat boards were supported on joists, and underfloor ventilated gaps kept everything dry and soil-free, the new coverings of rugs, carpets and linoleum were no longer suitable for the displaced *Aglenus*. It does sometimes occur out of doors, very rarely, in leaf litter, compost or manure heaps, but its previous association with human houses is at an end.

Likewise, the large ground beetle *Sphodrus leucophthalmus*, the specialist predator of cellar beetle (*Blaps mucronata*) larvae, was always rare in Britain, but as underfloor and cellar hygiene has improved, *Blaps* has declined to the point where there are just not enough populations to provide hosts for *Sphodrus*. Unlike its close relative *Laemostenus terricola*, which attacks insect larvae in fox, badger and rabbit burrows, *Sphodrus* is unable to cope with life out of doors; it was last seen in Britain nearly 40 years ago, and is now probably extinct here.

THE CLEAR-UP BRIGADE

Beetles dominate the dung fauna, and this is an important habitat for them the world over. In Britain, farm animals provide much of the dung through which coleopterists can search, and this is a rewarding pastime once any self-consciousness is overcome. Our largest dung beetles are the dor beetles, also called dumbledors (Geotrupidae), named for their heavy, droning flight at dusk on

FIG 113. A well-burrowed cowpat. The shiny black tail end of *Teuchestes* (formerly *Aphodius*) *fossor* is just visible near the bottom of the picture.

warm summer evenings. These broad, domed, shining blue-back beetles usually work in male/female pairs to excavate a deep burrow under or beside the dung, then stock the ends of the tunnels with balls or sausages of dung in which their eggs are laid. A spoil heap of soil is often a giveaway that they have been active under a cowpat. In sandy areas, the minotaur beetle (*Typhaeus typhoeus*) digs a tunnel up to 2 m long, and is equally happy with sheep nuggets or rabbit crottels.

Unfortunately, we have none of the exotic dung-roller scarabs in Britain, although they do occur in southern Europe. These famous beetles, worshipped by the ancient Egyptians, sculpt a large ball of dung, then roll it well away from the crowded dropping and bury it secretly, often many metres distant. Instead, as in many cool-temperate zones, British dung is mostly occupied by what I call *Aphodius*, which have now been divided into 29 smaller genera. These stout cylindrical beetles range in size from *Esymus pusillus* (2.5–3.7 mm) to *Acrossus rufipes* (10–13 mm), and it is not uncommon to find 10 or 12 species together

FIG 114. One of the beautifully detailed paintings of roller scarabs by E.J. Detmold, from Fabre (1921). Note the lack of tarsi, a strange adaptation to walking backwards found in *Scarabaeus sacer* and other species.

in a single dropping. Many are prettily marked or handsomely ribbed. Dung beetles fly readily, and can be seen zigzagging their way upwind to locate a fresh pat, minutes after it has been extruded. There are well-documented accounts of tropical dung beetles homing in on a recent pile of elephant dung in their thousands – 16,000 beetles spiriting away 1.9 kilograms of excrement in under two hours (Anderson & Coe, 1974). Such statistics are scant in Britain, but I have seen a cowpat heave as upwards of 200 *Nimbus contaminatus* writhed beneath a thin rind of sun-dried dung a few hours old.

Such large numbers of individuals and species, all vying for a piece of the same precious dung cake, have led to them dividing up the spoils, physically, geographically or temporally. The rare *Sigorus porcus* is reputed to be a cuckoo parasitoid, descending into *Geotrupes* burrows to lay its own eggs on the dors' stores. *Limarus zenkeri* is often reported to favour deer fewmets, but in reality it seems just to prefer tree cover; it is seldom found with its meadowland congeners in the middle of an open field, but readily takes to cow or horse dung where these animals are grazed in woodland. Some (e.g. *Agoliinus lapponum*) are northern and western hill or mountain species, while others (e.g. *Melinopterus consputus*) avoid the crush by being more active in autumn and winter.

Herbivore dung provides the majority of dung beetles with ample livelihoods, and there is a good argument that their dung is just partly processed plant material anyway. There is much nutritional content still available, and it has the added bonus of being served up in a partially digested, bacteria-rich stew. Searching horse and cow dung is relatively easy on the nose, but the riper droppings of dog, cat and fox should not be passed by. They are more tolerated

FIG 115. The common *Melinopterus* (formerly *Aphodius*) *sphacelatus*. Its stout cylindrical form, broad shovel head, strongly toothed front legs and lamellate antennae clearly identify it as a scarabaeid dung beetle.

FIG 116. The broad, powerful form of *Onthophagus coenobita*. This specimen has been hard at work, pushing through the dung and digging the deep tunnel (right) for its brood masses.

by *Onthophagus*[21] species; like *Geotrupes*, these make small tunnels in the soil under the dung and bury small dung morsels for their eggs. We have a rather impoverished fauna in Britain, but even in my garden, cat droppings regularly house the attractive bronze-and-straw *O. coenobita* (Fig. 116), and there is a tale, now slightly confounded in the retelling, of the very rare *O. nuchicornis* turning up under dog (or maybe it was human) excrement on the dunes of Camber Sands in Sussex in 1972 (Jones, 2017). It is still a mystery why the only British specimen of the Franco-Iberian *O. furcatus* was found under a goose dropping at Kew.

Carrion, too, is a bountiful supply of beetle nutrition. There is nothing quite like turning over the festering remains of a dead sheep at the foot of Grasmoor, south of Cockermouth in the Lake District, and seeing a hundred of the large, hairy grey *Creophilus maxillosus* rove beetles scuttle off, all waving their tails in defiance. There were a dozen burying beetles too – three different *Nicrophorus* species – but I suspect that no matter how hard they dug, their combined efforts would have been insufficient to inter the bloated woolly corpse. Large bodies are uncommon in the countryside, though dead foxes can be found; those hit by cars sometimes have the strength to crawl a few dozen metres before finally expiring. One of the first serious articles on the timeline of a corpse was that by Easton (1967), who studied the beetles on a dead fox as it decomposed over 18 months on Bookham Common in Surrey. His paper, in which he recorded 87 beetle species,

---

21 The name *Onthophagus*, by the way, derives from the Greek ονθος, or *onthos*, meaning 'slime', and φαγειν, or *phagein*, meaning 'eating' – particularly apt for a group of beetles often found under the riper droppings of omnivores and carnivores.

FIG 117. Dung also attracts predators. The beautiful *Ontholestes murinus* is a rapid, darting denizen of very fresh dung, snatching greenbottles and other flies from the moist surface while somehow managing to prevent its delicately tessellated, pubescent coat from becoming mired.

including two new to Britain, is now widely quoted in the science of forensic entomology, where the succession of different insect species is often a key factor in determining time and place of death in homicide investigations.

Carrion is not, like dung, all putrescent decay, although it can be pretty ripe during the first deliquescent fly-maggot invasion phase. Fresh, it is little different from meat, and is scavenged by what might be considered traditional omnivores and meat-eaters – in Britain, animals like foxes, badgers and stoats, but also squirrels, rodents, birds of prey, corvids and woodpeckers. The beetle fauna must wait for these larger, more aggressively active creatures to have their fill first – Easton was lucky that his experimental corpse was not dragged off by vertebrate scavengers, including dogs being walked on Bookham Common; as a teenager, I had to skewer my own modest test samples (a couple of dead sparrows in my parents' garden) into the ground to prevent them being removed by the neighbour's cat. It's no wonder that burying beetles work so assiduously to remove their prize from view. In North America, the once widespread *Nicrophorus americanus* is critically endangered, having disappeared from more than 90 per cent of its original range through the eastern United States and Canada since the 1920s. Habitat loss, through intensified agriculture and human encroachment, has been blamed, but mainly because this encourages vertebrate scavengers like skunks, opossums and raccoons, which remove the carrion first.

There is a fledgling carrion beetle recording scheme in Britain, and time will tell how species are doing. There is certainly death going on all around in the countryside, but one aspect of carrion is much less obvious now than previously – the gamekeeper's gibbet. As a boy, I remember often finding lines of dead animals strung up on a fence, usually on a private estate through which I was trespassing. Although one traditional take was that this supposedly frightened off further vermin, its most important function was probably to show the lord of the manor that the man was doing his job properly. In times past, certain creatures had a bounty on each head, so gibbets were a sort of tally for payment. Being air-dried, they were less inclined to reek, but still produced plenty of interesting beetles, including many *Dermestes* species, histerids, rove beetles and the attractive nitidulids *Omosita colon* and *O. discoidea*.

ANIMAL NESTS

Perhaps one of the most specialist habitats that beetles have successfully invaded is the nests of other animals. Here, they feed on droppings, moulted fur and feathers, spilled food, the occasional dead nestling and general warm, dank, fermenting decay. During the first part of the twentieth century, there was a sudden interest in animal nest beetles (Joy, 1906a,b, 1908, 1930), and some species previously considered among 'our rarest Coleoptera' were suddenly found to be relatively widespread in this hidden niche. Books and articles of the time gave detailed descriptions of how to find and dig up mole nests (a much larger molehill surrounded by smaller satellites) to unearth the strange ridged histerid *Onthophilus punctatus* or the rove beetles *Quedius puncticollis* and *Q. nigrocaeruleus*. Rabbit, fox and badger holes were plundered for the ground beetle *Laemostenus terricola*. Owl nests in hollow trees produced specialist histerids like *Dendrophilus punctatus* and *Gnathoncus nannetensis*, and hide beetles (Trogidae) like *Trox scaber* (Fig. 118).

For all of these, the 'best' time of year to find the beetles was when the nests were actively occupied by their hosts, usually when their young ones were in evidence. Perhaps this was all very well in an era when bird-nesting was considered an acceptable rural pursuit, and many wild animals were regarded as vermin, or at least fair game, but nowadays most nesting birds and animals

FIG 118. *Trox scaber* is a scavenger in animal nests, carrion and other forms of decaying organic matter. This one turned up under richly aromatic fungoid bark at Lurgashall, West Sussex, 27 December 1984.

are protected from disturbance by law. Opportunities still arise though. After the infamous gale of 15–16 October 1987, I trotted the 150 m from the south-London fixer-upper house I was living in down to the meagre utility open space at Nunhead Green to examine a large limb broken from an old black poplar tree at the edge of the children's playground. The cavity exposed could have housed woodpeckers, squirrels or owls, and although it had probably been disused for some weeks, it was still home to the scarce rove beetles *Quedius truncicola* and *Q. brevicornis*, the first time I had found either in the metropolis.

The knowledge that beetles lived in animal nests was a revelation to coleopterists, and nowhere more so than when they started to find them in ant nests. Donisthorpe's (1927) book is still my go-to reference, although there is a more up-to-date list of 369 ant-associated beetles (73 categorised as true myrmecophiles) given by Päivinen *et al.* (2002). Unlike the temporarily established breeding safe houses constructed by birds and mammals, ant nests are long-lived, continuously occupied for decades at least, and potentially surviving for centuries, with a continual turnover of the many thousands of sterile worker females and easy replacement of the fertile egg-laying queens. They are complex structures, invertebrate cities, with elaborate passages, cavities and specialist chambers used for storing food, rearing young and sheltering the occupants against inclement weather. In Britain, the largest nests are the heaped mounds of pine needles and other leaf litter, up to 1 m high and 2 m across, constructed by the wood ant (*Formica rufa*), but the distinctive domed earth-pile anthills erected by the yellow meadow ant (*Lasius flavus*), standing out from sheep-grazed hillsides of the South Downs, are equally impressive.

Beetles can invade the ant nests just as they would any other mouldering pile of plant or animal material, but because ants aggressively defend their territories, ant-nest beetles have usually evolved some way of hiding, blending in or otherwise avoiding the belligerent attentions of their hosts. The tough, rounded, seed-like shape of the histerids *Myrmetes paykulli* and *Hetaerius ferrugineus*, with their neat legs folded away into underside grooves, makes them difficult for ants to bite. The curl-tailed rove beetle *Drusilla canaliculata* is a rapid mover and can skitter away in the tunnels. To the untrained human eye, *Drusilla* looks remarkably like an ant when it jerkily twitches its way through them if you turn over a board where ants are living. This is unlikely to fool any ants, though, which are adept at feeling their way in the darkness and famously use chemical scents to recognise each other, find the forage trail leading to food sources and raise the alarm if the colony is under attack. If the ants discover *Drusilla* they actively attack it, perhaps instinctively recognising that it is an ant predator. *Drusilla* produces defensive chemicals (Brand *et al.*, 1973) and there is ample evidence to show that many other ant-nest beetles produce chemicals which either confuse their hosts, or actively encourage them to befriend their beetle guests.

The pretty pot beetle *Clytra quadripunctata* hangs from a twig near an ant nest and coats each egg with its own frass, patted into shape with its broad hind tarsi (scatoshelling and coprocoating are both suggested names for this process). It then drops them to oblivion below, where they are picked up by ants and taken back to the nest. Quite what the ants think they have collected is unknown, but the larva then feeds on dropped litter, dead plant material and ant faeces, all the time enlarging its frass pot. Whether it gives off ant-placating chemicals is as yet undiscovered, but almost certainly it escapes notice because it simply smells and tastes, to the ants, like just any other part of their convoluted nest. Conversely, the rather inadequately named scarce seven-spot ladybird (*Coccinella magnifica*) appears to be immune from attack by wood ants because of its smell. Although it does not live deep in the nest, it feeds on the ant-tended aphid colonies nearby, with apparent impunity, while the virtually identical 'common' seven-spot ladybird (*Coccinella septempunctata*) is vigorously assailed and repelled. Taste experiments (Sloggett & Majerus, 2003) show that *Coccinella magnifica* adults and larvae alike are more or less ignored by the ants, but that the ladybird chemicals are ant-repellent rather than ant-tricking.

I've never found *Myrmetes* in any ant nest, but one landed on my rucksack as I sat down for a sandwich lunch in Bedgebury Forest, Kent, last year (2016). It obviously moves from one colony to the next and is able to exist as a free-living beetle out in the open. At the extreme, though, some beetles are so tied in to their ant-nest niche that they are unable to survive outside the closeted confines of the ant lodgings. The bizarre rove beetle *Claviger testaceus* is the same size and pale straw colour as its yellow meadow ant benefactors; it is blind, its antennae are reduced and stumpy, it cannot fly, and its short legs and stunted tarsi allow little more than a waddle, yet it is treated by the ants as one of their own. As the ants run past *Claviger*, they often stop to examine it, caressing it with their antennae and licking it with their mouthparts. Special glands, near the base of the beetle's abdomen, are tufts of modified trichomes, hairs arising from pores, which release specific ant-appeasing pheromones. The ants lick at these secretions, then regurgitate food into the beetle's mouth, just as they do to each other. Similar tasty trichome glands occur in the rare staph genera *Lomechusa* and *Lomechusoides*.

Although we have no termite nests in Britain, elsewhere around the globe they are equally long-lived and equally invaded by beetle lodgers. Being concrete-hard, they are a tad more difficult to examine, but every year new species are being discovered, living in strange places, making their strange ways in a hostile but complex world. Most recently, Maruyama (2012) found a small (3 mm), dumpy scarabaeid, *Eocorythoderus incredibilis*, living with Cambodian forest termites; it has a scruff-of-the-neck carrying-handle node at the base of its elytra, which the termites use to move it from chamber to chamber along with their brood. Utterly weird.

In Britain, there are a few scavengers in bumblebee nests – *Antherophagus* species invade the jumbled wax cells of colonies, eating whatever they can find in the way of nectar, pollen or detritus. *Quedius* (formerly *Velleius*) *dilatatus* haunts hornet colonies in hollow trees, scavenging on the remains of the dead insects the large wasps have brought back for their brood but dropped into the depths of the papery carton nests. Solitary-bee tunnels, dug into a sandy bank or dead tree trunk, are sometimes frequented by spider beetles (*Ptinus*), which lurk in the burrows. As yet, no beetles are regularly found invading the British colonies of that most social of social insects, the honey bee. However, a small black southern African nitidulid, *Aethena tumida*, has been spread to eastern Canada and the United States (1996), Mexico, Hawai'i, the Caribbean, Australia (2002) and the Philippines, and is now a considerable international hive pest (Neumann & Elzen, 2004). Its larvae devour the bee brood, honey and stored pollen, causing extensive burrowing damage to the combs, messing everything with their excrement, causing the honey throughout the combs to ferment, and eventually destroying the colony. It reached southern Italy in 2014, presumably with the international commercial trade in queen bees, and despite active eradication procedures was still present there at the end of 2016. Beekeepers in Britain are on high alert should it arrive here.

CHAPTER 7

British Beetle Families

I T SEEMS AN INEVITABLE REQUIREMENT that whenever a new book on beetles emerges, its author promotes a new arrangement of families, superfamilies or suborders, with knock-on effects for the wholesale scattering of subfamilies, tribes and genera into some newfangled classification scheme, wholly different from the one produced only a decade or so before. I found it terribly frustrating when I started out. It seemed change for change's sake, and I used to think that this was the vain posturing of highfalutin entomologists wanting to make their mark, by declaiming a more or less random new blueprint as superior and modern, compared to all the obsolete fanciful clutter that went before. I now humbly offer my apologies to those I have wronged.

In truth, the vast diversity of the Coleoptera means that even with the familiar wild enthusiasm of coleopterists, it is immensely difficult to establish just how the different groups have evolved, how they have diverged and yet how they are related to each other, perhaps more so than in almost any other order of insects. The discovery of a vaguely intermediate form from the other side of the planet can throw previously held local beliefs out of the window. To a certain extent, trying to arrange this four-dimensional evolutionary complexity into a standard linear sequence so that checklists and monographs can follow some sort of reasoned order, from start to finish, is an artificial construct anyway.

The species is still the universal currency of the entomologist; gathering these into cohesive groups may work at the level of the genus, but for many genera the allocation into a family or higher grouping can be changed with any new appraisal. With each advance in microscope clarity, larval discovery, biochemical sensitivity and now DNA-sequencing technology, the sometimes hidden relationships are easily reinterpreted. New rearrangements of once sacrosanct lineages are suddenly thrown up, and groups once thought distantly separated are shown to be more intimately connected. It is no wonder that taxonomic lists transform so drastically, so often. It's still frustrating though, and can seem like the reshuffling of a pack of cards.

The following list of 102 families (plus two not very British families) follows the taxonomic order set down in the latest British checklist by Duff (2017), which in its turn follows the mammoth work by Bouchard *et al.* (2011) on 211 families, 541 subfamilies and 1,663 tribes, of all known groups of extant and fossil beetles, and formalised in the *Catalogue of Palaearctic Coleoptera* (Löbl *et al.*, 2003–16). I have no doubt, though, that this will be only a temporary respite, and that 20 years from now yet another arrangement will be in vogue. I've already rejigged this entire chapter once during the writing, after starting it using an earlier version of the British checklist (Duff, 2012a). Andrew Duff alerted me to the imminent 2017 edition almost the moment I had finished that first draft, and he kindly supplied an advance copy. Phew. Oh, and please note, species and genus numbers are also subject to change without notice.

The thumbnail illustrations are mostly from Reitter (1908–16), and have been selected to show typical species from each family. Identification references are not exhaustive, but are mostly to recent major works in English, concentrating on the British, or at least appropriate European fauna.

SUBORDER MYXOPHAGA CROWSON, 1955

Superfamily SPHAERIUSOIDEA Erichson, 1845

1. **Sphaeriusidae** Erichson, 1845; 1 genus, 1 species
A single rare species, the minute (0.6–0.8 mm), shining, convex, almost globular, semi-aquatic *Sphaerius acaroides*, found in the muddy edges of fen water (Huntingdonshire) or crumbling maritime cliffs (Dorset), where it feeds on microscopic algae. Sometimes given the cobbled English name minute bog beetles. Abdomen with only three visible sternites, and tarsi 3-3-3, making this the only British representative of the obscure suborder Myxophaga (about 65 species worldwide). Secretive and easily overlooked. Identification: Duff (2012b).

Sphaerius

SUBORDER ADEPHAGA CLAIRVILLE, 1806

2. Gyrinidae Latreille, 1810; 2 genera, 12 species
Small (3.5–7.5 mm), distinctive long-oval body shape, with
smooth lines, gently convex, some sub-cylindrical, shining
black above (except *Orectochilus villosus*, dull grey due to
dense, short hairs), elytra delicately striate. Front legs long
for clasping prey; middle and hind legs short paddles;
tarsi 5–5–5. Antennae short, 8–11 segments, first segment
small, second large and thickened, third inserted at side
of second and the remainder forming a dense club. Eyes
completely divided, giving the appearance of having four
of them, so that the upper portions look up to the sky and

Gyrinus

the lower sections look into the water. The larvae and adults are active predators.
The whirligigs, so called for the mad gyrating gangs that swirl and circle on
the surface of ponds, lakes and ditches, particularly on sunny days. The furious
swimming of squadrons of *Gyrinus* is a superb predator- and entomologist-
avoidance tactic, the glints of sun on their bodies combining with the reflections
in the rippling water to create a confusing maelstrom of flashing lights for the
onlooker attempting the snatch-and-hope netting technique. Identification:
Holmen (1987), Friday (1988), Foster (2008), Foster & Friday (2011), Duff (2012b).

3. Carabidae Latreille, 1802; 86 genera, 364 species
Very small (1.6 mm, *Trachys bistriatus*) to very large (35 mm,
Carabus intricatus). The ground beetles (also including the
tiger beetles, subfamily Cicindelinae), perhaps the most
typically beetlish group, with long running legs, tarsi 5–5–5,
and smooth, elegant, sometimes shining body forms.
Antennae relatively long, filiform, 11 segments, inserted
between eye and mandible, most segments with hairs,
setae or pubescence, although the basal 3–4 often smooth
and glabrous. Most have the distinct handsome carabid
body form: large head with prominent eyes, narrower than
the rounded or cordate pronotum, which itself is narrower

Blethisa

than the long-oval or sub-parallel elytra, which are usually striate or linearly
sculptured. The imposing members of the genus *Carabus* ably demonstrate
this. The violet ground beetle (*Carabus violaceus*) is typical in its sleek lines and
smoothly wrinkled elytra, their edges decorated with a band of rich purple

sheen. Likewise, the genera
Pterostichus, Leistus, Calathus
and *Agonum* are unmistakably
carabids. A few are broadly
rounded; the globose
Omophron limbatum looks
remarkably like a *Haliplus*
water beetle, the large genus
(28 species) *Amara* comprises
broadly oval metallic species,
sometimes called sun beetles
for their habit of dashing
about in the open. The
narrow *Odacantha melanura* secretes
itself in the leaves and sheaths of
reeds and waterside grasses, and the
cylindrical members of *Dyschirius*
and *Clivina* burrow into the soil, and
have short toothed, fossorial legs.

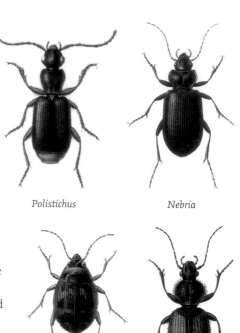

Polistichus

Nebria

Omophron

Panagaeus

Typically found running about
in the open, on mud and sand
at the edges of ponds, or hiding
under logs and stones (the all-
black *Nebria brevicollis* and the
sometimes red-legged *Pterostichus madidus* are two
of the commonest), although a few are arboreal
(the rare *Calosoma inquisitor* preys on *Tortrix* and
other moth caterpillars). Most ground beetles are
fast and active, with long, slim legs. This, together
with large eyes and sharp jaws, makes them expert
predators, although a few (e.g. *Zabrus tenebroides*,
Curtonotus aulicus) are seed-eaters. The larvae are,
likewise, mostly predators, but those of *Lebia* and the
bombardier beetles *Brachinus* are ectoparasitoids,
attacking the pupae of other beetles. A distinctive
feature of all carabids is the large trochanter, bulging
out at the base of the hind femur. This articulation,
and increased musculature of the back legs, allows
the beetles to wedge-push. Forcing itself head first

Carabus

into a crevice in the root thatch, the beetle is then able to raise its body at the rear end, effectively opening up the gap into which it can push deeper. *Aepus* take this to the extreme of living in silt-filled cracks in foreshore rocks that are covered by the sea at high tide.

Ground beetles are generally smooth, black or shining metallic, but many are attractively patterned, brightly coloured or bizarrely sculptured. There is a long tradition of featuring pictures of them in beetle books and they have been the regular subject of useful identification guides. They make an excellent group for someone wanting to get into beetles, offering a combination of easily distinct species and more challenging genera like *Bembidion* (56 species), *Harpalus* (22 species) and *Ophonus* (14 species). Identification: Lindroth (1974, 1985–86), Trautner & Geigenmuller (1987), Forsythe (2000), Luff (2007), Duff (2012b).

Rhysodes

–. **Rhysodidae** Laporte, 1840; 0 genera, 0 species
Small (6–8 mm), narrow, elongate, parallel-sided, reddish brown, body heavily grooved, with pronounced elytral striae, furrows on head and pronotum. Antennae relatively thick, 11 moniliform antennomeres. Legs short, slim, tarsi 5–5–5. Technically not a true British species, and never recorded here during recent historical times, however *Rhysodes sulcatus* is known from subfossil remains from about 10,000–3,000 years ago. Widespread but very scarce in Europe. In rotten wood, an Urwaldrelikt (wildwood relic) from a time of dense primeval forest cover. Identification: Freude *et al.* (1971), Lindroth (1985–86).

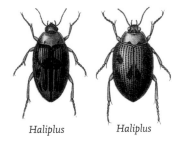

Haliplus   *Haliplus*

4. **Haliplidae** Aubé, 1836; 3 genera, 19 species
Small (2.5–4.5 mm), broad, semi-globular body, smooth lines, various yellows flecked with patterns of black, scutellum invisible, prosternum with a keel. Elytra striate, often very strongly punctured (e.g. *Peltodytes caesus*) or interstices raised into ribs (e.g. *Brychius elevatus*). All legs slender, mid- and hind pairs with hair fringes for swimming, tarsi 5–5–5; underside of abdomen almost completely hidden by the large flat coxal plates of the hind legs. Antennae filiform, often apparently 10 antennomeres, but there are 11. Rather dumpy but attractive crawling water beetles, which swim using alternate legs like a runner, rather

than synchronised sculling like true water beetles. Larvae feed on algae, adults predatory. Breathing air is stored under the elytra and large coxal plates. Charming and distinctive beetles, mostly (17 species) in the large genus *Haliplus*, which is divided into subgenera on the basis of punctuation and sculpturing. They use their crawling gait to swim/climb through semi-submerged vegetation and thick algal mats at the edges of ponds, lakes and streams. Identification: Balfour-Browne (1953), Holmen (1987), Friday (1988), Foster & Friday (2011), Duff (2012b).

5. **Noteridae** Thomson, C.G., 1860; 1 genus, 2 species
Hemi-ovoid, smoothly rounded, oval, blunter at head, sharper at tail, convex above, scutellum invisible, body flat beneath, dull orange-brown. Legs short, tarsi 5–5–5. Antennae slim, but male with fifth antennomere enlarged. Sometimes called burrowing water beetles, they swim, crawl or burrow in mud and vegetation at pond edges. In the past, the two widespread and common species have both, confusingly, been called the same names due to repeated misidentifications. There is some overlap, but

Noterus

Noterus clavicornis is generally larger (4–5 mm) than *N. crassicornis* (3.5–4.0 mm), and coleopterists still refer to 'bigger' or 'smaller' *Noterus* to avoid confusion. So much for scientific nomenclature. At least mostly predatory. Identification: Balfour-Browne (1953), Holmen (1987), Friday (1988), Foster & Friday (2011), Duff (2012b).

6. **Hygrobiidae** Régimbart, 1878 (1837); 1 genus, 1 species
Hygrobia hermanni, distinctive, large (8.5–10.0 mm), stout, bulging, yellowish orange and black. Relatively long legs, tarsi 5–5–5, hind tarsi very long. Antennae filiform. This widespread and sometimes common species, the screech beetle, is named for its ability to squeak loudly, a defence mechanism, if picked up. It rubs a stridulatory file on the underside of the elytral apex against the sharp edge of abdominal sternite 7. The strong vibrations are clearly palpable if the beetle is held between finger and thumb, and must feel like an electric shock in the mouth of a fish or beak of a bird. Predatory among vegetation at edges

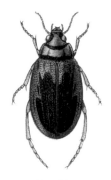

Hygrobius

of ponds, lakes and ditches. Identification: Balfour-Browne (1953), Friday (1988), Nilsson & Holmen (1995), Foster & Friday (2011), Duff (2012b).

7. **Dytiscidae** Leach, 1815; 29 genera, 118 species
Very small (1.5 mm, *Bidessus minutissimus*) to very large (38 mm,
Dytiscus dimidiatus). Smooth oval shape, broad and round
to nearly parallel-sided, shining, slightly convex to rather
globular. Body almost devoid of the fine touch-sensitive
setae found on almost all other groups of beetles. Legs
relatively long, especially hind pair, flattened, and broadly
fringed with hairs to create swimming paddles. Hind coxae
large and flat, covering much of the first abdominal sternite.
Tarsi 5-5-5, but in the subfamily Hydroporinae (mostly small

Acilius

species, 1.5–5.8 mm), tarsomere 4 of
front and middle legs is so small that
formula appears to be 4–4–5. Antennae
filiform, 11 antennomeres. Males
of the subfamily Dytiscinae (mostly
large species, 12–38 mm) have great
broad, round front tarsi, with sucker
pads to grip the backs of the females
during mating; some of the females
have strongly ridged, corrugated
elytra seemingly to counter male
grasping, the sexual dimorphism

Laccophilus

Dytiscus

having evolved in a traditional battle of the sexes. Thoracic glands, just behind the
head, secrete steroids that are distasteful to fish. The 'true' water beetles (sometimes
termed 'predacious' water beetles), streamlined, powerful swimmers. Although
predominantly black, brown or slightly greenish bronze, many are attractively
patterned or strikingly marked with pale streaks, stripes, dots or dashes. Adults and
larvae are predatory, larger species often attacking vertebrate prey like tadpoles or
small fish; they are also scavengers, eating dead fish or other carrion.

The Dytiscidae are distinctive and attractive beetles, and with several
identification guides available they lend themselves to study, even though there
are some challenging species groups. They fly well, and can colonise new ponds
very quickly; the almost ubiquitous *Agabus bipustulatus* is the most widespread
British species and has been known to arrive at garden ponds that are in the
process of being dug. Large dytiscids can be seen in clear water, hanging by the
tail end from the surface as they renew the air bubble trapped under their elytra,
but if disturbed by a sudden movement they dive down in a flash.

Both larvae and adults are fully aquatic, but the larvae heave themselves into
the damp foreshore to create a pupal cell in the muddy roots and stems, under a

log or other pondside object, just above the waterline. Most species overwinter as adults and although subdued, can be found even in the coldest months. Adults of the scarce northern and montane *Dytiscus lapponicus* have been seen swimming actively under thick ice. Identification: Balfour-Browne (1953), Friday (1988), Nilsson & Holmen (1995), Foster & Friday (2011), Duff (2012b).

SUBORDER POLYPHAGA EMERY, 1886

Superfamily HYDROPHILOIDEA Latreille, 1802

8. **Helophoridae** Leach, 1815; 1 genus, 20 species
Small (2.5–7.0 mm), oblong, flattened, yellowish, mottled to black, pronotum grooved or granulate, head with Y-shaped groove on frons, elytra smoothly striate or gently ribbed (more aquatic species), or sharply ridged (subgenus *Empleurus*, three mostly terrestrial species); underside of body covered with short silky pubescence. Legs slim, tarsi 5–5–5. Antennae with eight or nine segments, first large, last four forming a

Helophorus

club, slim but short, about equal in length to the relatively long, four-segmented maxillary palps. *Helophorus* are scavengers in muddy edges of ponds and ditches. They fly well, and are often found in herbage in marshes or stream banks, and come to moth traps and lighted windows. Identification: Hansen (1987), Friday (1988), Duff (2012b), Foster *et al.* (2014).

9. **Georissidae** Laporte, 1840; 1 genus, 1 species
Georissus crenulatus, 1.5–2.0 mm, compact, hunched, head hidden under front of pronotum, dumpy, shining black (although often encrusted with mud), elytra striate with rows of large, coarse dimples. Legs short but slim, front coxae flattened at the tip, forming two plates that more or less hide the prosternum; tarsi 4–4–4. Antennae short, nine segments, first two thickened, last three in a small club. Scarce, in aquatic vegetation, mud or sand at edge of ponds and streams. Identification: Duff (2012b), Foster *et al.* (2014).

Georissus

10. **Hydrochidae** Thomson, C.G., 1859; 1 genus, 7 species
Small (2.5–4.0 mm), narrow, sub-cylindrical, slightly flattened;
pronotum narrow, granulate or with uneven depressions; elytral
striae strongly punctate, black, shining or with slightly metallic
tints. Legs short but slim, tarsi apparently 4–4–4 because of
nearly obsolete basal tarsomere. Antennae short, with seven
segments, last three in a club. Maxillary palpi long. *Hydrochus* are
secretive, sluggish crawlers in aquatic plants and mud at edges of
ponds and streams. Identification: Hansen (1987), Friday (1988),
Duff (2012b), Foster *et al.* (2014).

Hydrochus

11. **Spercheidae** Erichson, 1837; 1 genus, 1 species
Spercheus emarginatus, small (5.5–7.0 mm), oval, foreparts
brown, elytra mottled. Legs slim, tarsi 5–5–5 but tarsomere
1 very small. Antennae short, stout, indistinctly clubbed,
seven-segmented, but antennomere 3 very small. In roots
and vegetation at muddy edges of ponds and streams. Last
recorded in Suffolk, 1956, probably extinct. Identification:
Duff (2012b), Foster *et al.* (2014).

Spercheus

12. **Hydrophilidae** Latreille, 1802; 18 genera, 73 species
Very small (1.3 mm, *Chaetarthria
seminulum*) to extremely large
(48 mm, *Hydrophilus piceus*), mostly
round or oval, smooth, shining,
convex or domed above, flat
underneath. Elytra striate, or not,
or with limited number of narrow
grooves. Meso- and metasterna
with a continuous keel, sometimes
prolonged into a sharp, backwards-
directed spine. Legs either slim,
fringed with hairs, or short, stout and blade-like;
tarsi 5–5–5 or 5–4–4 (*Cymbiodyta marginellus*). Antennae
with seven to nine segments, last three in a club,
shorter or not much longer than the slim maxillary
palpi. A diverse family with some sleek water-dwelling
swimmers and some more convex species that burrow into dung and other

Enochrus

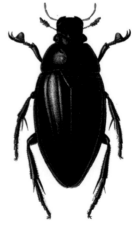

Hydrophilus

putrid, rotting matter. One of Britain's largest insects, the great silver water beetle (*H. piceus*), occurs in slightly brackish ditches around the coast. In the water, it appears silver underneath because of the air bubble (plastron) encasing its body on minute hydrophobic hairs; it replenishes this by arriving at the water surface head first and breaking the surface tension with its antennae. The swimming dung beetles (*Sphaeridium*) land on fresh cowpats and push through the semi-liquid excrement with their broad paddle legs, along with the slower, more domed but more numerous (22 species) *Cercyon* species. The larvae are mostly predators, but the adults are scavengers or graze in decaying matter. Identification: Hansen (1987), Friday (1988), Duff (2012b), Foster *et al.* (2014).

13. **Sphaeritidae** Shuckard, 1839; 1 genus, 1 species
Sphaerites glabratus, 4.5–6.0 mm, broad, oval, moderately convex, shining black, with a strongly greenish brass lustre, very finely striate. Legs slim, spinose externally, tarsi 5–5–5. Antennae with 11 segments, first long, a scape, last three in a tight club. Very rare, northern Britain. Associated with rotting birch trees and the birch polypore fungus (*Piptoporus betulinus*). Identification: Duff (2012b).

Sphaerites

14. **Histeridae** Gyllenhal, 1808; 21 genera, 52 species
Minute (0.8 mm, *Aeletes atomarius*) to medium (11 mm, *Hister quadrimaculatus*), very stout, compact, dumpy, mostly smoothly rounded and convex, shining black (a few marked with reddish spots) or slightly metallic bronze or green, prosternum expanded into a keel or prominence. Pronotum broad, often with grooves or striae at sides. Larger species especially have the elytra short, clearly exposing the last two segments of the abdomen, striate, although striae sometimes shortened or obscured by other confused punctuation. Legs broad, flat, tibiae toothed for digging into decaying wood, dung, carrion or fungi. Tarsi 5–5–5 or 5–5–4 (tribe Acritini). Antennae with 11 segments, apical antennomeres forming a club (usually abrupt, three segments), and basal segment often longer, a curved scape. If disturbed, they will draw legs and antennae tight in under the body, where they sit almost flush. Although they are general decay scavengers and predators, some

Hister   Onthophilus

species have become adapted to very specific habitats, including mammalian dung (most *Hister, Margarinotus, Atholus, Saprinus* and *Hypocaccus*), ant colonies (*Myrmetes paykulli*), stored products (*Carcinops pumilio*), bird and animal nests (*Gnathoncus, Dendrophilus*), and the burrows of bark beetles (*Plegaderus*). Despite their penchant for putrid decay, these are attractive and charismatic beetles. The minute (1.2–1.6 mm) *Kissister minimus* occurs in large numbers in grass thatch, and can be collected from playing fields using a suction sampler. The rather slim, long-oval *Paromalus flavicornis* occurs under fungoid bark. The bizarre sulcate *Onthophilus striatus* is local but widespread, under horse dung or in compost, but its congener *O. punctatus* appears to be confined to mole nests. The peculiar *Hololepta plana*, discovered in Britain only in 2009, is extremely flattened, to push under the decaying bark of dead trees. Sometimes called clown beetles for some not very clear association with the Latin word *histrio* meaning 'actor', and from which we also get the word histrionics. Identification: Halstead (1963), Duff (2012b).

Superfamily STAPHYLINOIDEA Latreille, 1802

15. Hydraenidae Mulsant, 1844; 4 genera, 32 species
Very small (1.1–2.5 mm), long-oval or narrow, slightly flattened beetles, mostly with a narrowed waist between a rounded (e.g. *Ochthebius*) or heart-shaped (e.g. *Hydraena*) pronotum, and rounded or parallel-sided striate elytra. Legs relatively long and slim, tarsi 5-5-5, but basal segments 1–3 sometimes minute or very small. Antennae eight- or nine-segmented, apical five forming a club, shorter than the long or very long maxillary palps. Secretive adults live in damp moss or waterlogged vegetation at the edges of ponds and streams, generalist feeders on algal growth and microorganisms. Identification: Hansen (1987), Friday (1988), Duff (2012b).

Ochthebius

16. Ptiliidae Erichson, 1845; 19 genera, 76 species
The feather-wing beetles, minute (0.45–1.20 mm), oval to elongate, slightly flattened, smooth, shining, although many with sparse but long, outstanding pubescence. Mostly dirty yellowish or brown, sometimes with short elytra (*Ptinella*) exposing several segments at tip of abdomen. Legs slim, tarsi 3-3-3. Antennae long and slim, hairy, 11 segments,

Ptilium

apical three antennomeres forming a vague club. Named for the wings, which are simple membranous lobes fringed with long hairs, giving a feather-like appearance. Mould-feeders, eating fungal hyphae and spores, under bark or in leaf litter, compost, dung, manure and wood mould. *Baranowskiella ehnstromi*, at 0.45–0.55 mm Europe's smallest beetle, was discovered in Britain in 2015 on its fungal foodstuff, the bracket fungus *Phellinus conchatus*. Study of this family requires a much higher (at least ×200) compound microscope power than the normal ×10–30 magnification of a stereomicroscope. Collection also requires a delicacy unnecessary elsewhere in the Coleoptera: lick the tip of a soft blade of grass, then drag this gently over the beetle, which is lifted up by the spittle; place the beetle and grass into a small glass tube with a shred of paper tissue to avoid fouling with condensation. Identification: Darby (2012).

17. **Leiodidae** Fleming, 1821; 22 genera, 94 species
Minute (1.5 mm, *Liocyrtusa vittata*) to small (7 mm, *Leiodes cinnamomea*), oblong to oval. A diverse group, previously divided into separate smaller families.

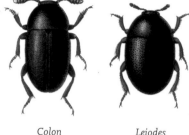

Colon   Leiodes

Subfamily Leiodinae are round, domed, convex, some striate, mostly shining black or reddish-yellow fungus-feeders, many able to roll into a sphere; tarsi 5–5–5 or 5–5–4; 11-segmented antennae, last antennomeres forming a loose club, often with segment 8 much smaller than 7 or 9. Mostly secretive, scarce species, with little of their life histories known, although there are some associations with subterranean fungi and slime-mould fruiting bodies. They are often found by evening sweeping, and some come to moth traps and lighted windows. The convex, glossy *Anisotoma* and almost hemispherical *Agathidium* are sometimes found under loose fungoid bark.

Subfamily Cholevidae are long-oval or teardrop-shaped, matt because of dense micropunctures and fine pubescence; tarsi 5–5–5; slim or clubbed antennae; feed in carrion, dung and other decay.

Subfamily Coloninae are intermediate in form, with a compact, five-segmented antennal club; their larvae, biology and food are unknown, and almost nothing is known about their life history, although they may be mostly subterranean and are usually found by evening sweeping.

The bizarre subfamily Platypsyllinae contains two tiny (2.0–2.2 mm) species. The blind, pale greyish-yellow *Leptinus testaceus* lives in the nests of mice and other small mammals, and on the animals themselves; it is thought to scavenge on their dung and dropped food morsels. The obligate beaver parasite *Platypsyllus castoris* lives in the nests and in the dense pelt of the animal, eating skin particles, fatty secretions and blood. It was discovered in Britain in 2012 on a trapped beaver kit during monitoring, after the hosts were reintroduced here from Norwegian stock in 2009. Identification: Cooter (1996), Duff (2012b).

18. **Silphidae** Latreille, 1806; 9 genera, 21 species
Medium (7 mm, *Thanatophilus dispar*) to very large (25 mm, *Necrodes littoralis*) carrion beetles, mostly stout, oblong, somewhat flattened; elytra usually with raised ribs, sculpture or dimples, and often shortened to expose last one to three abdominal segments. Legs long and slim, although tibiae often spinose, tarsi 5–5–5. Antennae with 10 (*Nicrophorus*) or 11

Nicrophorus   *Silpha*

antennomeres, with a gradual or abrupt four-segmented club. Most are carrion-feeders, although the pretty yellow-and-black *Dendroxena quadrimaculata* attacks moth caterpillars on trees, some *Silpha* species may be predatory on slugs, and *Aclypea opaca* eats beet plants and other Chenopodiaceae. The handsome and distinctive *Nicrophorus*, black or banded with orange bars, are the well-known sexton or burying beetles, which work in sexual pairs to excavate the soil from under a small animal corpse, eventually interring it, a hidden food source for them and their offspring. They fly readily and are attracted to moth traps and lighted windows, or are seen flying about in bright sunshine, looking for all the world like rather dopey bumblebees. Identification: Joy (1932), Duff (2012b).

19. **Staphylinidae** Latreille, 1802; 275 genera, 1,129 species
The rove beetles. Minute (e.g. *Oligota parva*, 0.7 mm) to very large (*Ocypus olens*, the devil's coach-horse, 28 mm), generally slim, narrow, cylindrical or flattened beetles with a prominent head, round, barrel-shaped or square pronotum, and distinctively short elytra, exposing typically six, sometimes five or fewer, abdominal segments, the highly flexible tail or hind body that can be held curled up in life. However, a few are broad, oval or convex, or with normal elytra and only the tip of the abdomen visible. Legs usually medium length, slim, various tarsal formulae. Antennae usually slim, sometimes slightly serrate or clubbed. This is

Ocypus

a huge and highly diverse family, and following analysis of larval characters, several previously separate groups have been brought in under one family name. The shortened elytra of the majority of species allow even greater flexibility to push into small crevices, under bark, into grass-root thatch or any other tiny space, and yet still allow flight wings to be stashed tightly away ready for the insect to take to the air. The staphylinid wing has reduced cross-veins, or enclosed cells, and this seems to have allowed a complex double-concertina origami type of wing-folding. Because of their short wing-cases, rove beetles have a familiar generalised form and are mostly immediately recognisable, although many of the smaller species

Hygronoma

form very difficult groups often separable only by high-power examination of dissected genitalia preparations. They occur in a huge range of habitats, some are detritivores in dung or other decay, some are scavengers, a number live with ants, many are predators. The exposed, flexible hind body can contract significantly at death, as the abdominal segments telescope in on each other (subfamily Tachyporinae in particular), so body lengths of museum specimens can vary quite considerably. Some keys get over this by giving lengths from tip of head to hind margin of the elytra. Sizes here and elsewhere in this book are total lengths, though, of live or freshly killed specimens.

Subfamily Omalinae; 28 genera, 71 species. Small (1.2–4.0 mm), compact, robust, often flattened, black or brown, elytra comparatively long, reaching half the length of the abdomen, and distinguished by having two ocelli towards the back of the head. Mostly secretive, slow-moving, under bark or in decaying organic matter, carrion, dung, bird nests, etc. The bizarre *Micralymma marinum* occurs in silt-filled rock cracks on the seashore, and is often covered up by water at high tide. Identification: Joy (1932), Tottenham (1954), Freude *et al.* (1964), Lohse & Lucht (1989).

Subfamily Proteininae; 3 genera, 11 species. Small (2.0–6.5 mm), broad, flattened. Dirty pinkish *Metopsia gallica* has a single ocellus on head. Mostly secretive, slow-moving, in decaying organic matter. Identification: Joy (1932), Tottenham (1954), Freude *et al.* (1964), Lohse & Lucht (1989).

Subfamily Micropeplinae; 1 genus, 5 species. Minute (1.7–2.5 mm), black or dirty brown, short, broad, flat, elytra and hind body with raised keels. Antennae with nine antennomeres, last three forming a vague club. Tarsi 3–3–3. Slow-moving, in manure, rotting straw or decaying vegetable matter. Identification: Joy (1932), Tottenham (1954).

Subfamily Pselaphinae; 19 genera, 54 species. Previously considered a separate family, and convincingly distinct. Very small (0.7–3.5 mm), but relatively robust and compact, narrow (e.g. *Euplectus*, 15 species), or with bulbous elytra and hind body much broader than narrow, constricted pronotum (e.g. *Tychus, Bryaxis, Pselaphus*), black, reddish or brown, only five (or fewer) abdominal segments visible. Antennae clubbed, highly variable from two (in non-British species) to 11 antennomeres (six in *Claviger*); palps often very long, sometimes with modified enlarged segments, but very short, rudimentary and with just one segment in *Claviger*. Eyes small or very small, but with large granular-looking facets, or completely absent (*Claviger*). Legs long and slim, although femora bulbous, tarsi mostly 3–3–3. In decaying vegetable matter, rotting wood, moss, grass roots or ant nests, secretive but agile, and mostly uncommon. Identification: Pearce (1957, 1975), Freude *et al.* (1974), Lohse & Lucht (1989).

Claviger

Subfamily Phloeocharinae; 1 genus, 1 species. *Phloeocharis subtilissima*, 2 mm, narrow, flat, dark grey, elytra brownish, sluggish, common under bark, in moss. Identification: Joy (1932), Tottenham (1954).

Subfamily Tachyporinae; 13 genera, 68 species. Small (0.9–9.0 mm), with distinctive convex, rounded head and pronotum, tapered conical hind body, often with long setae projecting at sides of body. Sometimes shining, and many are brightly coloured black and reddish orange. In fungi, dung, decaying organic matter, grass roots, etc. Very active, rapid, fast runners, fly readily. Large numbers of the common *Tachyporus hypnorum* and *T. chrysomelinus* occur in long grass. Identification: Joy (1932), Freude *et al.* (1964), Booth (1984), Lohse & Lucht (1989).

Tachyporus

Subfamily Trichophyinae; 1 genus, 1 species. *Trichophya pilicornis*, 2.5–3.0 mm, dull yellow-brown to black, antennae long and thin, and covered in long hairs, small bulbous eyes. In moss, active, a fast mover, flies readily. Identification: Joy (1932).

Subfamily Habrocerinae; 1 genus, 1 species. *Habrocerus capillicornis*, 2.8–3.5 mm, dark brown, shining, antennae very long and thin, and covered in long hairs, hind coxae broadly triangular, concealing part of femora. In leaf litter, active and agile. Identification: Joy (1932).

Subfamily Aleocharinae; 127 genera, 455 species. Minute to small (0.7–9.0 mm), rather delicate, soft-bodied, black, brown or variously dirty yellow; antennae inserted between eyes, hind coxae expanded sideways under femora. Although containing a large number of difficult-to-identify groups, mostly separable only using genitalia dissection preparations, there are some very distinctive forms: the almost globular, shining *Cypha*, found in long grass; the parallel-sided, curl-tailed

Drusilla canaliculata, often found under
logs in company with ants, and running
with them in a mimicking gait; and the
short round, flat *Encephalus complicans*,
found in grass cuttings, compost
and leaf litter. Members of the genus
Aleochara are larger and stouter, with
characteristic broad, curled tails; they
occur in carrion, dung and putrid decay,
where their larvae are parasitoids of fly
maggots and puparia. Identification:
Freude *et al.* (1974), Lohse & Lucht (1989).

Drusilla   Encephalus

 Subfamily Scaphidiinae; 3 genera, 6 species. Previously considered a separate
family, and obviously distinct. Small (2–6 mm), shining, globular, smoothly
long-oval, elytra almost entirely covering abdomen, just the
pointed final segment exposed. Antennae with a gradual club.
Legs relatively long, fast runners. Black apart from distinctive
and widespread *Scaphidium quadrimaculatum*, which has four
bright red marks. Mostly in rotten fungus or decaying wood.
Identification: Joy (1932), Lott (2009).

 Subfamily Piestinae; 1 genus, 1 species. *Siagonium
quadricorne*, 4.0–5.5 mm, shining black and reddish brown,
parallel-sided, flattened, front coxae globose, male with long
mandibles and two short head horns, slow moving, common
under fungoid bark. Identification: Joy (1932), Tottenham

Scaphidium

(1954), Lott (2009).

 Subfamily Oxytelinae; 15 genera, 95 species. Small (2–7 mm), flattened or
sub-cylindrical, mostly black species, stout front tibiae often with one (*Anotylus*)
or two (*Bledius*) rows of spines, for burrowing in soil (especially
damp sand and silt near the coast), rotting vegetable matter or
dung. The genus *Oxytelus* has a strongly grooved and sculptured
pronotum. Males of some of the burrowing *Bledius* have a
long, pointed horn projecting from the front of the pronotum.
Identification: Tottenham (1954), Lott (2009).

 Subfamily Oxyporinae; 1 genus, 1 species. *Oxyporus rufus*,
6–9 mm, shining, robust, brightly reddish yellow, marked with
deep black on large head, apical half of elytra and tail tip of hind
body; long, prominent, curved jaws that cross each other; last
segment of palps broad, sickle-shaped. In fungi, often tucked into the parallel

Oxytelus

gills of rotten mushrooms. Active when
it gets going. Identification: Joy (1932),
Tottenham (1954), Lott & Anderson (2011).

Subfamily Scydmaeninae; 9 genera,
33 species. Minute (0.7–2.0 mm), rather
ant-like, smooth or shining, convex,
black, brown or reddish, with spindly
legs, tarsi 5–5–5; head broad with narrow
neck, pronotum narrow, elytra rounded
at humeri, and long, almost no abdomen

Oxyporus

Scydmaenus

exposed. Maxillary palpi long, four palpomeres, but first
segment very small. In decaying vegetable matter, moss and rotten wood. The
distinctive bulbous *Cephennium gallicum* (and probably other species) is a specialist
predator of oribatid mites and has suckers under the head to latch onto the
shining carapace of its victim while it scrapes a hole with its short mandibles.
Identification: Joy (1932), Freude *et al.* (1971), Lohse & Lucht (1989).

Subfamily Steniinae; 2 genera, 76 species. Small (2–6 mm), cylindrical, sleek,
with characteristic large domed eyes jutting out in front of narrow barrel-shaped
pronotum, mostly black or grey, but some yellow-spotted on each elytron, slim
legs and narrow antennae (inserted between eyes) sometimes lighter. The large
genus (75 species) *Stenus* often common in damp grassy places; runs fast and flies
readily. Despite some awkward, difficult-to-identify species groups, many are
pretty and distinctive. The elegant dark metallic blue-black *Dianous coerulescens*
occurs in wet moss in, under or beside waterfalls. Identification: Tottenham
(1954), Lott & Anderson (2011).

Subfamily Euaesthetinae; 2 genera, 4 species. Minute (1.4–1.6 mm), compact,
broad-headed, reddish brown, antennae with a two-segmented club, tarsi 4–4–4.
Damp places in decaying vegetable material. Identification: Lott & Anderson
(2011).

Subfamily Pseudopsinae; 1 genus, 1 species. *Pseudopsis*
sulcata, 3–4 mm, dirty brown, narrow, pronotum and elytra with
raised ridges, rare in decaying plant material. Identification:
Tottenham (1954), Lott & Anderson (2011).

Subfamily Paederinae; 16 genera, 62 species. Small to
medium (2.5–9.0 mm), slim, elegant, pronotum rather long
and head joined with distinct narrow neck, legs and antennae
slim. Members of the genus *Paederus* are prettily marked with
yellowish red and blue-black. In damp grassy places, active and
relatively fast moving. Identification: Lott & Anderson (2011).

Paederus

Subfamily Staphylininae; 33 genera, 186 species. Small to large (3.5–28.0 mm), parallel-sided, slightly flattened, front tarsi usually very broad. These are fast-moving and active beetles, and the typical view of them is the curled tail disappearing into the grass-root thatch, or down a crack in the soil. Here are the largest and most obvious British rove beetles, and although this group contains the large and tricky genera *Quedius* (46 species) and *Philonthus* (46 species), many are strikingly coloured and distinct enough to be identified from pictures. Technically, only the giant (24–28 mm), matt black *Ocypus olens*, usually found under logs and stones, should be called 'devil's coach-horse', but this term becomes blurred for some other equally impressive beasts. Most are predators, living in dung, compost, carrion or other decaying organic matter, under bark, or in bird and animal nests. Usually black or dark brown, they vary from shining metallic bronze to bold black and red. The very rare *Quedius* (*Velleius*) *dilatatus* (blackish brown, 15–24 mm) breeds in hornet nests. The delicately powdered black-and-grey *Creophilus maxillosus* (14–20 mm) occurs under carrion. The bumblebee-mimicking black-and-yellow furry *Emus hirtus* (18–26 mm) is very rare under fresh cow dung in north Kent, but the equally pretty *Ontholestes murinus* and *O. tessellatus* are widespread and common, darting about on very fresh droppings, attacking and eating blow-flies. Identification: Lott & Anderson (2011).

Creophilus

Superfamily SCARABAEOIDEA Latreille, 1802

20. **Geotrupidae** Latreille, 1802; 5 genera, 8 species
Large (7–26 mm), broad-oval, glossy, domed, stoutly built, black, sometimes with brownish, metallic violet, greenish-bronze or bluish tinges. Legs stout, broad, toothed, for digging, tarsi 5–5–5. Antennae with 11 antennomeres, and stout three-segmented club. Apart from the small (7–10 mm), rare, subterranean fungus-feeding *Odontaeus armiger*, these are mainly dung beetles. Called dor beetles (or dumbledors), for their loud, humming flight. These are our largest dung beetles, and quite a sight (and sound) flying over grazing pastures on a summer evening. They bury a pear-sized node of dung in the soil and lay an egg on it. They have stout, spined legs, powerful enough to push out through clenched fingers, and are often quoted, by coleopterists, as the childhood inspiration behind an interest in beetles. *Geotrupes spiniger* and the

Geotrupes

very similar *G. stercorarius* both widespread under horse
and cow droppings, burying a bolus of dung for each
egg. *Typhaeus typhoeus*, especially active in winter, and the
males armed with three sharp horns on the pronotum,
buries dung (especially rabbit crottels) deep in sandy soil.
Identification: Joy (1932), Britton (1956), Jessop (1986).

21. **Trogidae** MacLeay, 1819; 1 genus, 3 species
Medium (5–10 mm) convex, squat, broad-oval, sub-
cylindrical species, black, brown or grey, spiny, elytra
striate or granular, often encrusted. Broad, stout, toothed
legs for digging, tarsi 5–5–5. Antennae with nine or 10
antennomeres, and stout three-segmented club. In
carrion, old bones and owl nests. Former minor pest
of tanneries, glue factories, leather stores; *Trox scaber*
sometimes in garden compost where inappropriate
meat/bone kitchen waste is dumped. Identification: Joy
(1932), Britton (1956), Jessop (1986).

22. **Lucanidae** Latreille, 1804; 4 genera, 4 species
Large (*Sinodendron cylindricum*, 10–15 mm) to huge
(*Lucanus cervus*, the stag beetle, 20–66 mm), black or
dark brown, stout, broad, chunky, with strong legs,
tarsi 5–5–5. Antennae of 10 segments, with first long, a
scape, remainder at elbow angle, ending with a strong
four-segmented pectinate club. Fat C-shaped maggots
in rotten, mostly subterranean wood, *Lucanus* taking
three to seven years before metamorphosing into short-
lived (a few weeks) adult. *Lucanus* males have famously
developed long curved, antler-like mandibles, with
which they fight over females, and despite their size they
fly readily on warm June and July evenings, sounding like
model aeroplanes. Male *Sinodendron* has erect, curved horn
rising from head. Commonest British species is broad,
flattened black lesser stag beetle, *Dorcus parallelipipedus*.
Slightly metallic blue-black *Platycerus caraboides* not
recorded for many years and almost certainly extinct.
Identification: Joy (1932), Britton (1956), Jessop (1986).

Typhaeus

Trox

Sinodendron

23. **Scarabaeidae** Latreille, 1802; 55 genera, 84 species

Small (*Oxyomus sylvestris*, 2.0–2.5 mm) to very large (the rare vagrant *Polyphylla fullo*, up to 38 mm), handsome, stout, convex, thickset or cylindrical species; pronotum broad, often convex; elytra can be striate, sometimes ribbed, granular, or smooth. Legs strong, often with tibiae toothed, tarsi 5–5–5 (although front tarsi absent in some exotic species). Antennae with three to six final antennomeres, flat, forming a stout club, the segments of which can be articulated, open or closed. The family fairly neatly falls into two groups. The dung beetles (subfamilies Aegialiinae, Aphodiinae and Scarabaeinae) mainly breed in dung, but also rotting fungi, decaying vegetable matter and rotten wood, and generally have a squatter form and broader digging legs. The Aphodini (46 relatively stout, sub-cylindrical black, brown or mottled species) dominate the British dung fauna; 10–15 species can occur together in a single pat. Largest British species in this tribe (11–12 mm) is *Acrossus rufipes*, deep chestnut brown, sometimes attracted to lighted windows. Scarce *Sigorus porcus* may be cuckoo parasitoid, laying its eggs in the brood dung balls of *Geotrupes* species. English scarab, *Copris lunaris*, domed, glossy and shining as if carved from obsidian, male with long back-curved head horn, has complex male/female pair-bonded burrowing and nesting behaviour, very rare indeed, may be extinct in Britain. The chafers (subfamilies Melolonthinae, Rutelinae and Cetoniinae) have plant-eating larvae, generally at roots, in leaf litter or wood mould, and day-flying leaf-nibbling adults are more elegant, convex, brightly coloured forms. Common cockchafer, *Melolontha melolontha*, also called maybug for its spring appearance, sometime agricultural pest with countless thousands feeding in orchards, now much reduced in numbers. Rose chafer, *Cetonia aurata*, strong metallic green, has an elytral notch just behind the humeri, so that it can close its wing-cases down over its abdomen, even when flying using its outstretched membranous hind wings. Very rare noble chafer, *Gnorimus nobilis*, elegant, dark metallic green, breeds

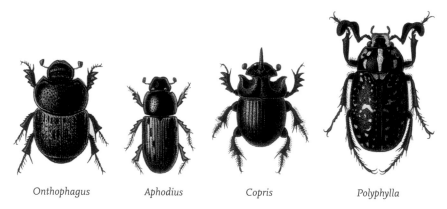

Onthophagus   Aphodius   Copris   Polyphylla

in rot holes in decrepit apple trees in old orchards. Identification: Joy (1932), Britton (1956), Jessop (1986).

Superfamily SCIRTOIDEA Fleming, 1821

24. **Eucinetidae** Lacordaire, 1857; 1 genus, 1 species
Nycteus meridionalis, 3.3–3.7 mm, teardrop-shaped, rounded at head, which is easily hidden under front of pronotum, contracted to elytral apex. Convex, black, vaguely reddish at head and tail tip, hind coxae enlarged, and oblique, concealing most of first abdominal segment; legs long, but stout, tibiae expanded at apex, able to make the beetle skip wildly, tarsi 5–5–5; antennae filiform, 11 antennomeres. Rare, under fungoid bark. Identification: Gardner (1969).

Nycteus

25. **Clambidae** Fischer von Waldheim, 1821; 2 genera, 10 species
Small (1.0–1.1 mm), convex, broad-oval, shining yellowish or black, head large and broad, and pronotum able to tuck under to form an almost complete sphere. Covered all over with short pubescence. Short, slim legs, hind coxae expanded into plates that partly conceal the femora, tarsi 4–4–4. Antennae with 10 segments, slim, with first and last two antennomeres enlarged. In decaying plant matter, under bark, secretive. Identification: Johnson (1966), Freude *et al.* (1971), Lohse & Lucht (1989).

Clambus

26. **Scirtidae** Fleming, 1821; 7 genera, 20 species
Small (1.5–5.5 mm), round or long-oval, moderately convex, but rather soft-bodied. Head usually hidden under front of pronotum. Smooth, or covered with short but dense pubescence. Legs slim, but the hind femora of *Scirtes* large, enabling them to hop, possibly aided by long spine at apex of hind tibiae, tarsi 5–5–5. Antennae filiform or serrate. Larvae are aquatic, adults usually on vegetation in marshy places. *Prionocyphon serricornis* breeds in flooded rot holes in trees. Identification: Joy (1932), Freude *et al.* (1979), Lohse & Lucht (1992).

Scirtes

Superfamily DASCILLOIDEA Guérin-Méneville, 1843 (1834)

27. **Dascillidae** Guérin-Méneville, 1843 (1834); 1 genus, 1 species
Dascillus cervinus, 9–11 mm, long-oval, nearly parallel-sided,
black or dark brown but appearing golden or olive grey
because of short, decumbent yellow pubescence, elytra very
weakly striate. Legs long, front coxae large, tarsi 5–5–5, with
segments 1–4 strongly bilobed. Antennae long and slim,
11 segments, first two short, third very long. Grassy places,
larvae at roots, adults usually on flowers, often orchids.
Identification: Joy (1932).

Dascilus

Superfamily BUPRESTOIDEA Leach, 1815

28. **Buprestidae** Leach, 1815; 5 genera, 17 species
Small (2–12 mm), slightly flattened, mostly sub-
cylindrical and elongate, some (*Trachys*) chunky
seed-shaped. Most are bullet-shaped, head short
and blunt, eyes prominent, pronotum broad,
elytra striate, edges characteristically sinuous
near middle, and narrowing to a fairly pointed
tip. Often brightly coloured, with high metallic
sheen, particularly on the underside. Legs slim,

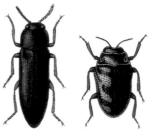

Agrilus   *Trachys*

tarsi 5–5–5. Antennae slim, filiform to serrate, with 11 antennomeres. Larvae
narrow, worm-like, but with a bulbous 'head' segment; mine leaves or burrow
in stems, or under bark of trees where the adults leave characteristic D-shaped
exit holes when they emerge. Called jewel beetles, and the shining, brightly
coloured elytra of exotic species were historically used to decorate rich fabrics,
jewellery and other items of clothing, although those of impoverished UK
fauna are rather demure. Mostly southern, scarce (barely into Scotland and
none apparently in Ireland), found by sweeping, or on flowers. The winding
burrows and exit holes of ruby-coloured *Agrilus sinuatus* occur widely across
England in dead hawthorn branches, although the adults are short-lived and
rarely seen. The once very rare, inky blue-green *A. biguttatus*, in thick, gnarled
oak bark, started to spread, or at least to be found much more widely, following
the windblown tree carnage of the powerful storm on 15–16 October 1987, and is
now known to be relatively widespread, at least in England. Identification: Levey
(1977), Bily (1982).

Superfamily BYRRHOIDEA Latreille, 1804

29. Byrrhidae Latreille, 1804; 7 genera, 13 species
Small (1.2–10.0 mm), round, convex, domed, often nearly
spherical, head hidden under front edge of pronotum;
sensibly called pill beetles. Elytra shining, shallowly striate,
although this is sometimes obscured by covering of sparse
pubescence, short spines, dusting of coloured scales or
encrusting dirt. Legs short, flattened, held flush against
body, in special depressions, if disturbed, and tibiae grooved

Cytilus

along the edge to receive the 4–4–4 or 5–5–5 tarsi. Antennae gradually or abruptly
clubbed, or slim, filiform, 11 antennomeres. Secretive, under stones, in moss,
plant roots or grass thatch. The globose bronze-black, shining, but slightly
pubescent *Simplocaria semistriata* (2.5–2.7 mm) is probably the only common
British species, in moss, at grass roots, mostly in damp places, but the smoothly
variegated *Cytilus sericeus* (4–5 mm) sometimes occurs with it. Identification: Joy
(1932), Freude *et al.* (1979).

30. Elmidae Curtis, 1830; 7 genera, 12 species
Very small (1.2–4.0 mm), slim, elegant, slightly elongate,
sub-cylindrical species, called riffle beetles. Pronotum
often with stria or grooves down each side, or raised
granules. Elytra parallel-sided or broadened behind
middle, striate, ridged or granular. Legs long or very
long, especially tarsi (5–5–5), last segment of which is
long and broadened at grapple-like claws. Antennae
short, slim or slightly clubbed, mostly 11 segments, but
six-segmented in rare *Macronychus quadrituberculatus*.
Larvae and adults aquatic, most are scarce and secretive
in relatively fast-running fresh water, clinging to
underside of rocks, pebbles and logs. The best way to
find them is to stand in the stream bed, then kick up stones,
mud and silt so they are washed by the flow of water into the
net held immediately downstream. Identification: Holland
(1972), Friday (1988).

Stenelmis

31. Dryopidae Billberg, 1820; 2 genera, 9 species
Small (3.0–4.5 mm), slim, long-oval, sub-cylindrical,
parallel-sided species. Pronotum arched at sides, front

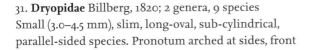

Dryops

angles protruding, head sunken within. Elytra shallowly striate, but surface covered all over with short, fine setae, giving beetles matt, dull grey, green or brown appearance. Legs long and slim, tarsi 5–5–5, long, especially last segment. Antennae short, first two of 8–10 antennomeres flattened into large lobe, remainder expanded sideways into a short, loose club. Larvae aquatic in mud or submerged wood; adults slow-moving, secretive in marshes or on riverbanks. Identification: Freude *et al.* (1979), Lohse & Lucht (1992).

32. **Limnichidae** Erichson, 1846; 1 genus, 1 species
Limnichius pygmaeus, 1.5–1.8 mm, round, oval, convex, black but appearing greyish because of thin covering of short decumbent hairs. Legs slim, yellow, tarsi 5–5–5. Antennae slim, 10 antennomeres, with last segment slightly enlarged into a small club. Secretive, under stones, in moss, at edges of running water, often in splash zone. Identification: Joy (1932), Friday (1988).

Limnichius

33. **Heteroceridae** MacLeay, 1825; 2 genera, 8 species
Small (2.5–5.5 mm), flattened, oval, sub-parallel, almost rectangular, head and pronotum broad; elytra squat, compact, vaguely striate, rather abruptly incurved at apex. Mandibles large and protruding, especially in males. Entire body covered with short, thick silky pubescence; appearance generally dark, with mottled paler pattern, often forming wavy, zigzag bars across the elytra. Legs short and broad, all tibiae toothed or spined, for digging, tarsi 5–5–5 but appearing 4–4–4 as first tarsomere is minute, capable of being folded back onto the tibiae when digging. Antennae short, segments 5–11 forming an oblong or serrate club. In mud and sand at edge of fresh water, and can be found by splashing the banks, causing them to exit their burrows. Sluggish on the ground, they can take to the air rapidly on hot days. The underside of the first abdominal segment has a curved, raised line on each side, topped with minute pegs (more deeply cut in males) – the stridulatory ridges. By rubbing the hind tibiae across the pegs, the males can make a distinct squeaking sound, thought to enable females to find them in the wet soil. Identification: Clarke (1973), Freude *et al.* (1979).

Heterocerus

34. Psephenidae Lacordaire, 1854; 1 genus, 1 species
Eubria palustris, 1.5–2.0 mm, broad, round, convex, shining
blackish brown, elytra striate only at base, and with
small prominence near scutellum. Legs slim, tarsi 5–5–5.
Antennae slim, serrate, segment 1 yellow. Rare. Flat,
round larvae (called water pennies) aquatic, clinging to
underside of pebbles or rocks, often in strong-flowing
currents; adults near the water, in moss, under stones.
Identification: Joy (1932).

Eubria

35. Ptilodactylidae Laporte, 1836; 1 genus, 1 species
Ptilodactyla exotica, 4.5–5.0 mm, long-oval, smooth lines,
slightly flattened, brown, scutellum heart-shaped. Body,
including limbs, covered with short pubescence. Legs slim,
tarsi 5–5–5, but segment 4 very small. Antennae serrate in
female, pectinate in male, each segment with long, thin
finger-like extension. A North American species, accidentally
introduced to Europe and surviving in heated buildings. In
Britain, only in glasshouses at Kew and Cambridge so far.
Identification: Mann (2006).

Ptilodactyla

Superfamily ELATEROIDEA Leach, 1815

36. Eucnemidae Eschscholtz, 1829; 6 genera, 7 species
Small (3.5–9.0 mm), elongate, very long-oval or sub-
cylindrical, sometimes slightly flattened, striate. Legs short
and slim, tarsi 5–5–5. Antennae slim, sometimes serrate or
pectinate, inserted at inner margin of eyes. Resembling click
beetles (Elateridae) in their smooth body lines, and also
possessing ability to hop by sudden jackknifing of the body.
Secretive, uncommon beetles that breed in rotten wood. The
most widespread is *Melasis buprestoides*, distinct for having
front edge of pronotum wider than hind edge. Identification:
Joy (1932), Allen (1969).

Melasis

37. Throscidae Laporte, 1840; 2 genera, 5 species
Small (1.5–3.5 mm), long-oval, slightly flattened, brownish or reddish black,
striate, covered with short pubescence, hind corners of pronotum extended
into short points. Eyes at least partly divided by a groove to receive the clubbed,

11-segmented antenna. Legs short, held flush to body if
disturbed (excellent seed mimics), hind coxae dilated into
plates that partly cover the femora, tarsi 5–5–5. Antennae
short with a three-segmented serrate club. Prosternum
with large forward-directed lobe that all but covers the
mouth. Larvae in rotten wood, adults on flowers and
herbage. Identification: Joy (1932), Freude *et al.* (1979).

Trixagus

38. **Elateridae** Leach, 1815; 38 genera, 73 species
Very small (*Panspaeus guttatus*, 2.8–3.5 mm) to
large (*Melanotus villosus*, 15–20 mm), elongate,
parallel-sided, narrow, slightly flattened; elytra
striate, elegantly curved towards pointed apex;
pronotum domed on disc, with characteristic
rounded front margin, hind angles often
drawn out into backward-pointing spines.
Mostly browns and black, but some (e.g.
Ampedus, mostly scarce) with vivid scarlet on
elytra. Others shining, dull with pubescence
or patterned with flattened scale-like setae.
Head usually inset into front of thorax.
Antennae inserted close to, but in front
of, eyes, 11 segments, sometimes filiform,
usually serrate, occasionally pectinate. Legs
slim, relatively short, tarsi 5–5–5. Underside
of thorax with long, backwards-directed
prosternal spine, which is housed behind in
a corresponding cavity in the mesosternum.
Rapid muscular flexion at this point suddenly
jackknifes the beetle, violently, altering its

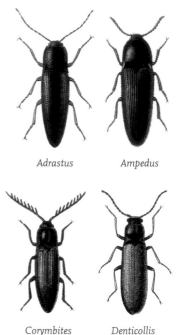

Adrastus   *Ampedus*

Corymbites   *Denticollis*

centre of gravity and flinging it with an audible click into the air, a defence escape
mechanism. Long cylindrical larvae (wireworms) and adults in decaying wood,
under bark, in wood mould and fungoid sap runs, although some (e.g. *Agriotes*)
at plant roots, giving them minor agricultural or horticultural pest status. Many
(possibly most) purportedly dead-wood feeders are at least partly predatory,
especially when reared in captivity, when cannibalism is also frequently observed.
Captive breeding is, apparently, enhanced by providing small portions of cheese
to supply important animal protein if other wood-boring maggots are in short
supply. Adults sometimes found flying, or on flowers. Many scarce species

of conservation concern are confined to truly ancient broadleaved woodland remnants, including the violet click beetle, *Limoniscus violaceus* (actually dull black with a purple tinge), and oak click beetle, *Lacon querceus*, at Windsor, and *Ampedus pomonae* around the New Forest area. Large, broad red-and-black *Elater ferrugineus*, once regarded as excessively rare, now found to be more widespread, but still very evasive, after a synthetic pheromone lure was developed to attract flying males for population studies. Some New World species are powerfully phosphorescent, having paired light-emitting organs near the front of the pronotum. Identification: Joy (1932), Leseigneur (1972), Freude *et al.* (1979).

39. Drilidae Blanchard, 1845; 1 genus, 1 species

Drilus flavescens, male a conventional beetle shape, 5–7 mm, long-oval or sub-parallel, black but elytra brownish yellow, rather soft, covered with pubescence; legs slim, black and yellow, tarsi 5–5–5; antennae long, black, 11 segments, strongly pectinate from fourth antennomere. Female is bizarre, 18–20 mm, larviform, rather caterpillar-like, lacking elytra or any tough sclerotised body plates, flabby, bloated, pale brown but with pairs of darker patches on each body segment, legs slim, antennae short and stout. Larva peculiar,

Drilus

soft, hairy, fringed with many finger-like growths; snail predators in chalk and limestone grassland. Uncommon, but males fly readily, females secretive and rarely found. Identification: Joy (1932).

40. Lycidae Laporte, 1836; 4 genera, 4 species

Small (5–9 mm), flattened, long, parallel-sided, soft-bodied, black, but elytra and sometimes also pronotum dull to bright red; elytra ribbed with raised lines and with numerous short cross-veins, sometimes giving appearance of rather square indentations and justifying common name net-winged beetles. Head concealed beneath front of pronotum, eyes larger in males than in females. Legs slim, tarsi 5–5–5. Antennae 11-segmented, long but robust, filiform or slightly pectinate, flattened. Larvae in rotten wood; short-lived adults flying, on herbage or flowers. Mostly scarce, only *Platycis minutus* is at all widespread in England. Identification: Joy (1932).

Platycis

41. **Lampyridae** Rafinesque, 1815; 3 genera, 3 species
The common glow-worm, *Lampyris noctiluca*, is the
only species regularly found (two others, doubtfully
British, may be extinct): medium-sized (5–13 mm),
male long-oval, flattened, parallel-sided, dull grey,
soft-bodied, head hidden beneath rounded front
of pronotum, elytra weakly ribbed towards base,
covered with short grey pubescence. Legs slim,
tarsi 5–5–5. Antennae short and slim. Female
larviform, lacking wings and elytra, and with only

Lampris   *Phosphaeus*

the nearly semicircular pronotal shield vaguely sclerotised. Legs and antennae
short and slim. The female (also larvae) glow magically, producing a wan green
light on the underside of the abdominal apex, to attract flying males. Larvae are
snail predators, usually in rough grassy places. Once widespread enough to earn
its common English name, now declining, with any spread or recolonisation of
sites hampered by flightlessness of the female. Elsewhere in the world, brighter,
often flashing species are known as fireflies. Identification: Joy (1932).

42. **Cantharidae** Imhoff, 1856 (1815); 7 genera, 41 species
Very small (*Malthodes lobatus*,
2 mm) to medium (*Cantharis rustica*,
13 mm), long, parallel-sided, rather
soft-bodied, elytra barely striate in
a few species, protonum rounded,
flattened, often keeled or wrinkled
along edges, sometimes (in difficult
genera *Malthinus* and *Malthodes*)
the elytra are shortened, exposing
wing-tips, abdominal segments
and frequently convoluted male

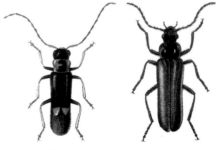

Malthodes   *Podabrus*

appendages, which are useful identification guides. Legs long and slim, tarsi
5–5–5, usually at least tarsomere 4 deeply divided, lobed, heart-shaped.
Antennae long, 11 segments, usually filiform. Called solider beetles in that their
predominantly black and red colours were frequently bright enough to resemble
military officers' dress uniforms. The very common red (tipped with vague black)
Rhagonycha fulva occurs in profusion, especially on hogweed flowers, from mid-
July onwards, frequently in mating pairs, male with clearly larger and more
bulbous eyes than female. Soft, velvety larvae are predatory in grass roots or
fungoid wood; adults often on flowers. Identification: Joy (1932), Freude *et al.* (1979).

Superfamily DERODONTOIDEA LeConte, 1861

43. **Derodontidae** LeConte, 1861; 1 genus, 1 species
Laricobius erichsonii, 1.5–2.5 mm, long-oval, head with large
eyes, and also with two pale ocelli at hind margin; black
pronotum broad, rounded or angular hexagonal; dark brown
elytra sub-parallel, strongly and coarsely striate, covered
all over with short, dense pubescence. Legs short, brown,
tarsi five- but appearing four-segmented as penultimate is
reduced. Short antennae, 11 antennomeres, with a loose club

Laricobius

of three segments. Resembles a member of the Latridiidae. A
recent arrival (1971) in Britain, a predator of aphids and adelgids on conifer trees.
Identification: Hammond & Barham (1982), Peacock (1993).

Superfamily BOSTRICHOIDEA Latreille 1802

44. **Dermestidae** Latreille, 1804; 13 genera, 41 species
Small (1.5–9.3 mm), short- to long-oval, usually compact,
convex or domed, not striate, black or brown, but many
covered or patterned with pale or coloured scales. Most
genera (not *Dermestes*) with a single ocellus on the head,
between the compound eyes. Legs slim, tucked up against
globose body if disturbed, tarsi 5–5–5, fifth segment long.
Antennae mostly with 9–11 segments, often clubbed. This
mixed family contains some of the most commercially
important pest species in the world. Many are cosmopolitan,
having been trafficked around the globe in traded goods. In
nature, their distinctive bristly larvae are carrion scavengers
at the dried sinew-and-feathers stage of decay, others occur
in bird, mammal or bee nests, but many have become pests
in stored products, especially attacking dried and cured meat
products (e.g. larder beetles, *Dermestes*), wool and silk carpets
(carpet beetles, *Anthrenus*), clothes, furs and hides (fur beetle,
Attagenus pellio), and stuffed animals in museums, including
insect collections (museum beetle, *Anthrenus verbasci*). Adults
sometimes occur on flowers. *Ctesias serra* scavenges on dead insect remains in the
webs made by crevice-dwelling spiders that live under the bark of old trees, and
although the smooth, shiny black adult is scarce (it may be short-lived), the bristly
larva is more often encountered. Identification: Peacock (1993), Háva (2011).

Dermestes

Attagenus

45. Bostrichidae Latreille, 1802; 4 genera, 5 species
Small (2.7–9.0 mm), cylindrical, head often concealed beneath
front of pronotum, mostly heavily punctured or granulate,
and elytra neatly or strongly striate. Legs short, slim, tarsi
5–5–5. Antennae short, 11 segments, with club discreet, or loose
pectinate. Wood-borers, narrow *Lyctus* species sometimes called
powder-post beetles for the powdery frass-filled voids they
create. Although the British fauna is seriously impoverished, 45

Rhyzopertha

other non-established species have been recorded as emerging
from imported wooden (including bamboo) furniture, ornaments, utensils, tools
and knick-knacks, the larvae having been gnawing away, unobserved, for months
or even years in the blocks and planks of wood from which the goods were
manufactured. Identification: Joy (1932), Freude *et al.* (1979).

46. Ptinidae Latreille, 1802; 29 genera, 57 species
Small (2–7 mm), short- or long-oval, mostly compact,
hunched, head more or less hidden under front of
pronotum. Shining, or covered in pubescence or scales,
smooth or striate. Legs slim, sometimes spider-like
(hence some are called spider beetles), tarsi 5–5–5.
Antennae mostly slim, 11 segments, sometimes serrate
(e.g. *Lasioderma serricorne*) or pectinate (e.g. male *Ptilinus
pectinicornis*). Previously divided into several smaller
families. The spider beetles (subfamily Ptininae), with

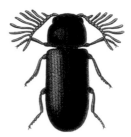

Ptilinus

a rather narrow pronotum ahead of somewhat bulging elytral shoulders, are
scavengers, under powdery dry bark and in animal nests (including those of
bumblebees and ants), or are stored-product
pests, sometimes found in mouldy food or
on mildewed walls of derelict buildings. The
widespread but local *Ptinus fur* shows sexual
dimorphism: male is rounded, globular and
dark with two rough bars of pale pubescence
(sometimes joined) across the elytra; female
is parallel-sided and uniformly light brown.
The Anobiinae, with last three segments of
antennae long, are rather cylindrical and
bore into dead heartwood; they include the
common domestic woodworm (*Anobium
punctatum*) and deathwatch (*Xestobium*

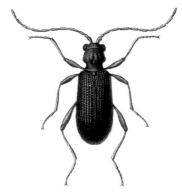

Ptinus

rufovillosum), the multiple generations of which can weaken or disintegrate seasoned wood of furniture or structural timber as they excavate burrows and voids over decades or centuries. Delicately patterned *Ochina ptinoides* occurs in ivy stems, and prettily patterned *Ptinomorphus imperialis* in old hedges, but often found by evening sweeping. Biscuit beetle, *Stegobium paniceum*, is a frequent nuisance in stored foods – cereals, flour, dried pasta, biscuits. The Dorcatominae, often cylindrical but some rather globose, with last three antennal segments in a large, loose serrate club, occur, rarely, in fungi and fungoid wood. Identification: Joy (1932), Freude *et al.* (1969), Zahradnik (2013).

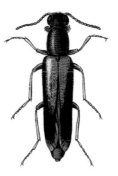

Ptinomorphus

Superfamily LYMEXYLOIDEA Fleming, 1821

47. **Lymexylidae** Fleming, 1821; 2 genera, 2 species
Medium (6–16 mm), narrow, parallel-sided, cylindrical; elytra not quite covering abdomen, soft, not striate but slightly ribbed; pronotum squarish, head bulbous. Legs slim, front coxae long, tarsi long, 5–5–5. Antennae with 11 antennomeres, slim, filiform or serrate. Maxillary palps of males expanded, flabellate or pectinate. Larvae are wood-borers, in heartwood of often dead standing trees, usually oak. Two British species, *Hylecoetus dermestoides* and *Lymexylon navale*, both very scarce. Identification: Joy (1932).

Lymexylon

Superfamily CLEROIDEA Latreille, 1802

48. **Phloiophilidae** Kiesenwetter, 1863; 1 genus, 1 species
Phloiophilus edwardsii, 2.0–2.6 mm, long-oval, broad, flattened, especially pronotum, pale brown, patterned with vague chevron or zigzag marks across elytra, not striate, covered with short pubescence. Legs slim, tarsi 5–5–5. Antennae short, filiform, with a loose three-segmented club. Scarce, on fungoid twigs and branches. Identification: Joy (1932).

Phloiophilus

49. **Trogossitidae** Latreille, 1802; 5 genera, 5 species
Small (4–11 mm), elongate to broad-
oval, flattened, flanged, to narrow
cylindrical, elytra more or less
striate. A mixed bag of variously
shaped species, previously spread
between several different families.
Tarsi 5–5–5, but first segment much
shorter than second, front coxae
strongly broadened. Antennae of
11 segments, last three forming a
serrate club. All British species are

Tenebroides   Thymalus

secretive and scarce. *Thymalus limbatus* strongly resembles a red tortoise beetle
(Chrysomelidae, subfamily Cassidinae) and occurs under fungoid bark. The
cadelle, *Tenebroides mauritanicus*, occurs in stored foods and is a predator of grain-
feeding beetle larvae. Identification: Joy (1932), Freude *et al.* (1967).

50. **Thanerocleridae** Chapin, 1924; 1 genus, 1 species
Thaneroclerus buquet, 5–6 mm, uniform orange-brown, elongate,
sub-cylindrical; pronotum finely punctured, rather globular,
constricted near scutellum to a peduncle; elytra coarsely
punctured, slightly pubescent. Legs and antennae medium
length, tarsi 5–5–5. A south Asian species, now occurring around
the world, first found in Britain breeding in Bombay ginger in
the Natural History Museum, London, where it was probably
predating other insect larvae. Identification: Gerstmeier (1998).

Thaneroclerus

51. **Cleridae** Latreille, 1802; 8 genera, 13 species
Small (3–11 mm), elongate, rather parallel-sided, sub-
cylindrical, covered with short, erect hairs, elytra striate
or not, pronotum rounded, cylindrical or heart-shaped,
head broad with bulging eyes. Legs medium-stout
to slim, tarsi 5–5–5. Antennae slim filiform, to short
serrate or clubbed. Larvae are predators, mostly of bark
beetles. Commonest is possibly metallic blue *Korynetes
caeruleus*, which can also occur in old houses because
of infestations by woodworm (*Anobium punctatum*) or
deathwatch (*Xestobium rufovillosum*). The pretty black,
white and red *Thanasimus formicarius* is also widespread,

Thanasimus

under bark of pine trees. Two rare *Trichodes* species (probably extinct in Britain) attack bee and wasp grubs, and *Necrobia* (three British species) feed in carrion, bones and sometimes stored dried meat products. Impoverished British fauna can be rather drab, but some brightly coloured exotic species give the family the name chequered beetles. Identification: Joy (1932), Gerstmeier (1998).

52. **Melyridae** Leach, 1815; 15 genera, 26 species
Small (2–8 mm), long-oval to linear parallel-sided, slightly flattened, soft-bodied; pronotum rounded, usually covered with short pubescence; elytra not striate, sometimes not quite covering abdomen. Legs slim, tarsi 5–5–5. Antennae slim, filiform or slightly serrate, 11 segments, basal segments expanded in males of some *Malachius* species. Several common species (notably *M. bipustulatus* and *Cordylepherus viridis*) are bright green, hence the name

Malachius

malachite beetles. Others are spotted or patterned with red. The red and golden-green scarlet malachite beetle, *M. aeneus*, was once widespread in Britain, but for no very clear reason has declined to the point of near extinction in the last 50 years and is now the subject of concerned conservation surveys. Several species have inflatable balloon-like red sacs under edges of pronotum and elytra, which are filled with fluid and expanded if attacked; they are thought to give off chemical odours to deter would-be predators. *Dasytes* species are deep leaden black and occur widely in rough grassy places. The metallic blue-green *Psilothrix viridicoerulea* is strictly coastal, on dunes and undercliffs, while *Dolichosoma lineare* occurs in or near saltmarshes. Larvae are predators in soil or rotten bark; adults found on flowers. Identification: Joy (1932), Freude *et al.* (1979).

Superfamily CUCUJOIDEA Latreille, 1802

53. **Byturidae** Gistel, 1848; 1 genus, 2 species
Small (3.5–5.0 mm), long-oval, sub-parallel, slightly flattened (especially the pronotum), pale brownish grey because of dense cover of short, pale pubescence. Legs slim, tarsi 5–5–5, fourth tarsomere very small. Head broad with eyes very convex (*Byturus tomentosus*) or very large (*B. ochraceus*), antennae of 11 segments, last three antennomeres forming a vague club. Both common, on flowers, especially bramble, *B. tomentosus* sometimes considered a pest of raspberries. Identification: Joy (1932).

Byturus

54. **Sphindidae** Jacquelin du Val, 1860; 2 genera, 2 species
Minute (1.3–2.4 mm), convex; globular, shining (*Aspidiphorus orbiculatus*); or parallel-sided, with very fine pubescence (*Sphindus dubius*); elytra striate, black. Legs short and slim, reddish, tarsi 5–5–5 (female) or 5–5–4 (male). Antennae short, with a gradual three-segmented club, apical segment very large in *Sphindus*. In fruiting bodies of slime moulds under bark or on rotten stumps. Both scarce. Identification: Joy (1932), Pope (1953).

Aspidiphorus

55. **Biphyllidae** LeConte, 1861; 2 genera, 2 species
Small (2.7–3.5 mm), long-oval, slightly flattened, smooth outline, pronotum with narrow raised ridge down each side, elytra striate, upper surface covered in short pubescence, all black with pale lunate mark across suture (*Biphyllus lunatus*) or brown with elytral shoulders obscurely paler (*Diplocoelus fagi*). Legs slim, tarsi 5–5–5. Antennae short, moniliform, with a club of two segments (*Biphyllus*) or three (*Diplocoelus*). *Diplocoelus* under fungoid bark, *Biphyllus* in fungal fruiting lumps of cramp balls (*Daldinia concentrica*), both scarce. Identification: Joy (1932).

Biphylus

56. **Erotylidae** Latreille, 1802; 4 genera, 8 species
Small (2.5–7.0 mm), mostly long-oval, rounded, convex, shining, black (or metallic) and pinkish orange, elytra sometimes striate, although the obscure *Cryptophilus integer* has short pubescence. Legs slim, tarsi 5–5–5, but first segment may be small, so appearing 4–4–4. Antennae short, 11 antennomeres, moniliform, with a three-segmented club. In fruiting bodies of bracket and other fungi, or under fungoid bark. Mostly scarce. Called pleasing fungus beetles in North America, because of their pleasing colours and shiny bodies. Identification: Joy (1932).

Triplax

57. **Monotomidae** Laporte, 1840; 2 genera, 23 species
Very small (1.9–4.8 mm), narrow, elongate, parallel-sided, flattened, pronotum square or elongate, head large-triangular, shining (*Rhizophagus*) or minutely pubescent (*Monotoma*); elytra striate, slightly shortened, leaving pygidium exposed. Legs slim or stout, tarsi 5–5–5, 5–5–4 or 4–4–4. Antennae short, slim, with 10 antennomeres, last one or two forming a club. *Rhizophagus*

Rhizophagus

species live under rotten bark, feeding on fungus, or are predatory. *Rhizophagus grandis* was recently introduced to Britain to combat the bark-boring forestry pest scolytid beetle *Dendroctonus micans*. *Rhizophagus parallelicollis* is probably subterranean, feeding on fly larvae in buried logs or decaying tree root systems; it has been called the graveyard beetle after being found in large numbers on interred coffins, where it is most likely a predator of scuttle flies (order Diptera, family Phoridae). *Monotoma* species occur in decaying plant matter like compost, grass cuttings and haystack refuse, or in ant nests. Identification: Peacock (1977).

58. **Cryptophagidae** Kirby, 1826; 11 genera, 101 species
Minute to very small (1.1–4.8 mm), short- to long-oval, rounded or sub-parallel, robust, moderately convex, mostly covered in short, often dense and silky pubescence, sides of pronotum often toothed, or with a prominent callosity near front angles (e.g. *Cryptophagus*), elytra not striate. Legs short, stout or slim, tarsi 5–5–5, or sometimes 5–5–4 in males. Antennae short, moniliform, often with a loose three-segmented club. Most are mould- or fungus-feeders, occurring in decaying plant material, but the rare (possibly extinct) *Hypocoprus latridioides*, and widespread but secretive and very tiny *Ootypus* and *Ephistemus*, can occur under dung; the three British species of *Antherophagus* breed in bumblebee nests and are sometimes found on flowers. The large genera *Cryptophagus* (36 small species) and *Atomaria* (44 very small species) are notoriously difficult to identify. Identification: Coombs & Woodroffe (1955), Freude *et al.* (1967), Lohse & Lucht (1992).

Cryptophagus

59. **Silvanidae** Kirby, 1837; 10 genera, 12 species
Very small (2–7 mm), elongate, parallel-sided, sub-cylindrical, flat or very flat, pronotum narrow, side margins with ridge, wavy or toothed, elytra striate, smooth or with short pubescence. Legs short, slim, tarsi 5–5–5 or 4–4–4 (*Uleiota planatus*). Head narrowed behind eyes, antennae 11-segmented, long filiform or short moniliform with a small, loose club. Mostly secretive and scarce fungus-feeders under bark (*Silvanus, Uleiota, Dendrophagus*, etc.), or in plant debris (*Ahasverus advena, Psammoecus bipunctatus*), although *Oryzaephilus surinamensis*, the saw-toothed grain beetle, occurs throughout the world as a stored-product pest, probably feeding on mould. Identification: Joy (1932), Freude *et al.* (1967).

Silvanus

60. **Cucujidae** Latreille, 1802; 1 genus, 2 species
Very small (3.5–4.5 mm), elongate, very flattened, pronotum
broad-rectangular, elytra parallel-sided, not striate. Legs short,
slim, tarsi 5–5–4 (male) or 5–5–5 (female). Head broad-triangular,
antennae short, slim, moniliform with a loose three-segmented
club. All pale brown (*Pediacus depressus*) or head and pronotum
darker (*P. dermestoides*). Under bark of fungoid trees. Both scarce.
Identification: Joy (1932), Allen, 1956.

Pediacus

61. **Phalacridae** Leach, 1815; 3 genera, 16 species
Very small (1.3–3.0 mm), short-oval, rounded, convex, domed
above, flat beneath, shining, smooth lines, elytra not striate
but with one (*Phalacrus, Stilbus*) or two (*Olibrus*) indented
grooves parallel to suture, scutellum relatively large. Legs
short, slim, tarsi 5–5–5, segment 4 very small. Antennae short,
slim, moniliform, 11 antennomeres, three-segmented club,
last segment the largest (distinctively pear-shaped in *Olibrus*).
Mostly on flowers. *Olibrus* species pollen-feeders, larvae
develop in the flowers. *Phalacrus* and *Stilbus* are mildew-,
ergot- and smut-fungus-feeders, hence their common names
flower beetles and smut beetles. Identification: Thompson (1958).

Olibrus

62. **Laemophloeidae** Ganglebauer, 1899; 5 genera, 13 species
Very small (1.3–4.8 mm), oval to elongate, flattened; pronotum
small, square or trapezoid, narrowed behind, usually with
thin groove or line on each side; elytra parallel-sided or
slightly rounded, more or less striate. Legs short, slim, tarsi
5–5–4 (female) or 5–5–5 (male). Head broad and flat, mandibles
large and prominent; antennae long, slim filiform, 10
segments. Although some are pests in stored grain, most are
scarce and secretive, under fungoid bark, where they may
be fungus-feeders or predators of bark beetles. The larvae
of *Cryptolestes* are apparently unique among beetles in having paired prothoracic
silk glands, with which they spin a cocoon to pupate in. Identification: Joy (1932),
Lefkovitch (1959), Freude *et al.* (1967).

Cryptolestes

63. **Kateretidae** Kirby, 1837; 3 genera, 9 species
Very small (1.8–2.5 mm), long-oval, pronotum rounded; elytra parallel-sided,
short and exposing two to three abdominal segments. Black, but covered with

short pubescence to give dull or grey appearance. Legs
short, stout, tarsi 5–5–5. Antennae short, slim, but with
last three segments forming a weak club. In flowers, often
each species associated with one particular plant, e.g.
some *Brachypterolus* on wild toadflax, others on garden
snapdragons, both common *Brachypterus* species on
stinging nettle (*Urtica dioica*); adults feed on petals and
pollen, larvae develop in seed capsules. Identification:
Kirk-Spriggs (1996).

Brachypterus

64. **Nitidulidae** Latreille, 1802; 16 genera, 92 species
Very small (1.4–6.5 mm), broad-oval, slightly parallel-sided,
flattened, pronotum often with flattened flange-like edges;
elytra sometimes not quite covering abdomen, leaving all
or part of pygidium exposed, or more (e.g. *Carpophilus*),
usually not striate, shining or dull. Legs short, stout, tibiae
sometimes broad and flat with minute teeth (*Meligethes*),
tarsi mostly 5–5–5, broad, segment 4 often very small.
Antennae with a more or less abrupt three-segmented
club. Several different life histories. Most, including

Meligethes

members of the large, difficult genus *Epuraea*, are fungus feeders (or possibly
predators), often breeding under bark but visiting flowers as adults. *Carpophilus*
are fungus-feeders, but some occur under bark and many cosmopolitan species
are worldwide pests in stored products like dried fruit. Members of the large
(38 species) and very difficult genus *Meligethes* breed in flowers, often with very
particular host-specific relationships, which can aid
identification. The very common greenish *Meligethes
aeneus* and *M. viridescens* are associated with wild charlock
(*Sinapis arvensis*), but fields of rape (*Brassica napus*) can host
countless billions of them. Others are carrion-feeders
(*Nitidula*, *Omosita*), live in ant nests (*Amphotis marginata*),
at fermenting sap runs on damaged trees (*Soronia*) or in
puffballs (*Pocadius*). The bizarre shining, globose, almost
hemispherical *Cybocephalus fodori* has a very broad head
and short broad-triangular scutellum, legs slim, relatively
long, tarsi 4–4–4; it was discovered in Britain only in 2001,

Omosita

and is a predator of scale insects (superfamily Coccoidea, family Diaspididae).
The sub-Saharan *Aethina tumida* is not yet known in Britain, but has spread to the
United States, Canada and Australia, where it is a pest in honey bee hives (and is

known as the small hive beetle), the larvae feeding gregariously and messily on stored honey and pollen in the combs. Identification: Joy (1932), Hinton (1945), Freude *et al.* (1967), Kirk-Spriggs (1996).

65. **Bothrideridae** Erichson, 1845; 3 genera, 5 species
Very small (1.3–4.0 mm), narrow, parallel-sided, sub-cylindrical, dirty reddish brown, pronotum long, partly covering head, elytra striate. Legs short, slim, tarsi 4–4–4. Antennae moniliform with a club of one segment (*Anommatus*, which also lacks eyes), or 2 segments (*Teredus* and *Oxylaemus*). Under fungoid bark, all scarce. Identification: Joy (1932), Buck (1957), Freude *et al.* (1967).

Anommatus

66. **Cerylonidae** Billberg, 1820; 2 genera, 5 species
Minute (1.0–2.5 mm); short-oval, convex, domed, pubescent and non-striate (*Murmidius*), or elongate, sub-parallel, slightly flattened, robust, shining and striate (*Cerylon*). Legs short, stout, tarsi 4–4–4. Antennae short, moniliform, 10 segments with last antennomere large, forming a club. Secretive under fungoid bark (*Cerylon*) or scarce in stored food products (*Murmidius*). Identification: Joy (1932), Freude *et al.* (1967).

Cerylon

67. **Alexiidae** Imhoff, 1856; 1 genus, 1 species
Sphaerosoma pilosum, 1.2–1.5 mm, round-oval, convex, domed, black to reddish, covered in long, sparse pubescence. Legs slim, short, tarsi 4–4–4. Antennae moniliform with 10 segments, last three forming a loose club. In moss, fungi, roots, etc. Identification: Joy (1932).

Sphaerosoma

68. **Endomychidae** Leach, 1815; 5 genera, 8 species
Small (1–5 mm), long-oval, convex, pronotum sometimes constricted in front of rather bulging elytra, smooth or covered in long, outstanding pubescence (*Mycetaea subterranea*). Legs slim, tarsi 4–4–4, but third segment is very small and hidden, so appearing 3–3–3. Antennae slim, moniliform, 11 antennomeres, but with a vague three-segmented club. Fungus-feeders, *Mycetaea* in mouldy hay or stored foodstuffs, *Lycoperdina* in puffballs, and the pretty spotted *Endomychus coccineus* (sometimes mistaken for a ladybird) under fungoid bark. Identification: Joy (1932).

Endomychus

69. **Coccinellidae** Latreille, 1807; 31 genera, 55 species
Minute (1.0–1.2 mm, *Clitostethus arcuatus*) to small (7–9 mm,
Anatis ocellata), round-oval, convex, domed above, flat beneath,
pronotum broad, elytra rounded, not striate. Legs slim, tarsi
4–4–4, but appearing 3–3–3 because third segment is very
small and hidden. Antennae inserted on the forehead, near
the eyes, short, slim, 11 antennomeres, with a loose three-
segmented club. The generally larger, smooth, shining,

Anisosticta

brightly coloured and spotted species (tribes Coccinellini,
Chilocorini and Epilachnini) are the familiar ladybirds, identified by their
patterns, the seven-spot (*Coccinella septempunctata*) and two-spot (*Adalia bipunctata*)
perhaps being traditionally most familiar, but with the several colour forms
of the Asian harlequin ladybird (*Harmonia axyridis*) fast becoming one of the
commonest and most widespread species in the country. The smaller, pubescent
Coccidulini and Scymnini include pretty but more subtle colour forms, but
are secretive and the genus *Scymnus* (11 species) can be difficult. Most adults
and larvae are predators of aphids, scale insects and other plant lice, although
some graze mildews from leaf surfaces and two species are herbivores – the
common 24-spot (*Subcoccinella vigintiquattuorpunctata*), mostly on false oat-grass
(*Arrhenatherum elatius*), and the scarce bryony ladybird (*Henosepilachna argus*) on
white bryony (*Bryonia dioica*), mostly in Surrey. Bright colours are a warning that
their bodies contain alkaloids, distasteful to predators; as a prelude, they are
able to reflex-bleed, exuding droplets of bitter-tasting, bright yellow or orange
haemolymph from the femoro-tibial joints. Identification: Joy (1932), Pope (1953),
Majerus & Kearns (1989), Roy *et al.* (2013).

70. **Corylophidae** LeConte, 1852; 5 genera, 12 species
Minute (0.4–1.2 mm), rounded, domed; pronotum broad,
mostly covering head, hind angles often swept back into a
sharp angle, but forming a uniform outline with the elytra,
which do not quite cover the last abdominal segments
and are not striate. Legs slim, tarsi 4–4–4. Antennae slim,
moniliform, with a three-segmented club. Wings short, but
fringed with long hairs; active flyers. Mould- and fungus-

Sericoderus

feeders, under bark, on fruiting bodies or in decaying vegetable matter, compost,
manure or grass heaps. Identification: Freude *et al.* (1971), Lohse & Lucht (1989),
Bowstead (1999).

71. **Latridiidae** Erichson, 1842; 13 genera, 56 species
Minute to very small (1.0–3.1 mm), elongate or long-oval,
convex or slightly flattened, usually covered with short,
outstanding pubescence. Latridiids have a distinctive shape,
with a large, frequently triangular head, small round or heart-
shaped pronotum, narrower than the relatively prominent
shoulders of the rounded or parallel-sided elytra, which are
striate, often heavily so (e.g. *Dienerella*). Legs slim, tarsi 3–3–3,
first tarsomere long. Antennae moniliform, 8–11 segments,
first antennomere enlarged, last three forming a loose club.

Stephostethus

Mould-feeders, under fungoid bark, in compost, leaf litter or slime moulds, or
generally found by sweeping in grassy places; a few are stored-product pests,
where forgotten sweepings have become dank and mouldy. The widespread
and prettily marked *Cartodere bifasciata*, pale with a darker chequer pattern on
elytra and flanged pronotal edges, and the similar but all-dark *C. nodifer*, are
Australasian species introduced to Europe and now common. Identification:
Hinton (1941), Freude *et al.* (1967), Lohse & Lucht (1992).

Superfamily TENEBRIONOIDEA Latreille, 1802

72. **Mycetophagidae** Leach, 1815; 7 genera, 15 species
Small (2–6 mm) and, apart from the latridiid-like *Berginus
tamarisci* (scarce on tamarisk), smoothly long-oval, slightly
flattened, mostly with covering of short, upstanding
pubescence, black or brown, striate elytra marked with
pattern of reddish or yellowish spots, bars or chevrons.
Legs slim, tarsi 4–4–4 (3–4–4 in males). Antennae with 11
antennomeres, the last four or five forming a gradual loose
club. Mostly in fungus and under fungoid bark, the prettily

Mycetophagus

marked (four red spots) *Mycetophagus quadripustulatus* being
our largest and commonest. Identification: Joy (1932), Buck
(1957), Freude *et al.* (1967).

73. **Ciidae** Leach, 1819; 7 genera, 22 species
Small (1.2–3.3 mm), long-oval, parallel-sided, cylindrical, convex,
shining smooth, or with sparse covering of thin pubescence or
short upright scales; pronotum relatively long, rounded barrel-
shaped; elytra not striate, rather abruptly rounded at apex. Legs
slim, tarsi 4–4–4. Antennae with eight antennomeres (*Octotemnus*

Cis

glabriculus), nine (*Ennearthron cornutum*, *Sulcacis nitidus*) or 10 (*Cis*, etc.), with last three forming a loose club. Males often with front of head and sometimes also pronotum raised into a ridge or sharp plate. In fungi, or under bark of fungoid logs, secretive, slow-moving. Identification: Freude *et al.* (1967).

74. **Tetratomidae** Billberg, 1820; 2 genera, 4 species
Small (3.0–4.5 mm), smoothly long-oval, slightly convex, head more or less hidden under front of broad, rounded pronotum, elytra sometimes striate (*Hallomenus binotatus*). Legs slim, tarsi 5–5–4. Antennae with 11 antennomeres, moniliform or slightly serrate, widening abruptly (*Tetratoma*) or gradually into a club at apex. Scarce, in fungus and under fungoid bark. Identification: Joy (1932), Buck (1954).

Tetratoma

75. **Melandryidae** Leach, 1815; 10 genera, 17 species
Small (*Abdera*, 2–4 mm) to medium (*Melandrya*, 9–15 mm), slim, long-oval to elongate, sub-parallel to cylindrical, elegant, with smooth lines, elytra sometimes striate (*Zilora*, *Melandrya*). Legs long, slim, tarsi 5–5–4. Antennae filiform, serrate or clubbed, 11 antennomeres, palpi long. In fungus or under fungoid bark. Mostly scarce species, but weakly ribbed, shining, slightly metallic greenish-blue-black *Melandrya caraboides* is widespread, and sometimes mistaken for a carabid ground beetle. The three smoothly rounded, long-oval *Orchesia* species are able to skip vigorously shrimp-like, much in the manner of the Mordellidae. Identification: Joy (1932), Buck (1954).

Conopalpus

76. **Mordellidae** Latreille, 1802; 5 genera, 17 species
Small (2–9 mm), long-oval, wedge-shaped, hunched, with a large triangular head on a narrow neck, which can be tucked right under broad, short pronotum; slightly shortened, non-striate, nearly parallel-sided elytra exposing long, sharp, spine-like pygidium. Upper surface covered with dense but short pubescence, sometimes forming shifting chequered or streaked patterns (*Tomoxia* and *Mordella*). Legs slim, hind legs long, tibiae and tarsi with microscopic but distinctive combs of tight bristles, which may

Tomoxia

aid the beetles in their rapid, erratic, skipping escape manoeuvres. Tarsi long, 5–5–4. Underside deep, convex, so natural resting position, legs pulled in, is on the beetle's side. Antennae short, with 11 antennomeres, more or less serrate. Larvae in plant stems (often forming a gall) or dead wood; mostly scarce adults on flowers, very active. Distinctive and attractive beetles, but *Mordellistena* (12 species) difficult. Identification: Buck (1954), Batten (1986).

77. **Ripiphoridae** Gemminger, 1870 (1855); 1 genus, 1 species
Metoecus paradoxus, 10–12 mm, highly distinctive, long, slim, hunched, wedge-shaped, large triangular black head tucked under bell-shaped black pronotum, scutellum invisible; elytra (black in female, yellowish in male) narrow, tapering, not meeting at suture and exposing wings and last abdominal segments. Underside deep, convex, almost keeled. Legs slim, long, tarsi 5–5–4. Antennae with 11 segments, pectinate in female, strongly bipectinate in male. Larvae in wasp nests, feeding on the brood; scarce (short-lived?) adults

Metoecus

flying, or laying eggs on dead wood, from which wasp hosts inadvertently pick up highly active triungulin larvae. Identification: Joy (1932), Buck (1954).

78. **Zopheridae** Solier, 1834; 9 genera, 13 species
Small (2–7 mm), long-oval to narrow-elongate, parallel-sided, slightly flattened to sub-cylindrical, smooth or pubescent, some sculptured and ridged, elytra striate. Legs short and slim, tarsi 5–5–4. Antennae slim with a small club, or highly compacted and fattened (*Orthocerus clavicornis*). A mixed family of scarce fungus-feeders or predators found under fungoid bark. Only the pretty black-and-red-marked *Bitoma crenata*, with deeply grooved pronotum and raised elytral lines, is widespread. The scarce flat, ridged, dirty pinkish *Langelandia anophthalma* is subterranean and lacks eyes. Identification: Joy (1932), Buck (1957).

Orthocerus

79. **Tenebrionidae** Latreille, 1802; 33 genera, 47 species
Very small (2.0–2.5 mm, *Eledona agricola*) to very large (18–25 mm, *Blaps mucronata*), highly variable in shape, from smoothly rounded long- or short-oval, to sub-cylindrical, to flattened, and although mainly with striate elytra, textures vary from smooth shining to roughly sculptured. Legs are long or short, stout or slim,

but all have tarsi 5–5–4. Antennae usually with
11 antennomeres, but can be moniliform, or
filiform, or serrate, or clubbed, attached ahead
of the eyes, under a relatively prominent
projecting frontal ridge. This family is well
known for being about the most heterogeneous
mixed bag it is possible to compile, and
although they are mostly large, distinct and
easily identified once the family is known, they
cause the beginner no end of problems for

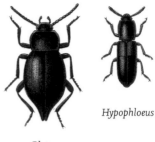

Hypophloeus

Blaps

the species' various general resemblances to other families. Some, like *Scaphidema
metallica* and *Diaperis boleti*, are round-oval, convex, globose and shining, like
chrysomelid leaf beetles. The heavy cellar beetle (*Blaps mucronata*), narrow *Helops
caeruleus* and elegant *Cteniopus sulphureus* resemble carabid ground beetles, with
their simple smooth or striate elytra, narrow
pronotum and even narrower head. The very long
antennae of *Pseudocistela ceramboides* make it look
like a longhorn cerambycid. The narrow *Corticeus*
and *Tribolium* species could be Zopheridae. Life
histories are equally varied. The aptly named cellar
beetle lives in houses, although cellars are less
often used for food storage nowadays and it seems
to be declining. Its nocturnal habits have given the
entire family the common name darkling beetles.

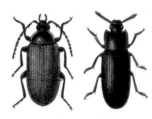

Pseudocistela   *Tribolium*

Pests in stored foodstuffs include the mealworm (*Tenebrio molitor*) and confused
flour beetle (*Tribolium confusum*). *Opatrum sabulosum* and *Melanimon tibialis* burrow
in loose dune sand at grass roots. *Eledona*, *Diaperis* and many others are fungus-
feeders, in fruiting bodies. *Prionychus*, *Nalassus* and *Helops* feed under fungoid
bark. *Corticeus* hunt adults and larvae in the burrows of bark beetles. Tenebrionids
are robust, adaptable and often long-lived beetles, and nearly 50 exotic species
have been recorded as introductions with imported goods; some may survive as
temporary colonies, but have not become established permanently. Elsewhere
in the world, tenebrionids are adapted to life in extremely arid environments,
including famously the Namib Desert, where they account for 80 per cent of beetle
diversity. Identification: Joy (1932), Brendell (1975), Novák (2014).

–. **Prostomidae** Thomson, C.G., 1859; 0 genera, 0 species
Small (5–6 mm), narrow, elongate, parallel-sided, somewhat flattened, reddish
brown, elytra striate, head and pronotum rather heavily punctured. Legs

short, slim, tarsi 4–4–4. Head broad, eyes small and protruding, antennae with 11 moniliform antennomeres, underside of head with two long, incurved projecting points, sometimes called the maxillary mala, over which the very large, broad, flattened serrate mandibles swing. Unmistakable. Technically not a true British group, and never recorded in recent historical times, but *Prostomis mandibularis* is widely found as subfossil remains in peat deposits from 10,000–5000 years ago. Like *Rhysodes sulcatus*, it is an Urwaldrelikt (wildwood relic) from a time of primeval forest cover. Identification: Freude *et al.* (1967).

Prostomis

80. **Oedemeridae** Latreille, 1810; 4 genera, 10 species
Small to medium (5–17 mm), slim, elongate, narrow, parallel-sided, head prominent and triangular, pronotum heart-shaped or contracted into a slight constriction near the middle, elytra long, rather soft, ribbed rather than striate, tapered and not meeting along suture in the common metallic green *Oedemera nobilis* and large brown *O. femoralis*. Legs long and slim, tarsi 5–5–4. Antennae long, slim, filiform, all antennomeres elongate, mostly 11 segments, but 12 in male *Nacerdes melanura*. Larvae develop in plant stems (*Oedemera*), or dead wood. The orange (elytral apex obscurely black) wharf borer, *N. melanura*, has a long, pale worm-like larva that bores into timber wharves below the high-tide mark, causing damage to piles and harbour fenders. The males of *O. nobilis* and *O. femoralis* have greatly swollen hind femora, and stout curved hind tibiae, the function of which is unknown. Identification: Joy (1932), Buck (1954).

Oedemera

81. **Meloidae** Gyllenhal, 1810; 3 genera, 10 species
Large to very large (7–36 mm), mostly black, head large and triangular, pronotum small, narrow, rounded; abdomen often bulbous, bloated (egg-laden females); elytra shortened, overlapping, splayed, barely covering half the extended abdomen. Legs relatively long, slim, tarsi 5–5–4. Antennae long, 11 segments, filiform, but in some *Meloe* middle segments modified, thickened and twisted into a kink, especially pronounced in males. Called oil or blister beetles because of the irritant chemicals (mainly the highly toxic cantharidin) exuded to deter predators, and which causes human skin

Meloe

to blister. Cantharidin was widely used until very recently to treat warts and molluscum, a viral infection causing verruca-like thickened skin lesions. In some parts of the world, cantharidin poisoning is known in horses and cows, which accidentally eat oil beetles, usually genera *Epicauta* in North America or *Mylabris* in Europe. In the case of beef cattle, the meat of animals so killed is toxic to humans. Bizarre life history starts with large numbers (many thousands) of minute but highly mobile, active, long-legged, pale, dirty yellowish triungulins, which climb herbage and flowers and grip passing insects. A few catch onto solitary bee species and, taken back to the small burrow nest, start to devour brood and food stores, transforming through a short-legged carabid-like larva, C-shaped grub and maggot-like prepupal larva, before finally pupating into nearly normal coleopteran pupa (hypermetamorphosis). Usually in warm, dry places – those favoured by their bee hosts. The metallic green Spanish fly, *Lytta vesicatoria* (a rare vagrant), and very rare brown-and-yellow *Sitaris muralis* are winged, more conventionally beetle-like. On the Continent, *Stenoria analis* is a widespread oil beetle parasite of the ivy bee (*Colletes hederae*); this bee arrived in Britain in 2001 and has quickly spread across southern England, so *Stenoria* should be sought out as a potential new British species too. *Meloe* species are great lumbering beasts, and lack wings; all are scarce, but because of their large size and strange form, they are often noted by general naturalists, much to the chagrin of coleopterists who hardly ever see them. Identification: Joy (1932), Buck (1954).

82. **Mycteridae** Oken, 1843; 1 genus, 1 species
Mycterus curculioides, 6–7 mm, long-oval, slightly convex, black, covered all over with grey or yellowish mottled pubescence, head projecting into a weevil-like snout. Legs slim, tarsi 5–5–4. Antennae slim, filiform, 11 antennomeres. Very rare, no modern records, probably extinct. Identification: Joy (1932), Buck (1954).

Mycterus

83. **Pythidae** Solier, 1834; 1 genus, 1 species
Pytho depressus, 9–13 mm, long-oval, flattened, shining, black or slightly bluish above and paler beneath, head large; pronotum small, rounded; elytra striate, slightly broadened behind. Legs slim, tarsi 5–5–4. Antennae short and stout, moniliform, 11 antennomeres. Scarce, Scotland, under pine bark. Flat larvae more often seen than presumably short-lived adults. Identification: Joy (1932), Buck (1954).

Pytho

84. **Pyrochroidae** Latreille, 1806; 2 genera, 3 species
Medium (7–17 mm), long-oval, slightly flattened, eyes
kidney-shaped; head large, attached by narrow neck to small,
rounded pronotum; elytra bright scarlet, soft, broadened
behind, finely pubescent. Legs slim, long, tarsi 5–5–4.
Antennae with 11 segments, pectinate, very strongly in the
scarce Scottish and Welsh *Schizotus pectinicornis*. Black-headed
Pyrochroa coccinea and red-headed *P. serraticornis* widespread,
often called cardinal beetles for their scarlet coloration.
Broad, flat, pale predacious larvae, with pronounced two-
pronged tail end, curved in *Schizotus*, straight in *Pyrochroa*,
under fungoid bark. Identification: Joy (1932), Buck (1954).

Schizotus

85. **Salpingidae** Leach, 1815; 6 genera, 11 species
Small (1.5–4.5 mm), long-oval to elongate, and apart from the
aberrant blind *Aglenus brunneus*, which is short cylindrical
with a large head and pronotum, they are sub-cylindrical or
flattened, parallel-sided, with small square or heart-shaped
pronotum and long striate or smooth elytra, often with
prominent shoulders behind the narrowed thorax. Legs slim,
tarsi 5–5–4. Antennae with 11 segments, abruptly clubbed in
Aglenus, more gently in other genera. Some are striking for their metallic bronze
colour (*Salpingus planirostris*), and for the head, which is extended into a shorter
(*Salpingus*) or longer (*Vincenzellus*), slightly bulbous rostrum, making them look
like weevils. Under fungoid bark or on dead twigs. Identification: Joy (1932),
Buck (1954).

Salpingus

86. **Anthicidae** Latreille 1819; 6 genera, 13 species
Small (2.5–4.0 mm), called ant beetles for their distinctive body
form of large square or triangular head on narrow neck, slim
heart-shaped pronotum and prominent rounded shoulders
to the long-oval non-striate elytra. Often prettily marked
with reddish spots or bars across the dark background. Legs
long, slim, tarsi 5–5–4. Antennae short but slim, filiform, or
very gently and slightly clubbed. Scavengers in various plant
materials, often found running actively on bare soil, or under
ground-cover herbage, in compost, grass cuttings or on flowers. The distinctive
Notoxus monoceros has a stout horn projecting over the head from the front edge
of the pronotum. Identification: Joy (1932), Buck (1954), Telnov (2010).

Anthicus

87. Aderidae Csiki, 1909; 3 genera, 3 species
Very small (1.5–2.0 mm), rather ant-like, large head with
relatively protruding eyes (especially *Euglenes oculatus*),
pronotum small and squarish ahead of the prominent
shoulders of the long-oval, non-striate elytra. Covered
with short pubescence. Legs slim, tarsi 5–5–4. Antennae
slim filiform or shorter, slightly pectinate. Last segment
of palps large, expanded cyathiform. All scarce, in dry
fungoid wood, or on flowers. Identification: Joy (1932), Buck (1954).

Aderus

88. Scraptiidae Gistel, 1848; 2 genera, 14 species
Small (2–4 mm), slim, parallel-sided or very gently rounded,
body slightly flattened (more so in the broader, and very
scarce, *Scraptia*), covered in fine pubescence, head rounded;
pronotum small, gently rounded at sides; elytra rounded at
tips, not quite covering pygidium. Legs long, slim, tarsi 5–5–4.
Antennae slim, 11 antennomeres, filiform, moniliform or
slightly pectinate. Larvae in rotten wood; adults on flowers,
often in great profusion. Most species black with red. The
striking *Anaspis fasciata* is dark, but has a squarish pale spot near each shoulder;
A. maculata is all orange-yellow with a dark scutellar area and spot-like blemish on
centre of each elytron. Identification: Buck (1954), Levey (2009).

Anaspis

Superfamily CHRYSOMELOIDEA Latreille, 1802

89. Cerambycidae Latreille, 1802; 51 genera, 71 species
Small (*Nathrius brevipennis*, 3–6 mm) to very large
(*Prionus coriarius*, 25–40 mm), long-oval to very
elongate, mostly parallel-sided, usually robust,
cylindrical or slightly flattened. Head prominent,
often with stout jaws, eyes large, often kidney-
shaped or indented, surrounding the antennal
bases. Pronotum small, sometimes armed with
a short, sharp spine on each side, or smoothly
barrel-shaped. Elytra not striate, but often ribbed,
usually with prominent shoulders behind the
narrower pronotum. The elytra are greatly shortened in *Nathrius*, *Molorchus* and
Glaphyra (all scarce), exposing the membranous flight wings and abdomen tip.
Body sometimes shining, but usually with very fine to rather dense dusting of

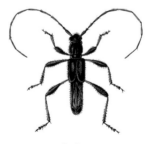

Glaphyra

short pubescence, sometimes coloured, or forming a
distinct pattern. Legs long to very long, femora often
bulbous at tips, tarsi 5–5–5, but appearing 4–4–4 because
of very small fourth tarsomere hidden between the
lobes of third; first three segments clothed with thick
brushes of short setae beneath. Antennae long, or very
long indeed, usually 11 antennomeres (12 in male *Prionus
coriarius*, and some exotic species with up to 20), filiform
(but serrate in *Prionus*, exceptionally so in male), never
clubbed, second segment usually the shortest. In
life, the long antennae are often held backwards,
over the body, reaching beyond the end of the
elytra; this gives the group their common name
longhorn beetles, and the former suborder name
Longicornia. Larvae in rotten wood, stems or plant
roots, adults often on flowers or flying, on cut logs.
Several very common species (e.g. *Rutpela maculata*,
Leptura quadrifasciata and *Clytus arietis*) are brightly
black and yellow, highly active wasp mimics, flying
about on flowers. The scarce (in fenland willows)
musk beetle, *Aromia moschata* (so called for smell of
defensive secretions), is bright metallic green or blue. The
small (4–7 mm) *Pogonocherus* are relatively short-legged,
mottled with a pattern of black, grey, white and fawn
hairs, perfectly camouflaged as they make their slow way
over lichen-covered bark. The male of the timberman,
Acanthocinus aedilis (scarce, Scotland mainly), has antennae
four or five times its body length, (twice body length in
female), and trails them like streamers when it flies. The
house longhorn, *Hylotrupes bajulus*, is the only known
British species to breed in well-seasoned wood, making it
a potential domestic pest in structural house timbers, but
it is very scarce, in Surrey. Many longhorns can also be
important for foresters, and elsewhere in the world many
are deemed significant pests. Nearly 75 exotic species
have also been recorded emerging from, or delivered
with, imported timber, including wooden boxes, pallets,
furniture and basketry. The British timber and forestry
industry is on high alert to prevent the establishment of

Asemm

Aromia

Pogonocherus

Rhagium

the notoriously invasive Chinese Asian longhorn, *Anoplophora glabripennis*. Since it arrived in North America in 1996, it has been the subject of a major containment exercise as thousands of trees are removed to prevent its spread. It has also been discovered in Austria, France and Italy. A large number of Cerambycidae species are of conservation concern because of their association with particular declining types of wooded habitat, and many have not been recorded for many years, so may already be extinct. This is a group of often striking, attractive and distinctive species, many of which can be identified from pictures, making them popular with beginners. Identification: Joy (1932), Duffy (1952), Bense (1995), Duff (2007, 2016).

90. **Megalopodidae** Latreille, 1802; 1 genus, 3 species
Small (2–3 mm), long-oval, head short; pronotum small, heart-shaped with a small tooth on each side; elytra not striate, rather cylindrical and parallel-sided, with prominent shoulders. Legs short but slim, tarsi 5–5–5, but appearing 4–4–4 since the fourth segment is small and hidden. All scarce, on poplars and aspens, larvae mining the leaves. *Zeugophora turneri* is entirely yellow, Z. *subspinosa* has elytra black, Z. *flavicollis* has elytra and rear of head black. Identification: Joy (1932), Hubble (2012), Cox (2016).

Zeugophora

91. **Orsodacnidae** Thomson, C.G., 1859; 1 genus, 2 species
Small (4–8 mm), long-oval, head short-triangular; pronotum narrow, heart-shaped, without lateral margin; elytra not striate, rounded or sub-parallel, shoulders prominent. Pale brownish, red or yellow, to blue-black, and with high colour variability within each species. Larval foodstuff uncertain, but they are probably plant-feeders, on buds or shoots. Adults on flowers, especially hawthorn blossom. *Orsodacne humeralis* is covered with short but dense pubescence, *O. cerasi* is nearly glabrous. Both scarce. Identification: Joy (1932), Hubble (2012), Cox (2016).

Orsodacne

92. **Chrysomelidae** Latreille, 1802; 63 genera, 220 species
Very small (*Mniophila muscorum*, 1.0–1.5 mm) to large (*Timarcha tenebricosa*, 11–18 mm), mostly shorter- or long-oval, convex, often shining, smoothly rounded, head rather hidden under front of pronotum, pronotum usually rounded, narrower at front edge, elytra striate or confusedly punctured, rarely smooth. Legs relatively short, stout tarsi 5–5–5, but appearing

Galerucella

4–4–4, fourth tarsomere hidden under third. Antennae generally short, slim, moniliform or filiform. This is a large group of often distinctive and attractive species, but it is also very diverse, with several very different body forms. Almost all, though, feed on living plant material, and many are identifiable from, or indicated by, their foodplants. Identification: Joy (1932), Hubble (2012), Cox (2016).

Subfamily Bruchinae; 4 genera, 17 species. The bean weevils. Short (2–5 mm), stout, chunky, convex (above and below), black or with mottled patches of pubescence, narrowed bell-shaped pronotum and large head connected by a flexible neck; elytra short, squarish, exposing at least the pygidium. Short, stout legs, hind femora enlarged, usually with teeth below. Eye indented, kidney-shaped, and partly surrounding base of serrate or clubbed antennae. *Bruchus* and *Bruchidius* are seed-feeders, mostly on Fabaceae, some are cosmopolitan stored-grain and seed pests. Fast-moving, agile, skipping gait, adults often on flowers. Additional identification: Cox (2001a).

Bruchus

Subfamily Donaciinae; 3 genera, 21 species. Reed beetles. Long-oval, 4.5–12.0 mm, or elongate and sub-cylindrical (*Plateumaris*), or rather flattened (*Donacia*), often metallic, but appearing wrinkled or dimpled, like beaten metal under a lens, head narrowed behind eyes; pronotum barrel-shaped, often sculpted, with ridges or depressions, narrowed ahead of broad elytral shoulders; elytra striate, each extended to an apical tooth in *Macroplea*; legs and antennae long. On waterside or aquatic vegetation. Additional identification: Menzies & Cox (1996).

Donacia

Subfamily Criocerinae; 4 genera, 7 species. Narrow, chunky, 2–8 mm, eyes notched, head hidden at front of pronotum, which is small, narrow, rounded, barrel-shaped, without lateral margins, or contracted before the broad shoulders of the long-oval or parallel-sided, sub-cylindrical, striate elytra. Legs short, tarsi broad. Brightly coloured, many metallic blue. The well-known bright scarlet (but black underside, legs, head and antennae) lily beetle, *Lilioceris lilii*, is a denigrated garden pest. The asparagus beetle, *Crioceris asparagi*, inky blue-black with red pronotum and elytral trim, and six large square cream elytral spots, is a riot of Mondrianesque bling.

Lilioceris

Subfamily Cryptocephalinae; 4 genera, 25 species. Short (2.5–11.0 mm), stout, squat, head hidden under front of domed pronotum; elytra broad, rather squarely contracted at apex, usually not quite completely covering the abdomen, striate or not; legs short; antennae slim. Mostly brightly coloured, some black (e.g. *Cryptocephalus moraei*, with small yellow markings), some strongly metallic (e.g. *Cryptocephalus aureoles, Cryptocephalus*

Cryptocephalus

hypochaeridis). Larvae terrestrial, feeding on dead leaves or lichens, and making a protective pot-shaped case from accumulated droppings, hence sometimes called pot beetles. Some (e.g. *Clytra quadripunctata*) drop frass-covered eggs that get taken back by ants to their nests. Mostly scarce.

Subfamily Lamprosomatinae; 1 genus, 1 species. *Oomorphus concolor*, small (2–3 mm), broad-oval, convex, hemispherical, shining metallic bronze-black; legs and antennae short, tibiae broad and flat. Usually in woodlands, in moss, although larvae reported to feed on ivy (*Hedera helix*), ground elder (*Aegopodium podagaria*) and other leaves.

Subfamily Eumolpinae; 1 genus, 1 species. *Bromius obscurus*, short (5–6 mm), squat, dull black, pronotum narrow, elytra almost square. On rosebay willowherb (*Chamaenerion angustifolium*), very rare, may just be vagrant.

Subfamily Chrysomelinae; 11 genera, 47 species. Typical leaf beetles, although highly variable in size (2–18 mm), convex, domed above, flat beneath, round, short (e.g. *Plagiodera*) or long (e.g. *Prasocuris*) oval, shining, many metallic (e.g. strongly punctured bronze *Chrysolina hyperici* on St John's-worts) or brightly and attractively coloured (e.g. black-and-red *Chrysomela populi*, on aspens and poplars); legs short, tarsi broad, antennae short but slim. The largest British species, the bloody-nosed

Chrysolina

beetle (*Timarcha tenebricosa*), exudes a drop of distasteful red haemolymph from its mouth if alarmed. The striped metallic rainbow leaf beetle, *Chrysolina cerealis*, is known from only a few patches of stunted thyme on the middle slopes of Snowdonia. It should not be confused with the striped metallic rosemary beetle, *Chrysolina americana*, a south European species accidentally introduced to Britain in 1995 and now a widespread garden species on rosemary (*Rosmarinus officinalis*) bushes. The international potato pest, the Colorado beetle, *Leptinotarsa decemlineata*, originally from the central United States, is part of this group. Although not established in Britain, it is now widespread in Continental Europe, having been accidentally introduced there during the First World War, and then widely spread in the chaos of the Second World War; its occurrence here is still a potential worry for farmers and smallholders.

Subfamily Galerucinae, tribe Galerucini; 10 genera, 19 species. Stout, broad- or long-oval, 3–12 mm, slightly flattened; head and pronotum small, rounded, margined at sides; elytra rather soft, usually broadest behind middle, pale browns, reds or black. Large, chunky *Galeruca tanaceti* jet black, on tansy (*Tanacetum vulgare*), female bloated with eggs, elytra barely covering swollen abdomen. Elongate, narrow *Luperus longicornis* often likely to confuse because of its cylindrical form and very long filiform antennae. Black-and-yellow-striped *Diabrotica virgifera*, an international pest of sweetcorn, originally from Central America but spread to North America and Europe, was recently discovered in Britain but seemingly successfully eradicated.

Subfamily Galerucinae, tribe Alticini; 20 genera, 128 species. The flea beetles, so called because the hind femora are greatly enlarged, allowing the beetles to jump wildly, often 30 cm or more, a predator-escape mechanism. All small (1–5 mm), mostly oval, domed, convex, shining. This section includes the large difficult genera *Longitarsus* (43 species, with long hind tarsus, at least as long as hind tibia, first tarsomere longer than remaining segments) and *Altica* (seven species, shining metallic blue), most of which require genitalia

Altica

dissection, but also some strikingly attractive species, like the metallic golden *Crepidodera*, blue-and-pink *Podagrica* and black-and-yellow-striped *Phyllotreta*. Many species are very abundant, and some are considered agricultural or horticultural pests.

Subfamily Cassidinae; 3 genera, 14 species. Small (4–9 mm), oval, flattened; the tortoise beetles, so called for their broad, smoothly rounded outline, flanged body edge, the fact that the head is entirely hidden under the front of the pronotum, and that they can retract legs and antennae and remain inviolate underneath the gently domed carapace if alarmed. Mostly dirty green (e.g. common *Cassida vibex* on thistles), but also speckled red and black (e.g. *C. murraea*, on fleabane, *Pulicaria dysenterica*) or streaked with metallic gold (e.g. *C. nobilis*, on goosefoots etc.). Larvae have a tail fork on which they construct a knobbly parasol or finger-like twigs of their own excrement and an assortment of moulted larval skins.

Cassida

Superfamily CURCULIONOIDEA Latreille, 1802

Note: there is a convention among coleopterists, which I continue here, for the following weevil families, that body length measurements *exclude* the snout or rostrum.

Division ORTHOCERI Schönherr, 1823

93. **Nemonychidae** Bedel, 1882; 1 genus, 1 species
Cimberis attelaboides, 2–5 mm, head and pronotum narrow ahead of slightly broader shoulders of long-oval, non-striate elytra. Black, but covered with short pale pubescence, appearing grey. Rostrum long, flattened, swollen at tip. Legs slim, tarsi 5–5–5, but appearing 4–4–4 as fourth tarsomere is very small, hidden under third. Antennae yellow, slim, 11 antennomeres, filiform, first segment not lengthened to form a scape, but rather stout, last three or four segments forming a slight club. Rare, on pines, larvae in male flowers. Identification: Morris (1990), Duff (2016).

Cimberis

94. **Anthribidae** Billberg, 1820; 9 genera, 10 species
Small (1.6–14.0 mm), short, squat or oval, head produced into a short, broad rostrum; pronotum broad, rounded or square; elytra striate, although sometimes only vaguely so, stout, rounded or squarish, not quite covering pygidium. Legs short but slim, tarsi 5–5–5, fourth tarsomere small, so appearing 4–4–4. Antennae with 11 segments, more or less filiform, first segment not forming a scape, but last three or four forming a slight club. Maxillary palps normal, flexible,

Platyrhinus

but small. Body usually covered all over with mottled pattern of pale hairs. *Bruchella rufipes*, 2.0–2.5 mm, black but uniform pale hairs making it appear grey, legs yellow, widespread on wild mignonette (*Reseda lutea*). *Platyrhinus resinosus*, 11–14 mm, handsomely marked with patches of bright white or cream, and grey and black pubescence, to give a chequered pattern that, combined with its cylindrical form, gnarled pronotum and ridged elytra, camouflages it as a piece of broken twig; scarce in cramp ball fungus (*Daldinia concentrica*). *Platystomos albinus*, 7–11 mm, similar, male has antennae very long, nearly as long as body, also scarce, under fungoid bark. The tiny *Choragus sheppardi*, all black or reddish brown, strongly punctured, elytra smoothly striate, on dead twigs, has the power of hopping. The two scarce species of *Anthribus* are highly unusual among weevils, in that they are predatory or semi-parasitic as larvae, attacking the egg-laden females of scale insects feeding on broadleaved trees. Identification: Morris (1990), Duff (2016).

95. Rhynchitidae Gistel, 1848; 9 genera, 18 species
Small (2.5–8.0 mm), stout, convex, head and short, round, strongly punctured pronotum narrow, contrasting with broad shoulders of rather square or rectangular, often powerfully striate, elytra. Lightly or densely covered in short pubescence. Rostrum, with jaws at tip, rather long and narrow, especially in females. Legs slim, tarsi 5–5–5, but fourth segment small, so appearing 4–4–4. Antennae with 11 antennomeres, slim, basal segment not lengthened into a scape, apical segments forming a loose club. Mostly

Rhynchites

black or dark metallic blue, green or bronze. Commonest British species is *Tatianaerhynchites aequatus*, on hawthorn blossom, elytra dull red, head and pronotum coppery. Identification: Morris (1990), Duff (2016).

96. Attelabidae Billberg, 1820; 2 genera, 2 species
Attelabus nitens short (4–6 mm), squat, hunched, shining; head narrow and protruding, with short rostrum and jaws at tip; legs long and slim, tarsi 5–5–5, but appearing 4–4–4; scutellum, underside of body and neatly clubbed antennae jet black; small rounded, domed pronotum and nearly square, vaguely striate elytra bright red. On oak trees, lays egg on leaf, then rolls leaf into tight, fat cigar-shaped packet, hence common name oak leaf-roller. *Apoderus coryli* similar, head longer and attached to thorax by narrow neck, colour slightly less bright, legs with some red, pronotum often with black, elytra more strongly striate. Makes short, neatly cylindrical leaf rolls on hazel (*Corylus avellana*), hence common name hazel leaf-roller. Identification: Morris (1990), Duff (2016).

Apoderus

97. Apionidae Schönherr, 1823; 34 genera, 87 species
Very small (1.2–5.0 mm), distinctively pear-shaped, with rounded (sometimes bulbous) striate elytra, narrowed in front to meet square or barrel-shaped pronotum, with relatively broad head, extended into long, narrow rostrum, with jaws at tip. Generally convex above and below, and often called seed

Apion

weevils because of body form, not food habits. Legs long, mostly dark coloured as body, but several genera with tibiae and/or femora yellow, trochanters long, widely separating coxae and femora, and easily visible in side view, tarsi 5–5–5. Antennae slim, 11 antennomeres, first segment (hardly a scape) about twice as long as next, finishing with a pointed three-segmented club. Body mostly dark, black, greyish or blue-green (sometimes metallic). The rare *Pseudaplemonus limonii* is a beautiful metallic purple-red jewel, on sea lavender (*Limonium vulgare*) in saltmarshes, where it is regularly covered by the sea at high tide. Some (e.g. *Apion*, on *Rumex* dock and sorrel plants) are a pinkish red, while others (e.g. scarce *Taeniapion urticarium*, on stinging nettles, *Urtica dioica*, or very scarce *Ixapion variegatum*, on mistletoe, *Viscum album*) are patterned with pale hairs or coloured scales. *Malvapion malvae* (common on mallows) grey and yellow. Females usually with longer rostrum than males, extremely so in *Rhopalapion longirostre* (where straight snout is almost equal to body length), which bites deep drill hole into large hollyhock (*Alcea rosea*) flower heads to lay eggs in developing seeds. Charming, attractive, very active and highly mobile little weevils, often occurring in profusion on foodplants. The majority occur on Fabaceae, particularly clovers, trefoils, vetches, broom and gorse. Identification: Morris (1990), Gønget (1997), Duff (2016).

98. **Nanophyidae** Gistel, 1848; 2 genera, 2 species
Tiny (1.5–2.0 mm), rounded, ovoid, convex, shining, but with sparse pubescence, pronotum strongly contracted to head, striate elytra with prominent shoulders, strongly contracted to apex, rostrum long. Legs slim, trochanters long, separating coxae and femora, tarsi 5–5–5, but appearing 4–4–4. *Nanophyes marmoratus* dark grey, legs mostly yellow, elytra with one strong and one faint chevron of yellow pubescence; widespread on purple loosestrife (*Lythrum salicaria*). *Dieckmanniellus gracilis* black, each elytron with orange splodge; rare, on water purslane (*L. portula*). Identification: Morris (1990), Gønget (1997), Duff (2016).

Nanophyes

99. **Dryophthoridae** Schönherr, 1825; 2 genera, 4 species
Small (2.0–3.5 mm), robust, rather narrow, slightly cylindrical, powerfully punctured pronotum as broad as strongly striate elytra. Legs short, stout, front tarsi with inward-projecting hook at apex. Antennae short, stout, with a scape, short funiculus and strong club. Three species of shining, rich chestnut-brown *Sitophilus* are cosmopolitan pests in stored whole grains, wheat, rice and maize, never found

Sitophilus

out of doors. *Dryophthorus corticalis*, dull brown, under fungoid bark, only in a few ancient woodlands, usually in company of brown tree ant, *Lasius brunneus*. Identification: Morris (2002), Duff (2016).

100. **Erirhinidae** Schönherr, 1825; 7 genera, 14 species
Minute (*Tanysphyrus lemnae*, 1.4–1.8 mm) to small (*Procas picipes*, 4–7 mm), rather slim and elegant, head narrow, pronotum rounded; elytra broad-shouldered, striate, long-oval. Legs slim, tibiae rather hooked at apices. Antennae slim, scape long, terminating with a tight club. Often on waterside plants. *Notaris acridulus* the only common species, brown and more or less shining, on sedges and rushes. The large, squat *Grypus equiseti* is attractively marked with blotches of pale and grey scales,

Grypus

has raised ribs along the short, broad elytra, and feeds on horsetails. The tiny *Stenopelmus rufinasus*, rounded, reddish because of a covering of scales, is a North American species that feeds solely on water fern (*Azolla filiculoides*). Identification: Morris (2002), Duff (2016).

101. **Raymondionymidae** Reitter, 1913; 1 genus, 1 species
Ferreria marqueti, 1.5–2.0 mm, narrow, brown; head hidden under front of globular pronotum, lacking eyes; elytra rounded long-oval, shining brown, thinly covered with short, sparse setae, upper surface strongly punctured. Legs and antennae short, stout. A scarce, secretive subterranean weevil, with poorly understood habitat requirements and life history. It has been found using deep (0.75 m) pitfall traps, and also under half-buried logs, mostly in cultivated areas like gardens. Identification: Morris (2002), Duff (2016).

Ferraria

Division GONATOCERI Schönherr, 1825

102. **Curculionidae** Latreille, 1802; 164 genera, 489 species
Minute (e.g. *Micrelus ericae*, 1.3–1.6 mm) to medium-large (e.g. *Liparus germanus*, 11–16 mm), and generally recognisable as the weevils, the head being drawn out into a longer or shorter rostrum (snout), jaws at the tip, and with elbowed, geniculate antennae, the long first segment forming a more or less slim scape, from which the small segments of the funiculus flex, usually ending in a terminal club; most have 11 antennomeres, but many have only 10. This is a large, diverse

and important family. Most feed on living plants, often on just a single plant species or genus. Because of their large species numbers, they are significant indicators of biodiversity, however some are considered important agricultural or horticultural pests. Many are attractively shaped or patterned, their markings and colours coming from a covering of scales or flat scale-like pubescence, and although some large genera can cause identification problems, this is a group well worthy of study by the novice. Nowadays, huge movement of

Trypodendron

horticultural produce means that plant-feeding beetles like weevils are quite likely to be introduced from the Continent, or further afield. Some will become established, but others are unlikely to succeed. The palm weevil, *Rhynchophorus ferrugineus*, from tropical Asia has been destroying palm trees in southern Europe and appeared (but was eradicated) in the United States. It was found in Brittany in 2013. Its larvae bore into the heart of the palm tree stem, causing damage to the growing zone, and allowing in fungal and other disease pathogens. It was recently found in an imported palm in a garden centre in Essex, but whether there are enough palms growing in Britain to support it has not yet been tested.

Subfamily Curculioninae; 29 genera, 110 species. A mixed group, but generally with pronotum narrower than the humeri of the elytra, and many are patterned with coloured scales, hiding the striae. The genus *Curculio* have extremely long rostrums, and gantry-like legs, to drill deep into acorns (e.g. the common reddish-brown *Curculio glandium*), hazel nuts (the similar *Curculio nucum*) or other tree fruit to make an egg hole. *Cionus* are short, squat, almost square, domed and convex, strongly chequered with scales, usually on figworts. By contrast, *Dorytomus* are parallel-sided, elegant, but also patterned with pale scales, mostly on trees. *Orchestes* and similar genera have enlarged hind femora, and can hop in the manner of flea beetles. Identification: Morris (2012), Duff (2016).

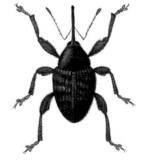

Curculio

Subfamily Bagoinae; 1 genus, 21 species. Small (2.5–5.0 mm), stout but cylindrical, pronotum barrel-shaped or slightly rounded, elytral shoulders prominent. The tibiae are characteristically curved into a hook on the inner side at the apex. All *Bagous* are aquatic, or sub-aquatic, on pond and stream plants, sluggish, often encrusted with mud and algae and difficult to find, if not genuinely very scarce. Identification: Morris (2002), Duff (2016).

Subfamily Baridinae; 5 genera, 7 species. Small (2.5–4.0 mm), slim, elegant, often smooth and shining, elytra barely broader than pronotum; legs short and stout, tarsi broad. Mostly scarce. Identification: Morris (2012), Duff (2016).

Subfamily Ceutorhynchinae; 30 genera, 94 species. Small (1.3–5.1 mm), broad, squat, hunched, some almost squarish, pronotum strongly narrowed in front; elytra usually with broad shoulders, gradually but evenly contracted to about two-thirds or three-quarters, then suddenly contracted to sutural angle; the whole subfamily have a 'look' about them. Legs short and stout. Antennae thin, scape long. Elytra of *Ceutorhynchus chalybaeus* (and others) are strikingly metallic blue; *Mogulones geographicus* (at roots of viper's-bugloss, *Echium vulgare*) is black, decorated with scales that form a road map of thin white lines, rivalled only by the black and white patterns of *Hadroplontus litura* and *H. trimaculatus* (on thistles); *Zacladus exiguus* and *Z. geranii* (both jet black, on cranesbills) have rows of raised tubercles on the elytral interstices. Identification: Morris (2008), Duff (2016).

Coeliodes

Subfamily Cossoninae; 13 genera, 16 species. Small (2.5–6.6 mm), generally elongate, parallel-sided, sub-cylindrical or slightly flattened, brown, pronotum barrel-shaped. Legs short, stout. Antennae short, thick. Mostly in or under fungoid wood. The hairy *Pselactus spadix* occurs in wooden pilings around harbours, or in driftwood in saltmarshes. The very narrow *Euophryum confine*, a native of New Zealand, is now very widespread in Britain, and the most common woodworm borer in household timber, but usually in damp rather than dry wood, where fungal rot has set in. Identification: Morris (2002), Duff (2016).

Cossonus

Subfamily Cryptorhynchinae; 4 genera, 5 species. Short (2.5–9.0 mm), stout, hunched, rather squat, with short legs. In dead wood, especially twigs and small branches. Secretive, mostly scarce. Identification: Morris (2002), Duff (2016).

Subfamily Cyclominae; 1 genus, 2 species. Both the scarce *Gronops lunatus* and rare *G. inaequalis* are small (3.0–3.5 mm), rather flattened, head and pronotum narrowed; elytra broad, with shoulders prominent and with several raised longitudinal ridges. Legs and antennae short and stout. Covered all over with mottled scales. Secretive, at plant roots. Identification: Morris (2002), Duff (2016).

Subfamily Entiminae; 28 genera, 112 species. Robust, often stout, rounded, rostrum generally short and broad, scape of antennae usually short, barely as long as rostrum is wide. Pronotum rounded, barrel-shaped or broad, but usually narrower than elytral shoulders. Often covered with coloured scales, or with outstanding setae. A large, important and varied group. The common vine weevil (*Otiorhynchus sulcatus*), reviled as a major garden pest (but not on vines), is dark brown, dusted with patches of paler scales, rather bulbous, active in a clockwork kind of way, one of many (24 species) in this diverse, mostly flightless ground-dwelling genus. The large genera *Phyllobius* (10 species) and *Polydrusus* (11 species) are among the most frequently encountered weevils; many feed on trees and bushes, and some have striking patterns of scales, frequently in metallic greens. Another large and frequently encountered genus is *Sitona* (15 species), which have a rather wedge-shaped head and thorax, and are mostly mottled or streaked with browns and greys; most feed on Fabaceae and some are considered agricultural pests. *Barypeithes*, though, are small, brown, shining, sometimes sparsely covered with long, thin hairs, and occur under logs and stones, at grass roots. Identification: Morris (1997), Duff (2016).

Phyllobius

Sitona

Subfamily Hyperinae; 3 genera, 18 species. Robust, stout, long-snouted; pronotum rounded, narrower than the broadly shouldered, parallel-sided or gently rounded elytra. Most are patterned all over with coloured scales (a strong green in the widespread *Hypera nigrirostris*, on clovers); these scales are often deeply divided into narrow horseshoe or eyebrow-tweezer shapes and are useful in identification. Identification: Morris (2002), Duff (2016).

Hypera

Subfamily Lixinae; 6 genera, 11 species. A mixed group of generally large (up to 16 mm) but rather slimly built, scale-patterned weevils. Most are scarce, but the mottled brownish *Larinus planus* and *Rhinocyllus conicus* occur sparingly in southern England on thistles, in the flower heads of which the larvae feed, and both seem to be having a resurgence. Identification: Morris (2002), Duff (2016).

Rhinocyllus

Subfamily Mesoptiliinae; 1 genus, 9 species. Robust, stout, rather long-snouted, pronotum broad; elytra rather parallel-sided, strongly striate. A single genus, *Magdalis*, of mostly scarce and secretive species breeding in the dead twig tips of various trees. The matt black *M. armigera* was once very common on elms, but populations crashed with the disappearance of large trees after the ravages of Dutch elm disease in the 1970s; trees and weevils are recovering now. Identification: Morris (2002), Duff (2016).

Trachodes

Subfamily Molytinae; 9 genera, 15 species. A mixed group, generally robust or very robust and stout, chunky, strongly punctured and striate (corrugatedly so in *Mitoplinthus caliginosus*), or covered in short, blunt, upstanding scale-like setae (*Trachodes hispidus*). At 9–15 mm, the handsome, smoothly shining black *Liparus coronatus* is one of Britain's largest weevils, and a personal favourite, it being the oldest extant specimen in my reference collection, from Eastbourne, 11 June 1967, when I was nine years old. The much smaller (2.4–2.8 mm) *Leiosoma deflexum* is also shining black, but very strongly punctured and striate. Both are usually ground dwelling, often found under planks of wood. The elegantly slim and prettily scaled *Hylobius abietis* and *Pissodes* species occur on pine trees. Identification: Morris (2002), Duff (2016).

Orobitis

Subfamily Orobitidinae; 1 genus, 1 species. *Orobitis cyaneus*, small (2.0–2.5 mm), shining blue-black, hunched and almost cartoonishly globular, head and pronotum small; elytra large, rounded and smoothly striate. Underside of body covered in pale scales. Legs and rostrum narrow and long. Scarce, on violets. Identification: Morris (2012), Duff (2016).

Subfamily Scolytinae; 32 genera, 67 species. Very small (e.g. *Phloeotribus rhododactylus*, 1.2–1.6 mm, in broom or gorse) to small (e.g. *Dendroctonus micans*, 6–8 mm, in spruce trees), compact, cylindrical, often short and stout, hunched, with head hidden at front of the stout pronotum, which is strongly or deeply punctured, or sculpted into a texture of fine cross-tuberculate protuberances; rostrum very broad and short, and not always apparent. Legs short and stout, tibiae usually with teeth. Most are shining or dull brown, sometimes with stout, outstanding setae along the striate elytra; *Hylesinus varius* is patterned with broad pale scales. In some species, the tips of the elytra are suddenly angled down at the apex, as if cut away – the declivity – where species- and sex-specific tooth-like spines or tubercles can project. Commonly called bark beetles, they burrow into mostly living timber to lay their eggs. The female chews a short, straight gallery just under the bark, laying eggs along its length, and the hatching grubs burrow out roughly at right angles to this maternal chamber, creating distinctive species-specific splayed patterns between bark and heartwood, which can remain long after the tree has died and the bark peeled away. Many are considered forestry pests for the burrowing damage they do to the trees, and for the pathogens they carry. *Scolytus scolytus* and *S. multistriatus* spread the spores of the fungal Dutch elm disease. A large number of exotic species have also been recorded, introduced from all over the world with imported timber, although few if any have been found breeding. Identification: Duffy (1953), Duff (2016).

Scolytus

Subfamily Platypodinae; 1 genus, 1 species. *Platypus cylindrus*, 5–8 mm, brown to blackish, cylindrical, head short and broad, pronotum barrel-shaped; elytra elongate, parallel-sided, striate, often chestnut coloured, shining, with brush of dense hairs at apex, and each with small, stout lobe-like tooth near tip. Antennae short, 6 antennomeres, scape short and thick, last segment forming a large round club. Bores directly into oak bark, pushing out a shower of fine sawdust behind it; sometimes called oak pinhole borer, scarce, but common enough where it occurs to be considered a minor forestry pest, and a mass attack by several dozen chewing their way into the bark leaves a striking cascade of pale wood powder down the tree trunk. Identification: Duffy (1953), Duff (2016).

Platypus

SIMPLE KEY TO THE COMMONER FAMILIES OF BRITISH BEETLES

As I have said elsewhere, most beetles can normally be identified to family because they have a 'look' about them. Trying to give substance to this nebulous concept in words can be a hiding to nothing (although I have tried on pp. 201–61), but if that is difficult, constructing an artificial dichotomous key to families soon becomes elaborate beyond comprehension. I have had to strike a compromise.

This key is largely based on that produced online in 1999 by Brian Eversham for the Wildlife Trust for Bedfordshire, Cambridgeshire and Northamptonshire, who has very kindly agreed to let me plagiarise it horribly. Like Brian, my stance must be clear from the start: the key is not meant to be comprehensive, but covers 47 (out of 102) families most likely to be seen by novices. There are 19 British beetle families that each contain just a single species; a further 23 families have between two and five species. In other words, 42 per cent of beetle families account for just 2.3 per cent of our species. Including all these obscure, rare or tiny beetles in a single key makes for a top-heavy complexity that may be appropriate in a specialist identification monograph, but is not useful here. On the other hand, our four largest families (Carabidae, Staphylinidae, Chrysomelidae and Curculionidae) together account for 2,202 species, more than half of the beetles occurring here.

This introductory key includes all British families with more than 100 species; many smaller families with large (>10 mm), distinctive or common species; families with a popular following in Britain and Europe; families that are prone to be found during pitfall trapping; and families for which I have a particular soft spot.

Full dichotomous keys down to family level are available by Joy (1932), Crowson (1956), Unwin (1984) and Duff (2012b), which I have also raided piecemeal. What follows here is a simplified form, from which to start. In the interests of brevity, I have slipped straight into some of the usual coleopterist jargon, but all terms are explained in the Glossary on pp. 409–15.

1   Hind coxae flattened, fixed immovably to metasternum, and appearing to divide the first visible sternite (S1) of the abdomen; antennae normally with 11 simple cylindrical or bead-like segments. Suborder Adephaga 2

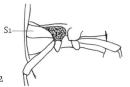
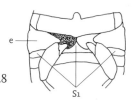

–   Hind coxae articulating with metasternum, moving when hind legs move, not dividing the first visible sternite (S1) of the abdomen, and instead sitting tight against the epimeron (e) of the metathorax; antennae variable, often thickened or clubbed. Suborder Polyphaga 8

2   Terrestrial beetles of various shapes. Legs not modified for swimming (not paddle-like, and tarsi without fringes of hairs); antennae relatively long, filiform . . . . **Carabidae**

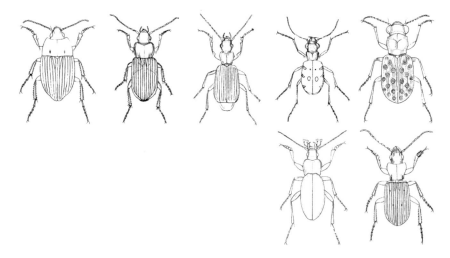

–   Aquatic beetles, mostly broader, rounded, streamlined, shaped for swimming. At least hind legs either paddle-shaped or with fringes of hair on the tarsi 3

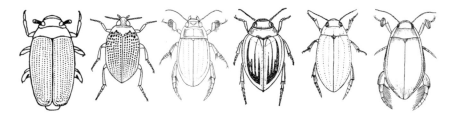

3   Middle and hind legs short and paddle-shaped; eyes completely divided (thus appearing to be four rather than two in number). Rather narrow oval beetles (3.5–7.8 mm long), which swim rapidly and erratically on the water surface; body with distinctive smooth outline; antennae reduced to short peg-like structures .... **Gyrinidae**

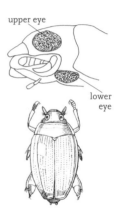

–   Middle and hind legs not paddle-shaped. Beetles of various shapes and sizes, which swim beneath the water surface 4

4   Hind coxae greatly enlarged, forming square or rectangular plates that cover the attachments of the hind femora and obscure much of the underside of the abdomen; antennae apparently with 10 segments, although under high power segment 1 is shown to be divided. Small (2.5–5.0 mm), rather squat beetles, yellowish brown, often patterned with vague stipples or chequered patches, and elytra with strong punctuation... **Haliplidae**

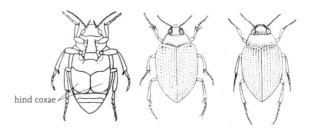

–   Hind coxae not expanded into large plates, articulating visibly with the bases of the femora. Beetles often larger or darker, and striae with smaller punctures 5

5   Head free and with distinct neck, not sunk into front of thorax, eyes strongly convex. Large (8.5–10.0 mm), distinctive stout and bulging form, dirty orange and black, hind tarsi very long, squeaks loudly when picked up .... **Hygrobiidae**

–   Head not free, more or less sunk into front of thorax; eyes less convex 6

6   Scutellum visible (left figure)......... **Dytiscidae** (part)

–   Scutellum hidden (right figure) 7

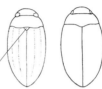

scutellum

7   Hind tarsi with two curved claws of equal length; abdomen with five visible sterna. Small (3.5–5.0 mm), dull orange-brown, hind margin of pronotum much wider than front edge, body distinctly broadest near base, just behind pronotum .... **Noteridae**

–   Hind tarsi each with one straight claw; abdomen with six visible sterna. Size variable, but usually more long-oval, widest near or behind middle of elytra; many species patterned with pale and dark marks .... **Dytiscidae** (Hydroporinae)

8   Elytra short, leaving at least three abdominal segments exposed 9

—   Elytra covering the entire abdomen, or if short, only last one or two
    segments exposed 11

9   Narrow, elongate, cylindrical or flattened, with three to six abdominal
    segments exposed. Usually small species – most under 5 mm, but
    largest to 28 mm .... **Staphylinidae**

—   Beetle large (usually 15–35 mm), broad and stout......................... 10

10 Antennae clubbed. Black, elytra long and parallel-sided with raised
    ridges, or strongly marked with orange bars . . . . **Silphidae** (Nicrophorinae)

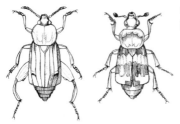

LEFT: Silphidae
(Nicrophorinae)

RIGHT: Meloidae

—   Antennae not clubbed, but may be kinked in middle. Black or with
    vague blue hints, elytra short and often splayed over relatively bloated
    abdomen . . . . **Meloidae**

11 Head produced in front of the eyes into a rostrum of varying length,
    with small jaws at the tip. Antennae often with first segment long, a
    scape, angled with funiculus and club. 12

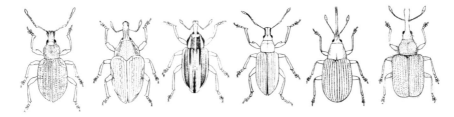

—   Head not produced into a rostrum 16

12 First antennal segment long, a scape, at least as long as next two
    segments together (usually much longer); antennae geniculate, angled
    at the funiculus and with a terminal club. . . . . **Curculionidae**

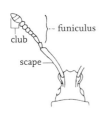

– First antennal segment short, at most twice as long as second segment, antennae not geniculate 13

13 Rostrum very short. Third tarsal segment (bilobed) partly recessed into second segment (also bilobed). Body form usually broad and stout; antennae often clubbed . . . . **Anthribidae**

– Rostrum longer 14

14 Head narrow, or very narrow with distinct neck, eyes well forward of pronotum; head and legs black, pronotum and elytra broad and stout, and strikingly bright red . . . . **Attelabidae**

– Head without such a narrow neck, form and colours otherwise. 15

15 Antennal club loose, individual segments distinct. Elytra relatively broad and parallel-sided, shoulders distinct. Mostly on broadleaved trees . . . .**Rhynchitidae**

LEFT: Rhynchitidae

RIGHT: Apionidae

– Antennal club compact, segmentation not so obvious. Elytra generally rounded at sides and pear-shaped. Mostly on herbs. . . . . . . **Apionidae**

16 Antennae clubbed, the last few segments clearly
    wider than the rest. 17

–   Antennae not clubbed, may be serrate, even pectinate, but segments
    generally of almost uniform width throughout (a few families with
    segments gradually thicker towards the tip key out both ways). 38

17 Maxillary palps long, at least three-quarters the
    length of the antennae (often much longer), tarsi
    with five simple segments. Water or dung beetles
    with flat underside and convex upper side. 18

–   Antennae much longer than palps. 21

18 Pronotum broadest at, or very close to, base; 1–50 mm . . . . **Hydrophilidae**

–   Pronotum broadest at middle and contracted to base. .19

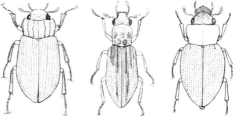

19 Of distinctive general shape, and pronotum with
five longitudinal furrows . . . . **Helophoridae**

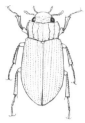

— Pronotum without long furrows, but may have dimples
or impressions 20

20 Underside with conspicuous round punctures; pronotum with distinct
arrangement of deep depressions . . . . **Hydrochidae**

LEFT: Hydrochidae

RIGHT: Hydraenidae

— Underside without round punctures (although other sculpturing may
be present); pronotum either with only vague shallow depressions, or
they form another pattern . . . . **Hydraenidae**

21 Front and middle tarsi with five segments, hind tarsi with only four
segments 22
— All tarsi with the same number of segments, commonly five, but
sometimes three or four. 25

22 Pronotum completely rounded at front, without protruding front
angles, broadest at base, and forming smooth outline with elytra;
middle and hind tarsi strongly tapered, and longer than tibiae 23
— Pronotum normally with front angles protruding, not completely
rounded, usually broadest at middle; tarsal segments of uniform
width, tarsi shorter than tibiae 24

23 Last abdominal segment produced into a point; underside of body very convex, beetle's natural resting position is on its side. Head broad, attached to thorax by narrow neck. . . . .**Mordellidae**

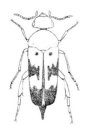

LEFT: Mordellidae

RIGHT: Scraptiidae

–   Last abdominal segment not produced into a point; underside of body not so convex. Head not so prominent . . . .**Scraptiidae**

24 Antennae inserted under a raised ridge on side of head, so that about the basal third of first segment is hidden . . . .**Tenebrionidae**
–   Antennae inserted at side of head so that its insertion, or a raised tubercle, can be seen from above . . . . **Melandryidae**

25 Antennal club lamelliform, widened on one side only, as a group of separate finger-like or plate-like segments 26

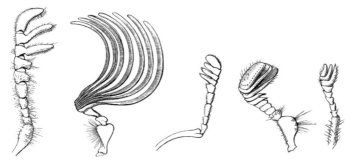

–   Antennal club more or less symmetrical, not made up of finger-like or plate-like segments 29

26 Antennae geniculate, first antennal segment long, a
    scape; final antennal segments forming a loose club,
    the antennomeres free and finger-like ......**Lucanidae**

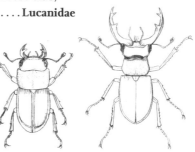

–   Antennae not geniculate, first antennal segment short, not forming a
    scape; final segments able to close to form a tight club, although they
    may relax and spread on death 27

27 Jaws projecting and visible when head viewed from
    above; antennae with 11 segments; eyes divided by a
    canthus into upper and lower parts. . . . . . . **Geotrupidae**

–   Jaws hidden; antennae with nine segments; eyes
    not divided 28

28 Epipleura broad, continuing to apex of elytra; upper surface granular,
    coarsely sculpted, with tufts of setae or scales. Dull black, often in
    carrion . . . .**Trogidae**

LEFT: Trogidae

RIGHT:
Scarabaeidae

–   Epipleura narrow, not continued to elytral apex; upper surface smooth
    or minutely hairy. . . . . **Scarabaeidae**

29 Body very rounded and convex, both above and below .30
– Body usually more elongate, and not convex below 31

30 Very hard, glossy, usually mainly black; elytra square or rectangular, short, leaving last one or two abdominal segments exposed; front tibiae usually broad, curved or toothed on outer edge; first antennal segment longer, a scape; antennal club usually short, almost spherical . . . . **Histeridae**

LEFT: Histeridae

Byrrhidae

– Less hard and glossy, usually hairy and dull brown; elytra longer, covering whole of abdomen; front tibiae less wide, straight; antennae without scape-like basal segment, club more oval and less distinct . . **Byrrhidae**

31 Overall shape broad-oval, pronotum wide, head very narrow (less than 40 per cent width of pronotum); apex of front tibiae oblique, with a long spur on inner side . . . .**Silphidae** (part)

– Various shapes; head broader in proportion to pronotum; apex of front tibiae less oblique, not produced on inner side 32

32 Antennae gradually thickened towards apex, so club is ill-defined; upper surface entirely covered with short pubescence; middle and hind tarsi as long as tibiae. . . . . **Leiodidae** (part)

– Antennal club distinct; middle and hind tarsi shorter 33

33 Tarsi all with five segments 34
– Tarsi all with three or four segments. 37

34 Antennal club tightly defined; elytra often rather
short, square-ended, exposing last one or two
abdominal segments . . . . **Nitidulidae**

– Antennal club of two to five loose segments 35

35 Pronotum much narrower than elytra; beetle rather long-oval,
sometimes parallel-sided or elongate; elytra usually with strongly
punctate striae, and with relatively prominent shoulders . . . .**Latridiidae**

LEFT:
Latridiidae

RIGHT:
Key entry 35,
part 2

– Pronotum almost as wide as elytra; rather broad and rounded beetles,
usually with non-striate and minutely pubescent elytra 36

36 Body with vague or pronounced pattern of hairs or scales; antennae
with club of two or three segments . . . . **Dermestidae**

LEFT: Dermestidae

RIGHT: Leiodidae

– Body with only short down; antennal club loose, often segment 8
smaller than either 7 or 9 . . . . **Leiodidae** (part)

37 Tarsi appearing to have three segments, but in reality having four, although segment 3 is minute; antennae with a distinct tight club; elytra without striae, usually smooth and glossy; rounded, often circular beetles with smoothly curved sides to the pronotum; often colourfully marked, red, black or yellow, with contrasting spots or marks (ladybirds) . . . . **Coccinellidae**

–   Tarsi with four simple segments; antennal club more elongate, or whole antennae thickened; parallel-sided, elongate beetles; pronotum usually with projecting front angles and often with serrations along the side margins; usually dull or yellowish brown . . . . **Cryptophagidae**

38 Front and middle tarsi with five segments, hind tarsi with only four segments 39
–   All tarsi with the same number of segments, commonly five, but sometimes three or four. 46

39 Pronotum clearly narrower at base than elytra at shoulders. 40
–   Pronotum not, or hardly, narrower than elytra at shoulders 43

40 At least pronotum and elytra bright red; antennae serrate or pectinate; 8–18 mm . . . . **Pyrochroidae**

–   Most of upper surface not bright red; antennae simple. 41

41 Elytra with three fine, raised longitudinal lines . . . . **Oedemeridae**

–   Elytra without three raised lines. 42

42 Head contracted immediately behind large eyes ....**Aderidae**

LEFT: Aderidae

RIGHT: Anthicidae

– Head contracted some distance behind small eyes ....**Anthicidae**

43 Pronotum completely rounded at front, without protruding front angles, broadest at base, and forming smooth outline with elytra; middle and hind tarsi strongly tapered, and longer than tibiae 44
– Pronotum normally with front angles protruding, not completely rounded, usually broadest at middle; tarsal segments of uniform width, tarsi shorter than tibiae 45

44 Last abdominal segment produced into a point; underside of body very convex, beetle's natural resting position is on its side........**Mordellidae**

LEFT: Mordellidae

RIGHT: Scraptiidae

– Last abdominal segment not produced into a point; underside of body not so convex ....**Scraptiidae**

45 Antennae inserted under a ridge on side of head, so that about the basal third of first segment is hidden ....**Tenebrionidae**
– Antennae inserted at side of head so that its insertion, or a raised tubercle, can be seen from above .... **Melandryidae**

46 Antennae filiform, long or very long, often longer than body; most
    antennal segments long, narrow and cylindrical, but segment 2
    much shorter than either 1 or 3; eyes notched at the front, where the
    antennae are inserted; elytra usually long and narrow; tarsi appearing
    four-segmented, but really five segments, the fourth hidden in the
    much-bilobed segment 3 . . . . **Cerambycidae**

  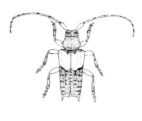 

–   Without these characters combined; eyes not notched (except in some
    Chrysomelidae, subfamily Bruchinae). 47

47 Tarsi like those of Cerambycidae, appearing four-segmented, but
    really five segments, the fourth being hidden in the much-bilobed
    segment 3. Generally broadly rounded, sometimes globular leaf
    and flea beetles, mostly glossy, often brightly coloured, metallic or
    patterned . . . . **Chrysomelidae**

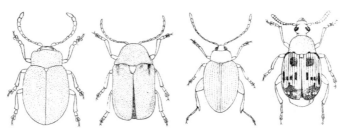

–   Tarsi otherwise 48

48 Tarsi five-segmented, usually segments 2–4 (but sometimes just 4)
    bilobed, rarely simple; antennae filiform, though basal one or two
    segments may be swollen; elytra elongate, parallel-sided, of a distinctly
    softer consistency. 49
–   Tarsi various, if five-segmented then segments 2 and 3 not bilobed;
    elytra not soft 52

49 Elytra bright red, with strongly raised longitudinal
ribs . . . . **Lycidae**

– Elytra various colours, but not with strongly
raised lines 50

50 Head completely hidden under front edge of
pronotum . . . .**Lampyridae**

– Head not completely covered by pronotum. 51

51 Broader beetles, elytra shorter; mostly dull or bright metallic green,
often with red marks; antennae usually black, marked with yellow . . . .**Melyridae**

LEFT: Melyridae

RIGHT: Cantharidae

– Narrower beetles, elytra longer; various shades of yellow, brown or
black, never metallic green; antennae usually a single colour . . . . .**Cantharidae**

52 Prosternum (p) fitting into mesosternum; beetles of
fairly similar distinctive narrow build and
smoothly elongate shape 53

– Prosternum not fitting into mesosternum; beetles
stouter, more solidly built 54

53 Pronotum with each hind angle produced into a tooth; beetles with a very long-oval or parallel-sided appearance; mostly dull browns, but some reds and blacks....................................**Elateridae**

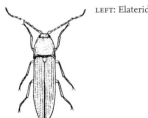

LEFT: Elateridae

RIGHT: Buprestidae

– Pronotum with hind angles not produced into a tooth; characteristic bullet-shaped beetles; often bright metallic greens, blues and golds...**Buprestidae**

54 Antennae simply filiform; pronotum much narrower then elytra... **Ptinidae**, Ptininae

LEFT: Ptinidae, Ptininae

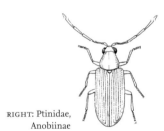

RIGHT: Ptinidae, Anobiinae

– Antennae filiform with last three segments much longer than the rest ....**Ptinidae**, Anobiinae

Skarabäus

In dem geheimnisvollen Käfer, der aus einer Kugel aus Mist zu entstehen schien, verehrten die Ägypter der Antike ihren Sonnengott Re. So wie Re den Sonnenball über den Himmel bewegte und abends in die Erde versenkte, so rollte der Käfer seine Mistkugel über den Boden und vergrub sie vor Sonnenuntergang in der Erde.
Und so wie die Sonne Leben keimen ließ, so ließ auch der Skarabäus aus Abfall wieder neues Leben sprießen.
Als Symbol der Unsterblichkeit, der Wiedergeburt und des Glücks wurde er in Siegeln, Amuletten, Schmuck und in der Schrift im Alltag der Ägypter allgegenwärtig.

Scarabaeus similis, Ägypten

Skarabäen aus Afrika

Tunesien

Ägyptische Skarabäen aus schwarzem Syenit. Nachbildungen

Transvaal　　Kenya　　Moçambique　　Simbabwe

In diesem Ausschnitt aus einem in Hieroglyphen geschriebenen Text wurde der Skarabäus als Schriftzeichen verwendet.

CHAPTER 8

The Human Significance of Beetles

SCARABS – THE FIRST BEETLEMANIA

Beetles were once among the most supremely recognised and revered of all creatures. In the third millennium BCE, what is now called the Old Kingdom of ancient Egypt was a flourishing civilisation and the icon of the age must surely have been the sacred scarab – *Scarabaeus sacer*. Scarabs occurred everywhere – in hieroglyphs, temple wall decoration, sculpture, seals and personal ornament. One of the most striking pieces of jewellery to be lifted reverently from the tomb of Tutankhamen (ruled *c*. 1332–23 BCE) was an exquisite necklace; the central chalcedony scarab is set around with wings, symbols and decorative motifs of gold inlaid with cloisonné, semi-precious stones and coloured glass. Sadly, no crown jewels that I know of in Europe ever featured a beetle. And where kings and pharaohs led, the fashion of the common folk followed too. Carved scarab amulets, talismans, rings, brooches, beads and tokens abound, and for at least a thousand years there was a Mediterranean-wide industry in creating them, from Sardinia to the Levant, from commissioned aristocratic high-status pieces of gold and gems, to roughly hewn stone and ceramic faience trinkets.

Received wisdom has it that the scarabs were worshipped by the Egyptians because their dung-ball rolling was seen as analogous to the sun god, Khepri

OPPOSITE PAGE: **FIG 119.** Almost every museum in the world has beetle exhibits, thanks to a fascination with ancient Egypt. Although small, the Naturmuseum in Augsburg, Germany, does not disappoint.

(often represented with a scarab for a head), rolling the glowing solar orb of life through the heavens. Perhaps the discovery of scarab pupae inside the hard, round dung shells was an inspiration, or justification, for mummification? New life emerging from the crisp pupal case, or the adult beetles erupting out of the hard dry ground from their subterranean interment come the wet season, offers as good a metaphor for life after death as any modern notion. The trouble is that the ancient Egyptian scarabs occur mostly as prehistoric artefacts; there is precious little in the way of primary historical (i.e. written) evidence of their significance. Certainly, many of the scarab talismans are inscribed on their flat undersurfaces with mottos like 'good health', 'true heart', 'good wishes' and 'life', but nowhere is there any explicit description of the cult's underlying doctrine. It was only with the Greek essayist Plutarch (46–120 CE), writing something like 2,000 years afterwards, that there comes any mention of sun or dung rolling. Plutarch (later followed by Aelian, Horapollo and Porphyry) also states that the Egyptians believed the beetles to be only male, but to be able to impregnate the dung ball with their seed to create new life. This male-only doctrine was oft quoted as the reason Egyptian soldiers wore scarab-decorated rings – a sign of their manliness perhaps. Great care must be taken when trying to interpret how the ancients thought, and what they believed. The Iron Age Greeks and Romans may have been writing earnestly and neutrally about what was to them moderately recent history, or they may have had their own agenda in interpreting the past for their own ends. It wasn't long before Christian scholars started to write scathingly about the folly of the Egyptian pagans, raising temples to cats, bullocks and dung beetles. There are some claims that images of dung beetles were representative of sinners in medieval Christianity, but I can't find much first-hand evidence to back this up. I wonder whether this was just a petulant rejection of the scarabs of the obviously heathen ancient Egyptians. Beavis (1988) gives a measured scholarly analysis of who said what and when about scarabs, sacred or otherwise.

Although it's difficult trying to understand exactly what our ancestors might have thought, back through the fogs of time, what is clear is that the artists and artisans who made the decorations were intimately familiar with the actual beetles themselves. William Matthew Flinders Petrie (1853–1942) made a valiant attempt (1917) to identify various original beetle genera, which scarab-carvers may have used as prototypes. His drawings of stylised genera *Hypselogenia*, *Copris*, *Catharsius* and *Gymnopleurus*, as well as the usual suspect *Scarabaeus*, next to examples of various carvings are very persuasive. Archaeologists still dispute whether, as Petrie suggested, these designs reflected a chronological change down the centuries (revisited by Ward, 1987). Some of the many thousands of scarab amulets in today's archaeological museums are little more than dome-shaped stereotypes, barely recognisable as beetles at all, but enough of them are delicately carved masterpieces, and I for one firmly believe that their creators were copying from life, or at least from a real dead beetle on the workbench beside them.

An obscure and now half-forgotten gothic novel, *The Beetle*, by Richard Marsh (1897), uses a beetle, sometimes described as a scarab, in a convoluted story of love, betrayal and weird Egyptian and Arab supernatural transmogrification. Set in mid-nineteenth century London, it follows several protagonists, including a Member of Parliament, a Sherlock Holmes-like detective, various wealthy aristocrats and an out-of-work office clerk, with some cult-of-Isis memes thrown in to catch the popular imagination of ancient Egypt that was burgeoning in Britain at the time. The beetle itself features little and is given no serious description, but this does not matter; the novel relies on the power of scarab imagery still deeply ingrained into the western psyche.

Today, dung beetles are, in my opinion, overlooked and undervalued. To vent my indignation, I've written a book about them (Jones, 2017). Quite apart from their handsome good looks, dung beetles perform one of the most important ecosystem services — recycling decaying organic matter. In Britain, urban humans mostly flush away their output without a second thought, but even in the wider countryside it is easy to ignore animal dung in the fields because it is cleared away by a quiet army of hidden recyclers, a significant portion of which are beetles. But elsewhere in the world, things do not run so smoothly. After 200 years of stock animal grazing in Australia, it finally dawned on the establishment that they were drowning in droppings. There came the realisation that native dung beetles, adapted to dry marsupial nuggets, could not cope with the copious much wetter output from what was now approaching 25 million cows, 70 million sheep and 25,000 horses. Various tabloid statistics abound: 2 million tonnes of dung dropped a day; the dung from 10 cows can smother a hectare of pasture in a year; each 2-kilogram dung pat produces 1,000 Australian bush flies, *Musca vetustissima* (which was *perfectly well* adapted to moist excrement, thank you); and 200,000 hectares (2,000 square kilometres) of grassland were rendered useless for grazing each year. Something had to be done about it all.

The Australian Dung Beetle Project, begun in 1964, imported eggs of 52 Eurasian and African dung beetle species better equipped for the higher water content of domestic farm animal dung, and captive-bred 43 of them for release. Between 1968 and 1984, 1.73 million dung beetles were set free across thousands of Australian pasturelands; 23 species became established. According to ongoing

FIG 120. We can sometimes only dream of a world where beetles are better understood. There is a message in my spoof sign: dog owners who clear up their pet's droppings into a poop bag, but then dump the bag in a hedge, are not only misunderstanding the ecology of dung recycling; they are actively thwarting it.

research (Edwards, 2007), the tide of ordure is being reversed, and every Australian dung beetle researcher already has a new wish list of extra species to import.

Dung beetle appreciation is creeping into our consciousness, and may yet give these creatures some of the praise they deserve. African game reserves have road signs warning drivers to give way to roller-dung scarabs trundling their brood balls across the tracks, much to the bemusement and pleasure of tourists, who broadcast pictures of them through social media. There is a large iron statue of a dung ball and two scarabs at London Zoo; the beetle on top of the ball is often joined by small children for visitor photo shoots. Britain's largest dung beetles are the dumbledors (Geotrupidae), named for their heavy, buzzing, bumbling flight, and although some dictionaries claim the name for bumblebees, no coleopterist would agree. Knowingly or unknowingly, that doyenne of Dickensian character names, J.K. Rowling, chose Dumbledore (the alternative spelling ending with an 'e') for Harry Potter's headmaster. When I caught a flying *Geotrupes* with my bare hands, on the banks of the Wolfgangsee near Salzburg in the summer of 2013, my Austrian hosts were immediately familiar with the name and listened politely enthralled while I pontificated on its various faecal burying tactics.

My favourite exhibit at the British Museum is a giant stone sculpture of a sacred scarab, 1.5 metres long (Fig. 121). Carved from green diorite sometime between 399 and 300 BCE, during the Ptolemaic era, it was made very late in the Egyptian fascination with dung beetles, and was possibly already a pastiche –

FIG 121. Giant carved stone statue of a sacred scarab in the British Museum, fondly admired by entomologists and non-entomologists alike.

a nostalgic half-memory, vintage or retro. Even so, it shows the power of the scarab beetle as an enduring symbol of veneration and popular affection.[17]

Affection is not just my hyperbole. Large, brightly coloured chafers (in the same family as scarabs) are regularly kept as pets. They are low maintenance (requiring just a few cut grapes every so often), and they are offbeat and exotic enough to appeal to oddballs, but they come without the danger of frightening or upsetting any snake- or lizard-averse relatives. And at the end of their lives they can become useful and interesting display specimens. Such was the destiny of

17 The only modern celebratory equivalent I can come up with (apart from the decorative statues of dung beetles at London Zoo, or similar adverts at other scientific institutions) is the boll weevil (*Anthonomus grandis*) statue in Enterprise, Alabama. It appears to be the only monumental statue erected to celebrate an agricultural pest. The boll weevil, a native of Mexico, arrived in 1915 and started to decimate the cotton crops. As a consequence, local farmers were encouraged to switch to farming peanuts and other crops, and this enforced change brought new prosperity to the area.

FIG 122. I can't help but think that more people should keep beetles as pets. As a child, I easily reared the oily blue-black grubs of bloody-nosed beetles (*Timarcha tenebricosa*) on bedstraw leaves growing in the garden. Even now, I can't resist picking up the adults for a quick tickle.

the Goliath beetle left in a tin can on the doorstep of the American Museum of Natural History in New York. It became a popular attraction in the entomology department, peeling its own bananas until it died 'at the venerable age of about seven months' in July 1959; it appears to be the only beetle to have had its own obituary published in the *New York Times* (Anon., 1959).

LADYBIRDS – WELL SPOTTED

Today almost all public affection for beetles has been monopolised by ladybirds. These colourful, spotted insects are everywhere prized for their cheerful decoration, their doggerel and nursery rhyme familiarity, and their aphid-devouring behaviour. Ladybirds are now seen as 'good' animals, helping the farmer or the gardener by destroying all those nasty plant lice. To the biologist, the notion of 'good' or 'bad' organisms is a nonsense, but humans have a tendency to classify things according to how useful, harmful or emotionally uplifting we might find them. Ladybirds are well placed to receive this adoration.

When California citrus farmers found their trees inundated with the powdery fluffs of the fluted cottony cushion scale (*Icerya purchasi*) in 1868, they were not, at first, uplifted. This Australasian plant bug had been accidentally introduced with imported acacias and by 1885 was so prevalent in the western United States that it threatened to decimate their industry. Instead, American entomologists went on the search for a predator, finding one in the Australian ladybird *Rodolia* (then *Vedalia*) *cardinalis*. It was released in California in

1888, and within a year the citrus crop had recovered, increasing threefold. This represents the first successful scientific biocontrol effort anywhere in the world,[18] and it is part of the folklore that has continued to heap praise on ladybirds everywhere. *Icerya* was found in Britain in the 1990s, and is now established in the London area. Less commercial panic surrounds its appearance here, and anyway *Rodolia* was found too, suggesting that both scale and predator had been accidentally introduced on the same batch of imported plants (Salisbury & Booth, 2004).

That ladybirds devour plant pests is a modern celebration – our love affair with them began through yet more religious iconography. The lady, after whom ladybirds are named, is none other than Our Lady, the Virgin Mary, mother of Jesus; medieval portraits often showed her wearing a rich red cloak, and the black spots of the common Eurasian seven-spot (*Coccinella septempunctata*) were said to represent the seven joys and seven sorrows of Mary (Majerus *et al.*, 2016). Elsewhere in Europe, similar pious names abound – in the German *marienkäfer*, Dutch *mariehone* and Spanish *mariquita* (all meaning 'Mary's beetle'), the French *bête à bon dieu* (good God's animal), and the Swedish *himmelska nyckla* (keys of heaven) or *Maria nyckelpiga* (Mary's key maid). In his curious and charming history of ladybirds, Exell (1991) lists 329 ladybird names in 55 languages, of which about a quarter refer to the Virgin Mary and another 50 mention God.

Today, ladybirds are seen as friendly and fun, happy characters in cartoons and storybooks, the subject of playground chants and nursery rhymes. They have also been misappropriated to become the brand logo for clothes, fried food, bicycles and books. There are Ladybird animation studios, fitness clubs, consultancies and vaping shops. The red-and-black spotty pattern is now used with vulgar abandon on buttons, badges, backpacks, beanbags, luggage, slippers, wellington boots, umbrellas, fancy-dress tutus, watches, earrings, socks, ties, rollercoasters, cupcakes, pens, pencils and rubbers. At Easter, there are even chocolate ladybirds. Why, I wonder, has there never been a ladybird-spotted red double-decker London bus?

Humans never worshipped ladybirds, although we once might have revered them. Now they are just an immediately recognisable and brightly coloured formula for anything cutesy. In an odd twist of fashion, no doubt borne out of

---

18 Although no beetle has been seriously released into Britain for biocontrol, plenty of common British species have been introduced elsewhere around the world. These have mostly been phytophagous species, introduced to combat European plants gone rogue on other continents, and include: *Galerucella calmariensis* and *Nanophyes marmoratus* against purple loosestrife (*Lythrum salicaria*) in North America; *Cassida rubiginosa*, *Larinus planus* and *Rhinocyllus conicus* against creeping thistle (*Cirsium arvense*) in Canada; and *Chrysolina hyperici* against St John's-wort in New Zealand.

FIG 123. An immaculate ladybird. Although most ladybirds are brightly spotted, there is sometimes great variation within a species. The 24-spot (*Subcoccinella vigintiquattuorpunctata*) can have fewer than its nominate number of spots, and may even lack them altogether. Barkingside, 22 September 1999.

a bitter-sweet nostalgia, the informative but mawkish children's literature once peddled under the Ladybird logo has now been supplanted by the company's spoof series of books for adults. Rather than reading about *British Wild Animals*, *William the Conqueror* or *Our Land in the Making*, we are offered *The Hipster*, *The Mid-life Crisis* and *The Zombie Apocalypse*. It's all very satirical, and many millennia from now it is certainly going to confuse future archaeologists.

A LIGHT IN THE DARK

In February 1965, when I was six, my family moved to a large but slightly rundown house in Newhaven. It had been unoccupied for many months, a burst pipe had brought down part of the kitchen ceiling and as well as open coal fires in several rooms, we used to have to light a coke-powered boiler every day to provide hot water. The late 1940s grey enamelled metal stove was labelled with a plaque announcing that it had been manufactured by the Glow-worm company, but the logo was a pale blobby, caterpillar-like maggot with big, lashed eyes, a smiling face and an incandescent lightbulb of a tail; I was not convinced.

In Britain, where the common glow-worm (*Lampyris noctiluca*), our only light-producing insect, is pale, muted and understated, it is very easily overlooked and there has never been much of a folk tradition celebrating its glorious power. It does appear, to good effect, through literature though. When Andrew Marvell (1621–78) wrote 'The Mower to the Glowworms' in 1681, he beautifully evoked their curious wonder by describing them as 'courteous lights' and 'living lamps, by whose dear light | The nightingale does sit so late'. Their gentle innocence is emphasised by:

> country comets, that portend
> No war nor prince's funeral,
> Shining unto no higher end,
> Than to presage the grass's fall.

William Cowper (1731–1800), perhaps unwittingly, inserts a bit of ecology into 'The Nightingale and the Glow-worm' (Cowper, 1860). The nightingale had

> The keen demands of appetite…
> And knew the glow-worm by his spark;
> So, stooping down from hawthorn top,
> He thought to put him in his crop.

The glow-worm, not surprisingly,

> Harangued him thus right eloquent
> 'Did you admire my lamp…
> As much as I your minstrelsy…
> For t'was the self-same power divine,
> Taught you to sing and me to shine.'

Cowper does not elucidate on the aposematic warning of the glow, or the beetle full of distasteful chemicals, but perhaps he had heard tell of birds rejecting them:

> The songster heard his short oration,
> And warbling out his approbation,
> Released him, as my story tells,
> And found a supper somewhere else.

Or it could just be poetic licence.

William Wordsworth (1770–1850) knew, in his 1802 poem 'Among All Lovely Things My Love Had Been', that his love had 'never seen | A glow-worm, never one', so when 'A single glow-worm did I chance to espy', he scooped it up and left it for her to see in the orchard – 'Oh! joy it was for her' (Wordsworth, 1904). The poet obviously had all the makings of a coleopterist; when I last came across a 'grass star', in the garden of a holiday gîte in France, I immediately took it inside, in my cupped hand, and woke my children to show them. Likewise, Alfred, Lord Tennyson (1809–92) contrasted the pinprick glow in the grass to the stellar pinpricks of the night sky in his 1847 epic *The Princess*:

> *Now poring on the glowworm, now the star,*
> *I paced the terrace, till the Bear had wheeled*
> *Through a great arc his seven slow suns.*

Lampyris still occurs on Tennyson Down, the chalk hill above Freshwater on the Isle of Wight, made famous by the poet laureate's regular promenades up there.

There is something quiet and secret, transient and ephemeral, about the glowing *Lampyris*. William Shakespeare (1564–1616) evokes its other-worldly quality in *Hamlet*, by having his father's ghost exit with the refrain:

> *The glowworm shows the matin to be near,*
> *And 'gins to pale his uneffectual fire.*
> *Adieu, adieu, adieu. Remember me.*

Most people, including the logo designers for the Glow-worm Boiler and Fireplace Company, forgot this long ago, and in any case have little idea exactly what *Lampyris* looks like. Let's not even go near the ludicrous illustration to grace the cover of a mid-twentieth-century version of Paul Lincke's piano sheet music for his song 'The Glow Worm', with its three(?)-winged, four-legged, phosphorescent-tailed, flying… somethings. Nevertheless, people are vaguely aware that *Lampyris* produces a dim green glow, and that has been enough for it to become the vague literary trope still, with which books and poems are regularly illuminated today.

At least fire-flies sound a bit more pyrotechnic. We have none in Britain, but they occur through much of the rest of the world. The Japanese rightly extolled the wonder of these much brighter beetles, and they act out a more vibrant part in oriental art and literature. There is still room for anatomical error though. The famous woodcut *Women Catching Fire-flies*, by Hosoda Eishi (1756–1829), shows his subjects wafting several rather moth-like insects into a paper lantern. Likewise, various prints by Utagawa Kunisada (1786–1865) show people with fans and paper lanterns surrounded by glowing blobs on insects of indeterminate body structure. Matsuo Basho (1644–94), perhaps the most famous poet of the Japanese Edo period, produced a haiku of such literary and observational simplicity that it sums up lampyrids everywhere:

Blade of grass
a fire-fly lands
takes off again.

Magical, and true.

The idea of bright fireflies giving enough light to see or read by is also very familiar. Disney's *A Bug's Life* (1998) depicts them as circus spotlights lighting up the performers in the ring – correctly beetle-like with glowing tails, thanks to the Disney art department's meticulous entomological field research. The brighter light of exotic beetles had been long reported. Italian historian Peter Martyr (1457–1526) claimed, in his various letters written during his travels in the New World (Martyr, 1555), that in the West Indies islanders would go out 'two Cucuji[19] tyed to the great toes of their feete: for the travailer goeth better by direction of the light of the Cucuji, than if he brought so many candles'. During his 1799–1804 expedition to the 'Equinoctial regions of America', Prussian explorer Alexander von Humboldt (1769–1859) often comments on the fiery stars in the herbage, and was one of the first to describe their use in lanterns: '

> *In the hut of the poorest inhabitants of the country, fifteen cocuyos, placed in a calabash pierced with holes, afford sufficient light to search for anything during the night. To shake the calabash forcibly is all that is necessary to excite the animal to increase the intensity of the luminous discs situated on each side of its body.*
>
> (Humboldt, 1852)

Again, a subliminal ecological message comes out here: if the beetle is alarmed by being bashed about in its calabash, its natural response is to glow brighter, reinforcing the deterrent message that its glow gives about its distasteful body chemicals, should an attacker try to eat it.

OTHER BEETLE MODELS

The stag beetle, *Lucanus cervus* (Fig. 124), has managed to insinuate itself into European culture by virtue of its large size and the male's monstrous jaws. Today, it is a flagship conservation icon, with its own Biodiversity Action Plan, public awareness campaigns, and citizen science monitoring and recording schemes. But like the ladybird, it also has religious and folk memories associated with it.

---

19 The click beetle *Pyrophorus noctilucus* was then often considered part of the family Cucujidae.

Some seem positively pagan: the beetle was once blamed for starting house fires, because it reputedly carried red-hot embers about in its antlers, or it attracted lightning. This last idea may be based on people seeing the beetles emerging from dead, possibly blackened and blasted, tree trunks and stumps. There also seems to be some form of Christian symbolism associated with the species, its horns being similar enough to those of the stags that were often used to represent Jesus Christ because of their ability to fight those tokens of satanic evil – snakes (Sprecher & Taroni, 2004). Sure enough, an identifiable stag beetle occurs in numerous devotional paintings, altarpieces and prayer books from the Middle Ages, including *Adoration of the Magi* (Stefan Lochner, 1440–45), *Jesus's Prayer in the Garden of Olives* (Michael Wolghemut, 1479) and *The Virgin Among the Multitude of Animals* (Albrecht Dürer, 1503). Dürer used the stag beetle in several monumental

FIG 124. Lillian and the stag beetle (*Lucanus cervus*). Today, the stag beetle is an icon of nature conservation, a dramatic but declining creature. It does have a place in the classroom, capturing the attention of children before they learn to be scared of insects.

FIG 125. In my house, everything can be beetle-themed, and anatomically correct at that – a magnificent birthday cake, but only one candle.

religious pictures, including *Adoration of the Magi* (1504), where it is depicted in careful three-quarters perspective. The same specimen, by the looks of it, was used as the model for his famous 1505 watercolour, often proclaimed as the first true-to-life beetle illustration in history. It is exquisite, rising lifelike off the paper, and would not look out of place in a modern identification guide. For the next 500 years, *Lucanus* was a popular adjunct, whether religious or merely embellishment, to still-life painting, decoration, ceramic art, jewellery, illustration, or cartoon, appearing on posters, postcards, stamps and coins (Sprecher & Taroni, 2004).

In Britain, the common cockchafer (no schoolboy jokes please), *Melolontha melolontha* (also known as the maybug for its spring flight; Fig. 126), once had a real place in the knowledge of a mostly rural population. On holiday in the Isle

FIG 126. Once derided as a noxious pest, the common cockchafer (*Melolontha melolontha*) is anything but common. It's certainly widespread in Britain, but flights of the beetles darkening the evening skies are a thing of the past.

of Wight in the early 1960s, I remember hearing someone my grandfather would have described as an 'old boy' recounting the din cockchafers made knocking on windows, and the tremendous rattle of clouds of them hitting the street lights. I was quite impressed, as I'd never seen one, and to date have only ever found the odd singleton. But back in the day they reached biblical proportions. The naturalist Thomas Mouffett (1553–1604) described how so many of them fell into the River Severn in 1574 that they clogged the waterwheels of the mills (Mouffet, 1634). During the eighteenth and nineteenth centuries, they were considered a serious pest of grasslands and orchards. Entomologist Eleanor Ormerod (1828–1901) recounts a harvest of 80 bushels of them from one farm (Ormerod, 1881). Her preferred method of control was to shake the fruit-tree branches over sheets and then feed the gathered beetles to pigs, or employ small children to go around squashing them.

The beetles are big enough and distinctive enough to feature in some of the same oil paintings as stag beetles. *Melolontha* was one of the many large, showy insects used again and again by Jan van Kessel the Elder (1626–79) in his still-life tableaux (Fig. 127). The clarity of his delineation is far better in *Melolontha* than in some of his slightly more fanciful insects, which still defy identification at close scrutiny. It's as if some of his creatures were copied from copies of copies, where an air of insectness has been retained, but specific identity clues have been blurred and lost. Painting to order probably meant that not all van Kessel's insect models were available alive or fresh for each commission. But the cockchafer is uniformly correct through all his iterations. The artist was obviously intimate with the common cockchafer – indeed, this beetle has long been a well-known image on mainland Europe and was widely represented in popular art right up to modern times. Swiss composer Josef Viktor Widmann (1842–1911) wrote an opera about it – *Maikäfer-Komodie* (1897). It appears on postcards, greetings cards and posters, in children's books and stories, and on beer-bottle labels. Every Easter, chocolate cockchafers appear in the shops along with chocolate eggs and rabbits. You can even make your own by purchasing the vintage tin moulds on eBay. I've yet to see a novelty chocolate stag beetle though.

Today, beetles appear on stamps – the mandibularly impressive *Chiasognathus granti* on a 1948 Chilean stamp is the earliest I can find – and they are incorporated into fantasy art, wrapping paper, clothing, luggage, greetings cards, buttons, crockery, cushions and cookware. These are frequently out-of-copyright images, filched from the great insect monographs of the past, and they certainly make attractive and unusual items. Often, though, beetle images tend to be rather generic. Writers, artists and commercial sales and branding teams just need to get across the basics of beetlery to an already confused and often entomologically

FIG 127. Details from one of Jan van Kessel the Elder's paintings. The cockchafer *Melolontha melolontha* is anatomically correct and easily identifiable, with its detailed colouring, lamellate antennae and downturned pygidium, although the tarsal segment numbers may be a bit confused. The other creature, however, is the stuff of nightmares – something like a dumbledor with crab's pincers.

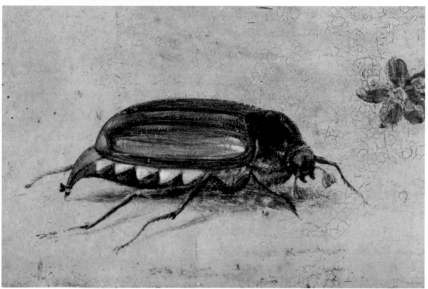

uneducated public. When Ferdinand Porsche designed a people's car, the Volkswagen was nicknamed beetle simply on the strength of its smoothly domed shape. The plastic constructions provided in the box game used for beetle drives look more like ants than beetles; they also have curled lepidopterous tongues and strange articulated tails, quite the chimera really. Elsewhere, beetle logos abound for companies selling shoes, bags, books, hardware and coffee. Some are obviously based on anatomically correct Coleoptera models, but many are simple shiny round or domed designs with varying numbers of legs, and sometimes with varying numbers of antennae. The beetle description used in Chapter 1 can be ultra-condensed to a stylised design, much like the ancient Egyptians' scarab amulets, and it is still enough to get the idea across. This is how marketing and brand awareness works.

FIG 128. Today, beetle motifs crop up in all manner of odd places, from fashion to soft furnishings.

A beetle theme can have negative, as well as positive, attributes. In his 1949 novel *Nineteen Eighty-four*, George Orwell's 'beetle-men' were nameless short, stout, scuttling figures in blue overalls that kept their heads down; they flourished under the dominion of the Party, and proliferated in the oppressing ministries. This is still an image that rings true with me whenever I meet mindless jobsworth bureaucracy.

Edgar Allen Poe (1843) went a little further in his short story 'The Gold-bug', which, in the same old confusion of things, was not a bug, but a beetle. It is definitely described as a scarab, but fancifully giving it two black spots on its wing-cases, like eyes, Poe's allusion is more to a skull than to any real insect. Some people have tried to identify an original species model for Poe's gold bug, and the astounding Central American *Chrysina aurigans* (pure gold) or *Chrysina limbata* (silver) might seem possible options. However, although the broader genus of shining green and metallic-tinged beetles, *Chrysina*, was known from 1828, the first truly brilliant burnished gold species were not described until the mid-1870s, fully 30 years after the mystery tale was published. The large (but not metallic) yellowish-orange *Cotalpa lanigera*, from the eastern United States, is another option, but I think it more likely that Poe just needed a large beetle-like beetle as a plot device to give an eerie weight to the mystery that unfolds.

There is nothing sinister in 'Forgiven' (1927). A.A. Milne's Alexander beetle is a delightful entomological aside, and well captures the absolutely typical fascination small children show towards small wildlife, before they learn to worry about dirt, creepy-crawlies or vermin. Although Alexander remains undescribed in the text, E.H. Shepard decided to illustrate the poem with something that looks very like a typical *Carabus*. Perhaps this was quite appropriate; *Carabus violaceus* is still typically a beetle of gardens, hedges and woodland edges, although it has probably declined since Christopher Robin's day. And although Mrs Tittlemouse was obviously offended by an uninvited beetle's 'little dirty feet', Beatrix Potter's painting (1910) shows another wholly unthreatening ground beetle, slightly stylised, but possibly *Nebria brevicollis* according to my eye.

BEETLES IN THE REAL HUMAN WORLD

On the whole, beetle references in art and literature are relatively benign, with appearances either being vaguely insect-like, comical, curious or just a bit weird. Nothing too unnerving. It is telling that among the great array of utterly dreadful science fiction horror B-movies pumped out over the last 70 years, there are plenty of radioactive wasps, mutated ants, killer bees, giant mantids, alien cockroaches and a half-human fly, but none that features evil beetles. Thankfully, Russell Hays' 'The Beetle Experiment' (*Amazing Stories*, 1929) never made it on to the big screen, although we should be impressed that the cover of the magazine shows a besuited, hatted scientist (?) with a double-barrelled shotgun facing up a giant, *but anatomically correct*, tiger beetle – the common North American *Cicindela togata* by my guess. Like the rest of the world, I will quickly gloss over *The Devil's Coach Horse* by Richard Lewis (1979), which, with its quack sci-fi mixture of genetic modification, human cadaver feeding and air disaster, could easily have been called '*Beetles on a Plane*'. Yes, best left well alone, that one.

If we think about this, beetles have come out a lot better than they ought, because in the real world some beetle species have had serious and devastating consequences for human beings. Perhaps it's to do with the fact that the adult beetles look so very different from the wyrms that are doing the destructive work: woodworms, mealworms, wireworms, etc. As ever, it is the larvae that do most of the feeding, and if it is human wares into which they are tucking, it is the larvae that do the damage. And beetles have actually done untold damage to humans.

There are endless beetle pests indoors, from woodworms in the floorboards to grain weevils in the larder, carpet beetles in the Axminster to bacon beetles under the kitchen cooker. Woodworm (*Anobium punctatum*; Fig. 130) and

Typhaeus
typhoeus

FIG 129. When one of Charles Darwin's sons, George, visited a friend, he naively asked, 'Where does your father do his barnacles?' (Darwin wrote important monographs on the group, 1852, 1854). My children might well ask the same about beetles. Our house is full of Coleoptera, and this spills over into our everyday lives. One of Verity's BTEC artworks explored mental illness through the internalisation of phobias. Calvin painted me a minotaur beetle picture for Christmas. And our Lego giraffe-necked weevil was a triumph.

FIG 130. Woodworm holes and sawdust frass – the all too familiar signs of unwanted guests in the understairs cupboard. My previous house, on Bellwood Road in Nunhead, was well occupied by *Anobium punctatum*.

deathwatch (*Xestobium rufovillosum*; Fig. 131) continue to be a major nuisance in all buildings where timber construction is used. This includes most conventional domestic houses. Although modern apartment and office blocks, skyscrapers and commercial buildings are now made of concrete, steel and glass, they contain much in the way of wooden furniture and fitments, all suitable food for woodworms. And, of course, many important historical buildings are timber-framed, at least in the roof. These need constant care and attention to stop them falling down around our ears when the frass-filled voids undermine the load-bearing beams. It was against the rabid attentions of deathwatch beetles in the roof of Westminster Hall (the oldest existing part of the Palace of Westminster) that entomologist Harold Maxwell-Lefroy was working when he founded Rentokil in 1925. Now an international pest-control conglomerate, probably dealing more with rats and mice, he originally wanted to call his company Entokil, emphasising its insect-based starting position, but this was turned down

FIG 131. My friends Jony and Fiona Russel were less than ecstatic when I found deathwatch beetles (*Xestobium rufovillosum*) crawling all over the floor in their lovely converted seventeenth-century abbey lodge house in the picturesque village of Banwell near Weston-super-Mare in 1994. I tried to console them with the news that as long as they checked the timbers every 50 years or so, they'd probably still get another few centuries of use from the timber-beam roof.

by the powers that be, because of existing trademarks. Chemical spraying and wood-injection are still the major treatments for domestic woodworm, although central heating and the vacuum cleaner have probably done much to combat the beetles too. Unfortunately, Maxwell-Lefroy accidentally gassed himself in October 1925 with the new poisons he was brewing in his poorly ventilated laboratory at Imperial College (Fleming, 2015).

No beetles are implicated in the spread of serious human diseases (they barely feature in textbooks of medical or veterinary entomology – that job is left to flies and cockroaches), but they do, I'm afraid, make up more than their fair share of domestic pests. Hinton (1945) wrote a most important monograph on the world's beetle pests of stored products, but he only got around to volume one in the series; there would certainly have been material for a further two or three. Throughout history, humans have had to cope with the vagaries of the weather when growing their crops, then with the devil-sent pestilences of grain weevils, bean weevils, flour beetles and mealworms in what is now termed, by the applied entomology fraternity, as post-harvest loss. In a developed world where we now regularly expect whole, undamaged, uninfested goods from the supermarket, it is difficult to remember that it was only a couple of generations ago, before the advent of serious crop-spraying, that almost every warehouse, shop and home in the country was plagued by beetles chewing away at things. In one nineteenth-century leather warehouse, the exasperated owners offered a £20,000 reward for any remedy against the larder and hide beetles (Dermestidae) ruining their stock (Butler, 1896).

One of the most important human pests is undoubtedly the grain weevil (*Sitophilus granarius*). Although flightless (indeed wingless), it has been carried across the globe in transported grain since the dawn of agriculture some 6,000–10,000 years ago. Archaeological digs confirm that it shadowed the expansion of grain farming and was known in Turkey around 5750–5500 BCE, the Levant in 5500 BCE, and Germany in 5140 BCE (King, 2013). It may have arrived in Britain with the Romans around 47 CE, first to Londinium, then to the garrison outposts of Luguvalium (Carlisle) and Aquae Arnemetiae (Buxton). Curiously, the beetle's native habitat and endemic origin are unknown. Today it only occurs in human grain stores, and has never been found 'in the wild' even in the Middle East or north-east Africa, where the original precursors of modern wheat, spelt, emmer and einkorn still grow wild. One appealing theory is that it started off infesting seed or acorn caches collected by an unknown hoarding rodent in the forests of the Indus Valley or Middle East. However, with climate warming after the end of the last ice age, altered weather patterns induced massive landscape change (subtropical forests became dry upland savannah). These changes were exacerbated by the arrival of sheep- and goat-herding, so much so that the hoarder declined, to the point of vanishment, and the weevil was forced to find new lodgings – grain stores in human habitations (Jones, 2015).

Sitophilus granarius was once an empire-threatening beast; during and after the First World War, the Royal Society reported that poor warehousing and high infestation rates made the British grain-storage situation perilous (Anon., 1918–21). Following similar food shortages during the Second World War, the United Nations Food and Agriculture Organization suggested that 10 per cent of all the world's cereal crops were destroyed by insect infestation, and *S. granarius* was the most serious pest by far. With changes in baking fashion, few domestic houses in Britain now have a grain store; instead, we buy plain and self-raising flour in packets. Consequently, the grain weevil's grip on our pantries has slackened, but elsewhere in the world it remains a terrible blight.

Pest beetles are still with us, and although we get used to treating some, or living with them, others keep coming along. The lovely scarlet lily beetle (*Lilioceris lilii*; Fig. 132) was very rare in Fowler's day (1890), known only from my part of

FIG 132. To the coleopterist, the lily beetle (*Lilioceris lilii*) is a scarlet jewel; to the lily gardener, it is spawn of the devil.

south-east London (Peckham, Deptford and Camberwell), and Swansea, but during the 1940s it turned up regularly around Guildford, Surrey, and started to spread (see p. 347). It seems likely that a hardier lineage was introduced here, one that could cope with Britain's cool, wet climate, and the beetle is now regularly listed among Britain's worst garden pests. I think the beetle is beautiful, but then I don't grow lilies in my garden. I miss the elms though. Growing up in East Sussex in the 1960s and 70s, English elms (*Ulmus procera*) were elegant and triumphant statements across the landscape of the Weald and river floodplains. These tall columnar trees, with broad domed canopies and deep green foliage, grew everywhere, but a shadow fell on them in the form of Dutch elm disease, caused by the fungus *Ophiostoma novo-ulmi*. This virulent disease hit hard and fast, and by 1980 more than 25 million mature elms had died and either fell or were cut down. The fungus infects the water-conducting tubular xylem cells in the tree bark, causing the tree to plug them, effectively cutting off water to the leaves. The spores are spread by the elm bark beetles *Scolytus scolytus* and *S. multistriatus*, when they chew into the bark to feed and to make their egg-laying burrows. The xylem plugging probably works to stop the disease getting into the tree's roots, because although the large old trunks die off, suckers soon sprout from the surviving rootstock. The elms were not killed; they still occur everywhere in hedges and woodlands, but these are now small scrubby growths. As soon as the tree trunks get to about 5–10 cm in diameter, they attract the

FIG 133. The elm weevil *Magdalis armigera* breeds in the terminal twigs of elm trees. Once common, it underwent catastrophic decline following the widespread removal of large elms from the English landscape during the outbreak of Dutch elm disease in the 1970s. Thankfully, the beetle is now making a comeback. (Penny Metal)

FIG 134. The Asian longhorn (*Anoplophora glabripennis*), photographed on a street in Beijing. (Nick Alexander)

beetles and the disease comes in again; there will be no more grand, handsome, towering mature specimens. The landscape is changed for ever.

To borrow a nautical metaphor, there are several other potential beetle pests in the offing, not quite in the harbour, but definitely this side of the horizon. The Asian longhorn (*Anoplophora glabripennis*; Fig. 134) is a large and attractive, shining black and white-speckled beetle. I was recently shown a photo of it on someone's phone and my anxious heart fluttered for a moment. But he had seen it on a recent visit to China, part of its native Far Eastern range; it was not yet in East Dulwich. This large longhorn somehow got into North America in 1996, possibly in wooden packing crates or pallets from China, and its convoluted burrowing under the bark and down into the heartwood of mainly poplar, willow, maple and birch trees soon left a trail of destruction in its wake. It is also established in central and southern Europe. There are regular reports of specimens emerging from wooden packing crates (Wright, 2000), so when a breeding colony (66 infested trees, mostly sycamore, *Acer pseudoplatanus*) was found at Paddock Wood in Kent in 2012, all hell broke loose (Straw *et al.*, 2015). More than 2,000 trees were felled to contain the outbreak, and although no further beetles have been sighted to date (2017), monitoring will continue until 2018 before any all-clear is sounded.

As an island we are protected from some unwanted invasions (more on how Britain got its beetle fauna in Chapter 10), but the huge transport of commercial goods, including forestry, agricultural and horticultural produce, means that invaders do not have to make their own way here – we bring them. Every year, scores of potential invasive pest species are discovered at British ports, and some inevitably become established here. I was surprised at the relative lack of furore on the discovery of the western corn rootworm (*Diabrotica virgifera*, Chrysomelidae) in several sweetcorn (*Zea mays*) fields in Surrey and Berkshire in 2003 (Ostojá-Starzewski, 2005). This is a beetle reputed to cause US$1 billion in losses in North America each year – US$800 million lost in destruction plus US$200 million for treating beetle infestations. It may just be coincidence that the localities seem to be near Gatwick and Heathrow airports. It is possible that the beetles arrived from North America (or central southern Europe, where it is now well established) on imported corn cobs but, with so much traffic passing through these international flight hubs, beetles could quite easily have arrived in innocent-looking luggage, crates or other freight packages. The beetles fly well, and being dependent almost exclusively on one foodplant, their entire evolutionary history has lined them up to be able to detect maize plant chemicals, at a distance, in the air. Several British colonies were found and were treated with chemicals and crop destruction; they persisted until 2007, but no further outbreaks have been reported since then. British maize farmers wait with bated breath though.

It was with bated breath that British potato farmers waited for the Colorado beetle (*Leptinotarsa decemlineata*) during almost the entire twentieth century. When it was first found, in 1811, this pretty domed, striped creature started out as a curious but rather insignificant insect, feeding on buffalo-bur (*Solanum rostratum*), a member of the nightshade family (Solanaceae), in the Rocky Mountains. But in 1859 it was found destroying potato plants (the closely related *S. tuberosum*) in Nebraska hundreds of miles to the east. This was the start of a rapid eastward spread, and by the time it reached the Atlantic coast, in 1874, it was already an internationally notorious agricultural pest. The beetle's arrival in Europe in 1877 was greeted with dismay; the 5-hectare plot at Mülheim, in Germany, was quickly cordoned off by troops, and the whole area was doused with petrol and burned. Unfortunately, quarantine procedures were relaxed when Europe descended into the chaos of war between 1914 and 1918. The next invasion of Colorado beetles was discovered in Bordeaux in 1922, and was widely blamed on the vast import of goods and materials with the US troops who had arrived in Europe a few years before. The subsequent spread was inexorable, through central Europe, Turkey, Kazakhstan, China and Russia to Kamchatka. It was about this time that wanted posters of the beetle started to appear, much in the style of those produced to find fugitive criminals. Showing the brightly coloured beetles, and the equally brightly coloured larvae, devouring potato leaves, these leaflets were pinned up in town halls, post offices and other municipal buildings around Europe. I have a clear memory of seeing one on the noticeboard of the old police station in Newhaven (demolished in about 1970), when my father and I went down to hand over a quantity of live machine-gun ammunition we'd dug up in the garden – our house was occupied by Canadian troops in 1944, prior to the D-Day landings.

Several times, the Colorado beetle has been intercepted in shipments at British ports (e.g. Liverpool in 1877, the first recorded), but it has never become established here. The climate may not be quite right for it in the British Isles, but nobody should be complacent. Vigilance is all. And we should add other beetles to that vigilance. The emerald ash borer (*Agrilus planipennis*, Buprestidae) is another Far Eastern species creating forestry havoc in the United States, killing ash trees as it goes, and it's spreading west through Eurasia – it's already on the outskirts of Moscow. Several species of *Agrilus* new to Britain have been discovered in the last 20 years, arriving either through their own flight across from the Continent, or with shipped goods. This one could be here soon.

The palm weevil (*Rhynchophorus ferrugineus*) is a large, sleek, striking reddish-orange and black curculionid from tropical Asia. There is no way it can move around temperate Europe except when its ornamental host plants are trafficked. This is exactly how it got to California in 2010 (although it was later eradicated, and the area was declared all clear as of January 2015), and how it has been moving around the Mediterranean, through Spain, Italy and southern France. It turned up in Brittany in 2013. The fat vermiform larvae bore into the heart of the palm tree stem, causing damage to the growing zone, and allowing in fungal and other disease pathogens. There was speculation that the beetle had been deliberately encouraged in some marginal areas of its native range because the cooked grubs were considered a delicacy; this strikes me as highly dubious, and more likely to be foreigner-blaming propaganda to divert attention away from the reality of lax biocontrol. It was found in an imported palm in a garden centre in Essex in 2016; there was some light media coverage. It seems doubtful that there are enough palms growing in Britain to support a viable invasion, but these trees are a noticeable feature in sheltered southern and western gardens, so that continued vigilance is still necessary.

At the same time, we should not panic unduly. Buglife recently drew up a list of 18 potential invasive invertebrates to threaten UK biomes, and only one of them was an insect, and not a beetle at that (the Asian hornet, *Vespa velutina*); the rest were freshwater or brackish crustaceans and molluscs. Despite the evidence of thousands of years of woodworms in architecture and grain weevils in archaeology, beetles are still seen as not very threatening. With this in mind, the future of human–beetle interactions looks bright.

WILL WE STILL LOVE THEM TOMORROW?

Apart from the occasional arrival of some new garden nuisance (several Balkan relatives of the vine weevil *Otiorhynchus sulcatus* have recently arrived in the London area (Barclay, 2003)) or more dire agricultural bogey, beetles seem set to maintain their good relationship with humans. Indeed, there are plans for us to exploit them commercially.

In the past, beetle exploitation was limited to Victorian collectors (I hesitate to use the word entomologists), paying inordinate sums of money for large and showy specimens at the regular natural history auctions held throughout the nineteenth century. It was not unusual for wealthy patrons to stump up five guineas – something like an average monthly wage in the 1850s – for a choice Goliath beetle. In 2003, I worked at the Horniman Museum in Forest Hill, researching the Horniman beetle *Ceratorrhina* (now *Cyprolais*) *hornimani*, a large green, yellow and red cetoniine chafer. An immensely rich shipowner, tea tycoon and Member of Parliament for Penryn and Falmouth, Frederick John Horniman had procured specimens from missionary correspondents in the mountains of Cameroon, and the traveller, naturalist and scientist Henry Walter Bates (1877) named the species in his honour. It was not unusual for expats of the

OPPOSITE PAGE: **FIG 135.** New beetle species continue to arrive in Britain with differing and sometimes unpredictable impacts. *Otiorhynchus armadillo* (top) is one of several south-east European vine weevils to become established in the London area in the last few years; it has yet to spread and torment many gardeners. Chelsea Harbour, 18 June 2004. Likewise *Henosepilachna argus* (centre) is still more or less confined to Surrey, where it grazes the leaves of white bryony (*Bryonia dioica*), though theoretically a jump to closely related cucumber, squash, pumpkin and courgette plants is not impossible (like bryony they are all in the Cucurbitaceae). Molsey, Surrey, 14 November 1997. Meanwhile the Asian harlequin ladybird, *Harmonia axyridis* (bottom), has spread widely and wildly since its arrival in Britain in 2004, and has been painted as a 'baddy' for its aggressive invasion of gardens and buildings.

time (soldiers, civil servants, adventurers and missionaries) to trade the natural history specimens they found on their travels, shipping them back home to eager patrons or crowded auction houses. Popular instruction manuals abound (e.g. Lettsom, 1774; Brown, 1833; Taylor, 1876) on how best to preserve anything from windblown seeds to whale skeletons. Chief items of interest were always brightly coloured exotic birds, butterflies and, of course, beetles. I could find no accounts of payment made by Horniman to the missionary station on the Mongo-ma-Lobah (God's Mountain) peak of Mt Cameroon, but on visits to the auctions of J.C. Stevens (an important London agent for numerous foreign travellers, including Alfred Russel Wallace) in Covent Garden, Horniman spent hundreds of pounds on insect cabinets, books and beetle specimens.

Collected museum-style specimens of large beetles still change hands for money, and there is an international trade in them, about which some entomologists (myself included) feel slightly uncomfortable. I regularly get random emails from sellers in China and South-east Asia offering me large and attractive ground beetles, jewel beetles, rhinoceros beetles, stags and chafers, and there are plenty of high street shops selling glass-framed beetles along with other insects, fossils, seashells and coloured rocks. Japan seems to have an inexhaustible thirst for beetle specimens (some kept alive as pets), and although tales of extra-large stag beetles being offered for £100,000 may be marketing hype, large numbers of live or dead beetles are widely available there in shops, online or even from vending machines.

The *ma'kech* is not yet available in vending machines. In Mexico and Central America, these bespoke jewel-encrusted ironclad beetles (family Zopheridae) are sold alive and tethered by a gold chain to a pin so that the living brooch can be worn openly on clothing. *Megazopherus chilensis* is a relatively long-lived (several months) tenebrionoid beetle with a thick, tough exoskeleton on which are glued polished gems or coloured glass beads. The *ma'kech* harks back to a Yucatán myth about the forbidden lover of a Mayan princess; he was turned into a beetle by the moon goddess so that he could visit her undetected by her guards.

Using real beetles as ornament and decoration has a long tradition. In 1888, when actress Ellen Terry (1847–1928) played Lady Macbeth in a dress and gown decorated with 1,000 iridescent green buprestid elytra (probably a *Sternocera* species from South-east Asia), she wowed Oscar Wilde and the West End, and was immortalised by the artist John Singer Sargent – his portrait of her still hangs in Tate Britain. But she was continuing a practice, the origins of which are lost in time. Along with bird feathers, mammal teeth and horns, whalebone, tortoiseshell and polished coral, animal products have long been used to ornament objects, clothing and the human body. Beetle elytra are tough, resilient,

FIG 136. Ivydale Primary School's answer to biscuit beetles – beetle biscuits, a useful and fun way to learn about bilateral symmetry, pronotal dominance, elytral structure and the correct number of eyes possessed by most beetles.

brightly coloured and patterned, and easily sewn or glued into artistic designs. The oldest recorded use may be the seventh-century Tamamushi Shrine, in Nara, Japan, named after the *tamamushi* (jewel beetle) *Chrysochroa fulgidissima*, with which it was once highly decorated. After 1,300 years, the brightly metallic green, gold and red elytra have all fallen away, but contemporary accounts confirm the design and elytral decoration. Buprestid and other brightly metallic elytra were also widely used in textiles in Mughal India (sixteenth to nineteenth centuries), and as personal adornment in Papua New Guinea and Amazonia. The technique, called beetlewing, still continues, although it is most usually seen in 'ethnic' tourist shops and street bazaars. There are plenty of objects available to buy online, including earrings, necklaces, hats and clothing. If you want to go into business yourself, 500 beetle wings, ostensibly farmed in Thailand according to the necessary Convention on International Trade in Endangered Species (CITES) regulations, will cost about £65.

I don't anticipate a branch of Beetles-Я-Us to open in my neighbourhood any time soon, but I would not be surprised to see a rise in mealworm-breeding facilities. These smooth, pale, tubular larvae of a once common larder pest, *Tenebrio molitor*, are widely available from pet shops and specialist stores, where they are sold as food for pet lizards and wild and caged birds (including farmed chickens), and as fishing bait. They are also increasingly being eaten by people.

There is a good argument that conventional meat-rearing agriculture is horribly wasteful when it comes to land use, water requirements, energy input, carbon footprint and greenhouse-gas emissions. Kilogram for kilogram, insect protein is much easier and more efficient to produce. All you need are tubs of appropriate carbohydrate feed kept at suitable temperature and humidity, and within 8–10 weeks you have a crop animal ready to harvest. The trouble is that an overcultured, squeamish western palate may not be up to maggots on toast, devilled larvae or barbecued grubs. However, mealworm larvae, ground down into a powdery meal of their own, make a useful ingredient in processed foods. Quorn, based on microorganism mycoprotein, sets a useful precedent here, and various start-up companies have tried to use insect protein in a similar way to make protein blocks that can be used in a variety of foods. To date, there are mealworm energy bars, cookies, flapjacks, bolognese sauce, 'sausage' rolls and 'meat' balls. The larvae can be fried whole and eaten like crisps or salted like nuts. They can be dipped in chocolate, scattered like croutons on salad, or stuck through with a cocktail stick along with cheese and pineapple. Or the ground flour can be used to enrich whatever cake, bread or bun you are baking.

At the moment, mealworms are a niche human food – more than a gimmick, but still some way off from mainstream. Apart from prudish revulsion at the idea of eating insects, there is also widespread ignorance and confusion about how they might ethically be introduced into our diet, or indeed what type of animal they might be anyway. When I innocently asked my local specialist food store whether they stocked any mealworm bars, cricket snacks or other insect-protein products, they replied that, no, sadly it went against their pescatarian policy.

CHAPTER 9

Evolutionary History of Beetles

BEFORE BEETLES RULED

Beetle exoskeletons are tough. This partly explains why beetles have been so successful on Earth, but it also explains why beetle remains (fossil or otherwise) are relatively well known. Unfortunately, it is the lead-up to beetledom that is not so clear from the fossil record; certain assumptions, extrapolations and educated guesswork sometimes have to be employed too.

Six-legged terrestrial animals, the epiclass Hexapoda, may be a genuine single-ancestor grouping, although there is still plenty of debate here. Further back in time, the Arthropoda are accepted as a natural group, but the split of the Myriapoda (millipedes and centipedes) and six-legged hexapods is still clouded in mystery. The monumental work by Grimaldi & Engel (2005) goes into exquisite but also excruciating detail about the evolution of insects; there have been some groundbreaking discoveries since, but their work still stands as the authoritative fount of fossil and phylogenic knowledge, and I admit to having drunk deep of it. This is not the latest word, though, as data and dating are reanalysed and re-reanalysed. When McKenna *et al.* (2015) published a dated phylogeny using molecular similarities and differences between beetle lines, it took only a few months before their data was reinterpreted by Toussaint *et al.* (2016), pushing back all the evolutionary dates again. To the non-palaeontologist me, it seems incredible that we are still fumbling 100 million years here or there, but the realist me recognises that, given we know so little about extant beetles, it is no wonder we know even littler about their long-gone ancestors. There is also a difference between a dated fossil, written in stone, and a suggested time that a lineage may have arisen or diverged, which will inevitably be an extrapolation backwards to some earlier point, and with all the uncertainties of error margins and assumptions about rates of evolution. The dates offered here are based mainly on those given by Toussaint *et al.* (2016), with the proviso that they, too, are likely to be revised at any time.

Extant hexapods have previously all been called insects, but recent studies identify the Diplura (two-pronged bristletails),[20] Protura (no common name) and Collembola (springtails) as a breakaway faction, the Entognatha; everything else is an insect though, sometimes also called the Ectognatha. What may be borderline non-terrestrial Hexapoda fossils occur in the earliest Devonian beds, around 400 mya. One of these at least (*Wingertshellicus backesi*, also named *Devonohexapodus bocksbergensis*) appears to have had six longer legs near the head, but numerous abdominal segments, each with short limblets. It looked a bit like a cartoon cross between a millipede and a damselfly nymph (Haas *et al.*, 2003; Kühl & Rust, 2009). More abstruse defining characters for modern Hexapoda include the second maxillae fused into the labium, loss of an articulated endite segment at the tip of the mandible, loss of all those short supernumerary abdominal legs and, crucially, only one pair of antennae. Latest thinking appears to suggest that the Hexapoda arose sometime in the Cambrian (500 mya) with the Insecta starting out in the Ordovician around 470 mya (Edgecombe & Legg, 2014). The early fossil traces on which these deductions are based are fascinating and tantalising, but a bit far removed from our central analysis of beetles, which we need to get back to now.

The next significant evolutionary event in the history of beetles was the appearance of winged insects. Wings are thought to have evolved by the various possible developmental pathways discussed in Chapter 6. All insects, except the almost antediluvian three-pronged bristletails, Archaeognatha, and equally prehistoric-looking silverfish, Zygentoma, have wings, although some like fleas and lice have subsequently lost them. It is clear that wings evolved in insects only the once, but relationships in some of the basal groups are still politely disputed by insect phylogenists. Diaphanous and fragile, wings are less fossilisable than other parts of a proto-hexapod's body, and although later insect wing imprints are frequently found (the giant dragonflies are always a wonder), there are few that offer much insight into when wings first arose. This probably happened some time before the late Carboniferous, maybe in the Devonian 400 mya, but fossils

---

20 Including *Campodea*, which gave its name to the campodeiform beetle larvae described on p. 72.

are scant and estimates are vague. There have been suggestions that metabolically expensive flight was able to arise in the Carboniferous/Permian 350–250 mya because of the hyperoxic atmosphere – there was about 31 per cent oxygen in the air compared to 21 per cent today (Dudley, 1998) – a theory also used to explain the existence of giant insects (Harrison *et al.*, 2010). However, flight probably started during the non-hyperoxic Devonian. The jury is still out.

The oldest known fossil insect, *Rhyniognatha hirsti*, from Early Devonian chert near Rhynie, in Scotland, is represented only by small portions of the head capsule, but the clearly triangular jaws, with their dicondylic (double-pivoted) construction, firmly imply that it belonged to an as yet unidentified group of winged insects. This would push the origin of insect wings back 80 million years.

Thankfully for this discussion on the Coleoptera, we can skirt around these non-beetle difficulties and address the undisputed fact that wings had already appeared in the Devonian. There are two major types of wing construction, one of which launched the beetle-dominated arm of the vast phylogenic tree of species into the formidable complexity we see today. Archaically, the first insect wings were stiff, held out sideways when at rest; the dragonflies (Odonata) still do this. Alternatively, they were flipped up to meet vertically over the body, as in mayflies (Ephemeroptera). Most other 'modern' insect groups, together the infraclass Neoptera, are able to flex the wings backwards, to come to rest horizontally over the abdomen, often the front wings resting on top of the hind wings, flattened or tent-like, and at least the rear portion of the wings folded along a crease. This mechanism hinges, quite literally, on a minute chitinous plate, the third axillary sclerite, at the side of the mesothorax, where a tiny thoracic muscle pulls, collapsing the wing membrane along the claval fold, between the cubital and anal veins. This seemingly small adjustment to the wing-base construction had momentous effects on insect evolution. Stowing the flight wings at rest like this enables insects to crawl into sheltered roosting spaces. This is exactly what cockroaches (Dictyoptera), stoneflies (Plecoptera), wasps (Hymenoptera), moths (Lepidoptera), grasshoppers (Orthoptera) and many other insects do, and which beetles have taken to a high art form.

Before beetles crawl off into the undergrowth, there is one more major evolutionary step to consider – the larva. Beetles, along with flies, bees, wasps, butterflies and moths (the Holometabola), have a wingless grub, maggot or caterpillar stage. This contrasts with the Hemimetabola (all other insect orders), which simply develop from miniatures of the adult, the nymphs growing larger wing buds with each successive moult until the final, fully winged mature is complete. As considered in Chapter 3, having a secretive, wingless larva living and feeding wholly separately and differently from the adult insect offers the possibility of occupying different niches, avoiding competition between parent and offspring. It also helps the growing insect avoid predators and parasitoids, which can much more easily track down the mobile flying adults. True, aquatic nymphs for some hemimetabolous insects then turn into terrestrial adults (dragonflies, alderflies, stoneflies, etc.), but these are small non-diverse groups; only the Hemiptera among the Hemimetabola offer up any serious numbers of species, but even an estimated world fauna approaching 100,000 species is dwarfed by the massed ranks of the holometabolous hordes. Holometaboly is unquestionably a highly successful strategy, used by at least 85 per cent of all extant insect species.

In the Holometabola, the pupa (or chrysalis) stage allows the replacement of the larval form by an adult cuticle of entirely different form. Throughout its larval life, the immature insect's body houses pockets of specialised epidermal cells (called imaginal discs for their approximately round shape), which start to develop only as the end of larvahood approaches. During, and in the run-up to, metamorphosis in the pupa, these discs proliferate and differentiate; they invaginate, then expand and flatten into wings, or telescope into legs, antennae and other structures. Their repression throughout the larval stage is under the control of juvenile hormone (see p. 79), which, as its name suggests, keeps the larva in a non-adult form. It is more than mere analogy to suggest that the larva is an extended free-living embryo. Comparing holometabolous beetle larvae to the hemimetabolous nymphs of plant bugs (Hemiptera), it is usually suggested that the larva is an extended pronymph. This stage occurs in the Hemimetabola, usually between hatching and the first-instar nymph. The pronymph usually has shorter or no appendages and no wing buds, does not feed, often lasts only a few hours, and in many cases remains mainly or entirely within the egg. It is still an embryo.

By extrapolation from phylogenic analysis, the Holometabola evolved in the Carboniferous, perhaps 390 mya. However, if fossil evidence for wing evolution is sparse, fossil evidence for larval evolution is verging on the mythical. A putative caterpillar fossil, *Metabolarva bella*, from the mid-Carboniferous about 310 mya, appears to be the earliest known immature holometabolous insect (Nel *et al.*, 2013). This relieves some of the scrutiny previously applied to another possible larva, *Srokalarva bertei*, from the same period, which was later reinterpreted as possibly a myriapod.

The Coleoptera were the first significant branch of the Holometabola to arise and found a dynasty that still exists into the modern day. Their closest relatives are the lace-winged Neuropterida, represented today by the lacewings (Neuroptera), snakeflies (Rhaphidioptera) and alderflies (Megaloptera).

This unlikely band of delicate, net-winged, fluttering insects has, on closer examination, clear similarities to beetles, mostly in sharing similar traits in ovipositor and wing articulation. Fossil lacewings and pre-beetles were so similar that they continue to confuse our modern eyes. The broad-winged *Adiphlebia lacoana* was described as the earliest known fossil beetle, from Carboniferous deposits in Illinois, by Béthoux (2009), but later redefined as a member of the extinct neuropteran family Strephocladidae (Kukalová-Peck & Beutel, 2012).

Molecular studies consistently show that beetles and Neuropterida shared a common ancestor in the Carboniferous, something like 375–350 mya. General similarities are still visible today. Most books on insects portray lacewings and alderflies with their wings splayed out as museum specimens, looking rather similar to dragonflies, but in life they flex their wings back, tent-like over the abdomen (they have that third axillary sclerite to thank for this). Internally, the primitive cupedid *Priacma* has the greatest number of thoracic muscles known in beetles; this follows a more basic construction (superseded by a reduced number of muscles in most modern beetle lines), and similar to that found in Megaloptera. Neuropterida larvae also look like those of beetles, and like beetle larvae they have six-segmented legs. Lacewing larvae (Fig. 137) patrolling an aphid colony mingle with very similar-looking ladybird larvae, and when I first found a snakefly larva (Fig. 138) under a piece of loose bark, it took more than a double take to realise that it was not that of a staphylinid.

FIG 137. Larva of a lacewing, showing its close resemblance to a ladybird larva. (Penny Metal)

FIG 138. The surprisingly coleopteron-looking larva of a snakefly, Rhaphidioptera.

BEETLES ARRIVE ON THE SCENE

Contrasting with the scant fossilisation of standard insect wings, with their pale, evanescent, lacy traceries, hard beetle elytra suddenly offer a wealth of fossils. It's not all rosy though; when beetles die they usually become dismembered, broken and fragmented, so it is common just to find isolated elytra. These can be examined and described, and sometimes (if they are similar to modern lineages) they can be recognised, but often they remain enigmatic. Isolated from other body parts and telling us little about the extinct beetles' overall form, they give no understanding of what carly beetles looked like, how they might have lived or what they were related to. The venation that offers so many identification clues to winged insects, living and fossil, is obscured or non-existent in elytra, so even the best-preserved specimens can remain tantalisingly disappointing. Hundreds of beetle fossils are described and given species names purely on the basis of loose elytra, without any clear idea of their overall classification. Nevertheless, a picture is emerging.

The Archostemata are widely considered to be the basal suborder of beetles. There are only a few surviving species alive today (including the biologically complex *Micromalthus* – see p. 76), but the distinctive square-punctured elytra of modern Cupedidae (none in Britain) echo the very earliest beetle fossils known. The oldest confirmed beetle fossils, with clearly developed elytra, are the Tshekardocoleidae, from eastern Europe and central Asia (Kirejtshuk *et al.*, 2013). Comprising 12 genera from the early Permian (about 295–270 mya), they are sometimes described as proto-Coleoptera or (and I love this one) beetloids, and had coriaceous, probably not fully sclerotized, elytra, which were narrow and pointed, and extended well beyond the end of the abdomen. The striae of the elytra correspond to a nearly full complement of standard insect wing veins, and the deep puncture-like depressions are clearly reminiscent of the multitudinous cells created by the complex lattice of cross-veins found in the Neuropterida, particularly the Megaloptera, alderflies. Modern cupedids often have the square elytral punctures floored with a clear (or at least translucent) membrane. More complete fossils show a distinct broad, punctured pronotum, almost typical beetle head with protruding rounded, multifaceted compound eyes, and antennae of 13 segments. The Archostemata are more diverse in fossils than in living species, and by the Late Triassic (210 mya), 20 fossil genera are recorded in the Cupedidae compared to about 10 extant today. Modern Cupedidae feed, as larvae, in fungus and rotten wood, and the adults occur under logs and loose bark. This fits intuitively with the idea that beetle diversity expanded as a direct result of the elytra giving protection to the membranous flight wings when the first proto-Coleoptera pushed into tight galleries under rocks and logs, or into foliage and soil, in fungus and under fungoid bark.

The Triassic was the age of beetle diversification. At the start of this geological period (about 250 mya), beetle fossils are barely represented. By 240–220 mya, some 250 fossil species are described, in 20 families, including 'true' beetles with hard, veinless elytra. Beetles were also becoming more common, with Coleoptera specimens making up about 20 per cent of all insects in the deposits. Jurassic fossils (200–145 mya) now account for more than 600 species, in 35 families, and easily identifiable modern families start to appear. A good breakdown of beetle evolution is provided by Lawrence & Newton (1982).

The Geadephaga (ground beetles) and Hydradephaga (predatory water beetles) appear in the Late Triassic (230–220 mya). One of the most interesting groups to appear in this fossil cornucopia is the Trachypachidae. Eurasian and North and South American remnants of this group number just six species, but about 20 are known as fossils. They resemble carabid ground beetles in their streamed lines, and molecular evidence based on extant species backs this up, but they have large, immovable coxal plates like the Haliplidae. Although not quite a missing link, they could equally be regarded as a sister group to the Hydradephaga. The Myxophaga do not seem to appear at all in the fossil record, but these are tiny insects, with just a handful of modern species worldwide that are difficult enough to find even when alive today.

The Polyphaga, the remaining beetle families, make up the vast majority (90 per cent) of modern beetles, and much of their astonishing diversity is linked to their evolution and interaction with flowering plants, which were themselves undergoing explosive radiation during the Permian and Triassic. To date, the oldest definitive polyphagan is a tiny (2.1 mm) but obvious rove beetle, *Leehermania prorova*, with typical shortened staphylinid elytra and conical tail segments, from the Late Triassic about 220 mya (Fraser *et al.*, 1996; Chatzimanolis *et al.*, 2012). Other families follow in the fossil record, but a detailed list is beyond the scope of this book. Every new discovery upsets, or at least reschedules, the previous time frame, but there are a few key observations to make.

Early Jurassic scarabs lack sufficient detail to make certain assumptions about their habits, but there has been a lively debate about whether there were dinosaur dung beetles to deal with dinosaur dung. The oldest specimen of the Scarabaeiformia appears to be *Holcorobeus nigrimontanus*, from the Late Jurassic, about 152 mya; it has typical scarab burrowing legs, and although these were probably for digging into the soil or humus layer, it is not biologically clear that dung was ever shovelled. An easy assumption is that large herbivorous animals like triceratops, stegosaurus and diplodocus extruded healthy quantities of richly aromatic, fibrous dung for a dung beetle fauna similar to that found in today's grazing meadows. But mammalian droppings (in which the vast majority of true dung beetles feed) are distinctly different from those of reptiles, birds or dinosaurs. In mammals, dung and urine are ejected separately, from separate orifices, at different times. Most other animals, however, combine gut contents ('dung') and kidney filtrate ('urine'), and store them together in a cloaca before passing them together. This accounts for the piebald splashes from birds, the high uric acid content of which is wholly unattractive to most dung beetles. Herbivorous dinosaurs were also mostly gorging on gymnosperms, the tough cycads and conifers that dominated at the time; this would have produced very different dung from that produced by today's angiosperm grazers. There is a good argument that modern dung beetles evolved only in response to the rise in modern herbivorous mammals, after the extinction of the dinosaurs (Arillo & Ortuno, 2008), and the evolution of scarab lines does seem to track the sequential rise of angiosperms and mammals (Ahrens *et al.*, 2014). Early scarabs were probably feeding in that catch-all substrate 'decaying organic matter' – mostly rotting plant litter in the humus layer – as do many modern Geotrupidae. Maybe it was dung flies that cleared up the sauropod scats.

The Phytophaga, together the longhorns (Cerambycidae), leaf beetles (Chrysomelidae etc.), and weevils (Curculionidae etc.), now constitute the largest lineage of plant-feeding animals after the Lepidoptera; almost all feed on living plants, 99 per cent on angiosperms (see below). The earliest known putative longhorn is possibly *Cerambycomima* from the Late Jurassic, 152 mya; its antennae are very long, with typical cerambycid antennomeres, but it is also very chrysomelid-like and has been interpreted as belonging to both families by different researchers. There appears to be a hiatus (at least in beetle fossils) during the Cretaceous (145–66 mya), but cerambycids are common in the Eocene (from 55 mya), particularly in Baltic amber – they may be preferentially trapped in the resin of the very trees into which they were boring. Chrysomelid leaf beetles follow a similar Late Cretaceous radiation with angiosperms.

Curculionids are well represented in the fossil record, but this may be because the rostrum snout makes otherwise slightly ambiguous compression fossils more easily identifiable. The Nemonychidae (just one rare British species) are usually considered to be the most primitive of the weevils; they mostly feed on the pollen of araucarian conifers in South America, Australia and New Zealand. The earliest known weevil fossils, dating from Late Jurassic deposits about 150 mya, are nemonychid-like.

Most modern beetle families appear in the next 100 million years, and by the time the crisp, clear fossils in amber start to appear about 130 mya, similarities with modern species are very close indeed. Some beetle fossils are known only from amber: Cantharidae in Baltic amber from the Eocene 40–50 mya, Bostrichiformia from Mesozoic amber 90–100 mya, and Lymexylidae from Burmese amber, also 100 mya. Care needs to be exercised in interpreting these finds. The Cantharidae may indeed be a young lineage, but their notoriously soft bodies may mean that earlier, non-amber fossils are just non-existent because fossilisation was such a rare event, not that the beetles themselves were non-existent. And it is not surprising that timber-boring beetles like bostrichids and lymexylids may be preferentially preserved in amber, since the resin-exuding pine-like trees producing the amber may well have been responding to attack from these very same insects. The commonest beetles in Dominican amber are from the weevil subfamily Platypodinae (Ross, 1998), pinhole borers well known for burrowing directly, perpendicularly, into living tree bark.

The remarkable thing about insects entombed in amber is the clarity and detail of the trapped animal. As well as offering a clear three-dimensional image, microscopic details like hairs, bristles and scales, as well as integument colour patterns, are available for examination. Soft bodies are well preserved, including larvae and triungulins. Delicate features like the multi-flabellate antennae of Rhipiphoridae and extruded ovipositors or half-expanded flight wings confirm similarities to modern groups. Behavioural artefacts are also preserved, such as sawdust excavated by those platypodids, still-coupled mating pairs or the self-constructed frass container of a *Cryptocephalus* 'pot beetle' larva.

By the Quaternary, 1.6–1.7 mya and especially in the last 500,000 years, the preserved fossil or subfossil Coleoptera remains are virtually identical to the modern fauna. Many modern species are obviously more than a million years old.

BEETLE–PLANT RADIATION

My sampling method of choice for beetles is to thrash around in the herbage with a sweep net. This finds flower visitors aplenty, but also seeks out the rich vein of plant-feeders among the hyperdiverse longhorns (Cerambycidae), leaf beetles (Chrysomelidae etc.) and weevils (Curculionidae etc.) – together the Phytophaga. These three groups of beetles have a rich history of close study; not only are they often distinctively and attractively marked, but their plant associations mean that many of the common species have come to the attention of farmers, foresters and gardeners who regard them as destructive pests. Monoculture farming is perhaps relatively pest-free nowadays because of pesticide spraying, but textbooks from the late nineteenth and early twentieth centuries offer the mustard leaf beetle (*Phaedon armoraciae*), the turnip flea beetle (*Phyllotreta nemorum*) and the bean weevil (*Sitona lineatus*) as particularly troublesome pests (Ormerod, 1881). And, of course, so many of the most devastating forestry pests are wood-boring bark beetles in the weevil subfamily Scolytinae – species like the great spruce bark beetle (*Dendroctonus micans*) and the famous elm bark beetles *Scolytus scolytus* (*Scolytus destructor* is one of its earlier names) and *Scolytus multistriatus*. Conversely, scarce species can be sought out, monitored and studied, by finding the marginally less evasive foodplants. The tiny (1.7–2.2 mm) but distinctive orthocerous weevil *Kalcapion semivittatum* was once regarded as very local in Britain, but seeking out its foodplant, annual mercury (*Mercurialis annua*), on brownfield sites has proved it to be quite widespread in south-east England. Finding the uncommon reed beetle *Donacia crassipes* at Wicken Fen was just a question of looking for its foodplant, *Nuphar* water lilies, on the floating pads of which it suns itself. There really was no need for my colleague Tony Drane to strip down to his Y-fronts and slip into the water after it; we just needed to look for the plant growing in a narrower ditch – simple.

Knowing the plant from which a beetle has been knocked is often also a useful help in identifying it. The ubiquitous *Trichosirocalus troglodytes* occurs everywhere on ribwort plantain (*Plantago lanceolata*), but its scarce congener *T. thalhammeri* (which can really only be separated by genitalia dissection), occurs sporadically around the coasts of Britain on the saltmarsh equivalent, sea plantain (*P. maritima*).

Today, the world Phytophaga, representing about 135,000 species (80 per cent of all herbivorous beetles and 50 per cent of all herbivorous insects), mostly feed on angiosperm flowering plants. However, it seems likely that plant-feeding arose early, about 50 million years after the appearance of beetles and before angiosperms themselves had evolved, and that modern beetle lineages that later evolved to feed on flowering plants were in already in place then. Permian fossils are entirely saprophagous Archostemata (Cupedidae, Tshekardocoleidae, etc.), but by the Middle Jurassic, herbivores made up about half of beetle fossil species, and by the Tertiary they overshadowed all the other guilds entirely.

DNA studies on modern beetles (Farrell, 1998) show a nice link to the parallel evolution of angiosperms. In the three phytophagous lines (longhorns, leaf beetles and weevils), the basal lineages, which can be usefully regarded as near approximations of the ancestral forms, are all associated with conifer or cycad gymnosperms, particularly the southern temperate relic floras of South America, South Africa, Australia and New Zealand that evolved following the break-up of Gondwana about 180 mya. In the longhorns, the tribe Asemini in subfamily Spondylidinae remain conifer-feeding; in Britain, six species – including the handsome *Asemum striatum* and two large *Arhopalus* species – are widespread but local under the bark of dead standing pine trees and stumps. Among the leaf beetles, most worldwide Orsodacninae are associated with gymnosperms, although the precise life histories of our two uncommon British species have not yet been determined. In the weevils, nemonychids are conifer- and cycad-feeders, and Britain's one very rare species, *Cimberis attelaboides*, is known only from scattered records on conifer trees. The implication is clear that these basal groups continued to feed on surviving gymnosperms, whereas the rest of their relatives went off to spread widely and wildly among the burgeoning angiosperms.

The numbers are inordinately different, with a current world total of basal gymnosperm-feeding Phytophaga hovering around the 225 species mark (Farrell, 1998), while the remainder of this great beetle grouping exceed 135,000 species, a 600-fold proliferation. Niche expansion has also occurred. Most of those basal phytophagous beetles feed on the male cones, pollen-bearing strobili to give them their technical name, whereas angiosperm feeders have developed beyond feeding in these nutrient-rich structures to eating foliage. The Phytophaga now encompass a vast range of strategies, from free-living caterpillar-like larvae eating the leaves, buds, flowers and stems, or becoming internal leaf miners, to seed-feeders, fruit-eaters, bark-excavators, twig-borers or root-burrowers. The clincher, if one were needed, is the group of leaf beetles, including the lily beetle (*Lilioceris lilii*) and reed beetles (*Donacia*); feeding on monocotyledons, these beetles appeared later, in the Palaeocene, with the rise of this relatively younger (120–100 mya) group of plants.

BEETLES' CLOSEST RELATIVES?

It is one of my greatest frustrations and regrets that I have never found a live specimen of the Strepsiptera. This peculiar order of insects has, in the past, been regarded as being closely related to Lepidoptera, Hemiptera and/or Hymenoptera, but for the last 125 years has traditionally been considered a sister group to the Coleoptera. Fowler (1891) included it under the suitably Victorian title 'abnormal Coleoptera' at the very end of his five-volume *magnum opus*. Sadly, the Strepsiptera are no longer considered beetles, but no book on beetles would be complete without them, so I've decided to slip them in here.

The Strepsiptera (in Britain comprising four families, seven genera and 10 species) are minute (1–3 mm long), highly sexually dimorphic insect parasites that may have diverged from the Coleoptera 300 mya or more (Toussaint *et al.*, 2016). Males are winged and free-living, while the females are wingless, larviform and viviparous, remaining permanent internal parasites inside the abdomens of their hosts. The host range is astonishing, including 35 families in seven insect orders – quite remarkable for an insect order of only 600 species worldwide – although in Britain species of *Elenchus* and *Halictophagus* attack only leafhoppers, while those of *Halictoxenos*, *Hylechthrus*, *Stylops*, *Paraxenos* and *Pseudoxenos* parasitise solitary bees.

Their claim to being beetles (or at least near) rests on the fact that, like beetles, the winged males have broad membranous flight wings arising from the third thoracic segment (hind-wing dominance in entomological parlance), and that ahead of these, on the second segment, are hard club-like flaps, analogous to elytra. These flaps are not flat, but buckled, and have given the group the sometime name twisted-wing flies. Also like beetles, the prothorax is free (although the upper carapace is nowhere near as developed as a beetle's pronotum), the underside abdominal sternites are more heavily sclerotised than the upper tergites, the antennae have a reduced number of segments (eight or fewer) that are strongly bifurcate or flabellate, and the mandibles of the pupa are immobile (they move in the Neuropterida). Adult males have bulbous, protruding eyes with large facets (often likened to raspberries), and look very like miniature males of the non-British beetle family Rhipiphoridae (next to which they were sometimes classified). The females, on the other hand, are very unbeetle-like, being little more than elongate bags of eggs. This is quite literally what they become, since during the final larval instar the ovaries autolyse and spill the eggs into the female's haemocoele (body cavity). The eggs, numbering 1,000–750,000 (an extraordinary quantity for so tiny an animal), hatch in the haemocoele, and are released live as microscopic triungulins. The triungulins also closely resemble those of rhipiphorids.

Unfortunately, there are also some non-beetle similarities. Male wings have only long, radiating veins; there are no cross-veins at all. Strepsipteran elytral knobs closely resemble the halteres (reduced hind wings) of flies; they may act like them, too, flapping out of phase with the flight wings, and possibly having some counterbalancing or motion-monitoring function. One suggestion is that the wings and 'elytra' of strepsipterans are truly homologous to the wings and halteres of flies, but that the thoracic segments from which they emerge have become switched by the action of a homoeotic gene, a not uncommon development in experimental insect models like *Drosophila* fruit flies, where four-winged or legs-for-antennae mutants can occur. In some species, the male strepsipteran pupariates inside the inflated and hardened skin of the final larval instar; this puparium (as opposed to a pupa or conventional chrysalis) is another feature shared with flies. There have even been suggestions that the Strepsiptera are not true holometabolans, since the larvae appear to have compound eyes (rather than mere stemmata), vestigial wing bugs appear in the second-instar stage of males, and the internal organs of the larvae are largely conserved on the change to adult. The Strepsiptera are very odd indeed. An interesting and useful review of the order is given by Kathirithamby (1989).

DNA studies of Strepsiptera are not always helpful either. Concentrating on one small gene (Strepsiptera have small genomes anyway), results have been equivocal, and studies linking them to the Diptera have been dismissed because large amounts of base substitution and insertions were regarded as convergent evolution – accidental artefacts rather than evidence for genuine shared heritage (Boussau *et al.* 2014). On the other hand, some DNA results seem to confirm the Strepsiptera as holometabolans. The fossil record of Strepsiptera is sparse – hardly surprising for such tiny insects – but some obvious male adults are known, delicately preserved in Cretaceous Burmese amber from 100–75 mya, as are some possible triungulins.

One of the reasons Strepsipterans are so difficult to classify is that their peculiar life history is almost unique among insects. While internal parasitoids (which kill their hosts) are widespread among various insect orders (although some might be termed obligate predators), the only other internal parasites are various bot flies (Diptera, family Oestridae) living under the skin or in the intestinal tracts of mammals. Like the bots, strepsipterans do not generally kill their hosts, although they may reduce their fitness by damaging them and neutering them. The mode of parasitism is also unique. Having located a suitable host larva (Hymenoptera) or nymph (Homoptera), the triungulin secretes enzymes that start to digest the host's integument, probably as the host moults and sclerotisation is at its least. As the larva digests its way in, the host cuticle and epidermis separate, and a sac of the host's skin (almost like scar tissue) forms around the strepsipteran larva, isolating it and avoiding all the usual antigen immune reactions. Eventually, this bag pinches off to contain the larva within it, thus the host cannot recognise that it even has a parasite because it feels as if the attacker is on the 'outside' of its integument, even though it is buried inside the animal's abdomen. This strategy probably explains why Strepsiptera have targeted so wide a range of host orders – they have not had to overcome a different set of physiological immune responses each time. Perhaps the acme of Strepsipteran oddity occurs in the pantropical Myrmecolacidae. In this group, males parasitise ants and females parasitise mantids, grasshoppers or crickets. Correlating males and females of the same species has been impossible, except for one case where DNA matching was able marry them up. Yes, the twisted-wing flies are very peculiar indeed.

The name Strepsiptera was coined by renowned entomologist the Reverend William Kirby (1759–1850) in 1813, when he realised the affinity of *Xenos vesparum* (described by Peter Rossi in 1793, and initially placed in the Hymenoptera, near the Ichneumonidae) and his own discovery (in 1802) of *Stylops melittae* (Fig. 139). Later, William Elford Leach (1791–1836) described *Stylops kirbii* in his honour. At the creation of the Entomological Society of London (later the Royal Entomological Society) in 1833, Kirby was made an honorary life president and *S. kirbii* was adopted as the society's official symbol. The journal *Stylops* was published from 1932, but was later incorporated into the society's proceedings, now *Systematic Entomology*. Images of the insect have appeared on the society's letterhead, on its publications, and on a grand dais, desk and presidential chair, presented to the society by one of its wealthy members, Robert Wylie Lloyd, in 1933. Sadly, *S. kirbii* (and several other extremely similar species) was subsequently synonymised with what later turned out to be a rather variable *S. melittae*, but the society still uses the logo.

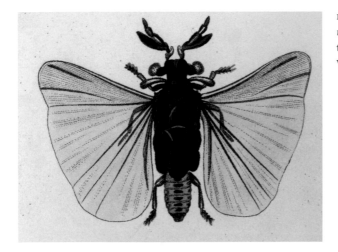

FIG 139. *Stylops melittae*, male, from the frontispiece of Westwood (1839).

Apart from sitting in the society's presidential throne (although I was never president of that august body), the closest I ever got to a Strepsipteran was on my first-year university field trip to Ashdown Forest (in the summer of 1977), when we undergraduates sweep-netted our way across the heather to collect insects for calculating diversity indices. One of my fellow students found a tiny winged creature that defied identification until someone (although not me) recognised it as probably *Stylops melittae*. I remember peering at it briefly down the microscope, but regrettably its significance as a bizarre evolutionary sister-group almost-beetloid was lost on me at the time.

JUST HOW MANY BEETLES (OR INDEED INSECTS) ARE THERE?

There runs an important aphorism about beetle diversity, which I originally heard from Peter Hammond when he gave a lecture to the British Entomological and Natural History Society in 1990. If an alien civilisation visited Earth and wanted to find out how life on this planet worked, with limited time and limited resources all they need do is study beetles, and dismiss everything else as mere sampling error. It has long been known that evolution has favoured beetles – they outnumber almost everything else here in terms of species, and coleopteran diversity in body structure, size, life history, ecological niche, specialist physiology and behavioural quirkiness is second to none. Much of this diversity is down to subjective interpretation, and I know plenty of dipterists who would argue that it is their favourite insects – the flies – that are really the most diverse in their life histories and behaviours. Species numbers, though, are at least quantifiable... or are they?

Traditionally, the total number of species of any given insect order has relied on estimates garnered from the numbers of species labelled and stored in the world's museums, usually backed up by descriptions published in the scientific literature of the last 250 years. It's amusing to read some of the early authors' suggestions that there may be an amazing 200 species of beetle in God's creation. John Ray (1691) states: 'The butterflies and beetles are such numerous tribes, that I believe in our own native country alone the species of each kind may amount to 150 or more'. Jan Swammerdam (1637–80) died before he could publish his research; it eventually saw the light of day in 1737–38, translated into the Latin *Biblia Naturae* and with an English translation (*Book of Nature*) in 1758. He reckoned on 203 beetles: 'nine of the largeſt kinds, twenty-one of the middling, thirty-ſeven of a ſmaller, and one hundred and thirty-ſix of the leaſt kind'. Those were certainly ſimpler times. Ironically, that same year, 1758, marked the 10th edition of Carl Linnaeus's *Systema Naturae*, in which the start of the modern system of binomial (genus and species) scientific names is decreed by international agreement. Linnaeus gave 654 beetles; his pupil Johan Christian Fabricius (1775) upped this to 4,112, and so started the grand scheme of finding and describing roughly four beetle species a day ever since. The first attempt at a world catalogue, by Gemminger & von Harold (1868–76), listed 77,000 species; that of Junk & Schenkling (1910–40) 221,500 species. We've never looked back.

Towards the end of the twentieth century, a consensus had started to appear that something in the order of 1 million arthropod species had been identified, described and named,[21] and that 40 per cent of them were beetles. For British coleopterists, a world beetle fauna of 400,000 species seemed an immeasurably bewildering number, especially since they were struggling to make sense of our 3,500 or so species. But if that were a staggering number, coleopterists (indeed biologists in general) were about to be knocked sideways by the next calculation.

In tropical Panama, Tony Erwin (1982) was using insecticide fogging machines hauled up high into the rainforest canopy, and retrieving the huge knock-down in collecting bins and on sheets spread out across the forest floor below. From one tree species (*Luehea seemannii*), he sorted 1,200 beetle species (including some new to science) into trophic groups and theorised that proportions of these

---

21 Everything else – plants, fungi and all other animals together – added maybe another half million species.

might be host-specific to that particular tree – he reasonably suggested 20 per cent of the herbivores, 5 per cent of the predators, 10 per cent of fungivores and 5 per cent of scavengers. This gave 162 beetle species specific to one tropical tree species. Given that there are 50,000 known species of tropical tree around the globe, and that beetles make up 40 per cent of all arthropods, and that there are twice as many species in the canopy as on the forest floor, a simple calculation gives a worldwide tropical rainforest arthropod fauna of around 30 million species (that would give 12.2 million beetle species). This was a truly frightening number, especially considering that at current rates most of these species would be extinct anyway before they could ever be found.

A subsequent remodelling using slightly different assumptions (Stork, 1988) gave a range of possible totals from 7.3 million to 81.4 million arthropod species (that would translate as 3–33 million beetles). This was crazy – rather than getting nearer to an answer, the numbers were getting wilder; biologists could not agree, even to within an order of magnitude, how many creatures they shared the planet with. The last 30 years have generated more calculation (Stork, 1993), estimation (Hammond, 1992), extrapolation (Colwell & Coddington, 1994) and random guesswork (Gaston, 1991). My apologies for a cheap joke – that last one is actually entirely unfair. Gaston (1991) estimated 5 million insect species based upon guesstimate responses from expert taxonomists, and although this was later ridiculed as unscientific, it was these self-same experts who were describing and ordering the species, and deciding what constituted a species anyway, so their combined judgements were just as valid as any supposedly more objective number-crunching method.

One of the latest reviews of several different methods (Stork et al., 2015) tries to get a bit closer to some overall agreement, and offers a measured (and pretty convincing, I think) exploration of the various previous reports and analyses. As well as looking again at other people's estimates, the authors also put forward some calculations of their own. Looking at the ratio of butterfly species to the rest of the British insect fauna (67:24,043), and then scaling up using estimates of 15,000–20,000 world butterflies, gave 5.4–7.2 million insect species (of which 1.0–1.2 million might be beetles). Another extrapolation was to consider the regional distribution of plants: if 2.1 per cent of the world's plant species occur in North America excluding Mexico, the continent's beetle fauna of 25,160 (anticipated to reach 28,000 when including undescribed taxa) (Marske & Ivie, 2003) could likewise be scaled up to 1.2–1.3 million species worldwide. These seem slightly less excessive numbers than those presented in Erwin's bombshell calculations. But we haven't finished yet.

For the British coleopterist, one measure produced by Stork *et al.* (2015) is of particular interest, because it relies on extrapolating up from our own relatively very well known beetles to give a measure of the still very hazy world beetle fauna. By recording the size of each British beetle species and noting the year in which it was originally named, it is possible to demonstrate that large, obvious species were described long ago (right back to Linnaeus's 1758, 10th-edition starting point of modern scientific names). As time went on, smaller, more secretive and more easily overlooked species were discovered and named. In effect, over time, the cumulative mean species body size of the British beetle fauna has reduced from about 7.0 mm (1758) to 3.5 mm (2011), and it does this in a fairly neatly descending line/curve as plotted on a logarithmic scale. The next stage was to take a random sample of 2,652 from the 179,649 named worldwide beetle species in London's Natural History Museum (thought to be about half the known world species) and find their mean size. It had to be a sample, because examining all 8,891 drawers of the museum's beetles, housed across 972 cabinets, would have taken an eternity. The sample was taken, and at 6.98 mm (and using 95 per cent confidence intervals) this corresponded to the mean size of British beetles known between 1758 and 1762, when 352 and 437 species had been recorded, respectively. By extrapolating up to 2011, when the British list stood at 4,069, the corresponding world beetle tally would be 1.7–2.1 million species – again, not quite Erwin's 12.2 million, but still likely to be four or five times our current running total.

This was only one of eight assessments examined, but was within the final suggested range of 2.6–7.8 (mean 5.5) million insect species worldwide, or 1.0–3.1 (mean 2.2) million beetles calculated overall. Arguments continue, but it is abundantly clear that there are an awful lot of beetles out there still waiting to be discovered. This means quite a lot of work for coleopterists (and entomologists), finding, describing and cataloguing them, for the next several hundred years.

CHAPTER 10

The Origins of British Beetles

WHY SO FEW BEETLE SPECIES IN BRITAIN?

Anyone who looks at beetles when they go on holiday to Europe, even after a short hop over the English Channel, is immediately struck by the far greater number and diversity of species over there. Estimates are thin on the ground because most European faunal studies are not based on sovereign nation states, but on overlapping geographic zones. However, Fairmaire (1902) reckoned on the possibility of 10,000 beetle species in France, compared to about 3,300 in Britain at the time (Sharp, 1909). When the monumental *Die Käfer Mitteleuropas* began publication 50 years ago, there were thought to be at least 8,000 beetle species (Freude *et al.*, 1965) in Germanic central Europe, compared to then 3,690 in the British Isles (Kloet & Hincks, 1945), although this had risen to nearly 11,000 by the time a full catalogue was published by Lucht (1987). Despite our green and pleasant land, the British Isles obviously have a seriously depauperate beetle fauna. We have the last ice age to blame for this.

At its maximum, something like 22,000 years ago, the Weichselian ice sheet covered almost all of northern Europe, completely engulfing Scandinavia, and covering significant parts of Poland, Germany, the Baltic states and northern Russia. In Britain, it covered Scotland and most of Wales, and extended down England to about the Humber–Severn line, and in Ireland it reached as far as a line from Dublin to Limerick. In places, the ice was 4,800 metres thick, and all life beneath it was squeezed out. Judging from similar ice-capped landscapes today, the rest of England, and extending well into mainland Europe, was a frigid arctic tundra – treeless, exposed, inhospitable. There would have been some plant and insect life, but nothing like the range that our modern balmy climate supports today.

As a consequence of so much of the world's water being solid mountainous ice, rather than liquid water flowing into the oceans, sea-levels were correspondingly much lower than today, sometimes 130 metres below their current station, exposing much of the continental shelf as dry land. What is now the North Sea was the open country of Doggerland, Ireland was joined to Britain by a continuous lowland (sometimes called Celticland), and the English Channel was merely a large river valley, the Fleuve Manche, stretching west to a vast estuary emptying half of northern Europe's rain and meltwater into the North Atlantic well beyond Brittany. It's difficult to picture exactly what the retreat of the ice might have been like, but retreat it did, starting about 13,000 years ago. And as it melted, the landscape changed, and with it the fauna. Beetles, by virtue of their preserved tough exoskeletons, provide a lot of clues, not just of their invasion and colonisation of post-glacial Britain, but of the landscape, flora and climate as these also changed.

William Sharp (1899) was one of the first to speculate on how Britain's modern coleopterous fauna was influenced by our previous glacial history. Some of his conjectures are interesting, if a bit dated. He rather simplistically tried to identify which segments of the beetle fauna either came north through France or from the east through Scandinavia. He recognised the importance of northern, montane beetles like *Dytiscus lapponicus*, *Pelophila borealis*, *Carabus glabratus* and *Miscodera arctica* in the puzzle. None of these now lives in lowland southern, eastern or central England, so how could they have invaded as the ice retreated and yet no longer occur there? Sharp's model of invasion seems to be as a gradually soaking stain of ink spreads across a large sheet of blotting paper.

Beirne (1952) and then Hammond (1974), armed with more geological, biogeographical and ecological knowledge, summarise a whole host of recent research into modern distributions and subfossil remains, and this has been updated and added to in the review by Kenward & Whitehouse (2010), from which much of the following discussion is taken. Several features emerge as the dominant ecological shapers. Importantly, even during the greatest snowy extent, much of the south of England and southern Ireland were ice-free, and supported a cold-adapted beetle fauna. The edge of the ice cap was not marked by a solid, immutable line of impenetrable permafrost; the ice edge came and went, oscillating over the decades and over the centuries, and just as in modern times, summer/winter fluctuations allowed some leeway for things to survive. There were always sheltered spots, especially near the warming Atlantic coast, and species were able to hibernate for many months of cold and snow cover. As the ice retreated northwards, so too did these tundra species. They advanced, colonising newly opened ground, but also disappeared behind, either because they could not cope with the increasingly clement climate, or because they were unable to compete with incoming competitive species.

Some beetles, like those mentioned by Sharp, were able to keep a toehold in Britain, but only in cooler northern latitudes and/or up cooler northern and western mountainsides. Lindroth (1935) identified 15 carabid ground beetles that were very clearly relics of a migrated community of tundraesque fauna; they showed similarities to the historical ground beetle fauna moving north through Scandinavia during the same thaw period. The rare montane carabid *Curtonotus* (*Amara*) *alpinus* occurred in north-east Italy 18,870 years ago, but is now limited to the mountains of Scotland and northern Scandinavia (Foddai & Minelli, 1994). Generalist predatory ground-dwelling beetles, it seems, have the advantage of pursuing a variety of prey, and are highly mobile and very adventurous. They cope well in areas of poor plant cover and bare ground, and are often among the first species to arrive at abandoned quarries, recent landslips, freshly dug pond edges and derelict brownfields. Plant-feeders, on the other hand, are less quick to colonise new territory – they have to rely on the plants migrating first. There is no point in being too adventurous if there is no foodplant available at a new location, so better instead to stay put and make use of the available on-site nutrition. So while some beetle species were able to move as the climate changed, others disappeared from Britain. This is perhaps a scenario that might come back to haunt us as modern climate change sweeps in.

The latest checklist of British beetles includes a supplement of more than 200 species known only from fossil or subfossil remains here (Buckland & Buckland, 2017). Forty-eight of these occurred in Britain during that last glaciation (115,000–15,000 years ago), but are now extinct. These include species like the ground beetles *Pterostichus middendorffii* (now mainly known from Russia), *Amara interstitialis* (Scandinavia, Siberia, Kamchatka, Alaska and Canada) and *Carabus cancellatus* (northern Europe and Siberia), and water beetles like the obviously named *Hydrobius arcticus* (Scandinavia, Siberia and Canada) and *Helophorus glacialis* (central and northern Europe). That they survive on mainland Europe (even some way south), but not in the British Isles, is partly a function of continental weather patterns: modern Europe is much colder in the height of winter than the British Isles, because our mild, damp, oceanic climate is fuelled by relatively warm sou'westerlies blowing up from the hot-water-bottle heat store of the Atlantic Ocean. Europe is also more mountainous, allowing cold-adapted beetles to retreat uphill to their comfort zones near the snowline.

Meanwhile, in glacial Britain, tundra was not as barren as it is sometimes portrayed in caveman films. Summers, though short and cool, would show a flush of plant growth, with flowers and seeds produced. Ground-feeding weevils were present; the likes of *Cathormiocerus curviscapus* (now limited to Iberia), *Otiorhynchus mandibularis* (eastern Europe) and *O. politus* (Scandinavia, Siberia and eastern Europe) were here, but are now extinct in Britain. This was also a time of mammoths, woolly rhinoceroses and giant elk, along with other surviving northern mammals like reindeer, bears and wolves. Numerous dung beetles known during the last glaciation no longer occur in north-west Europe, including *Aphodius holdereri*, once obviously widespread in Britain (Coope, 1973) but now confined to central Asia, 3,000–5,000 metres up in the foothills of the northern side of the Tibetan Himalayas, where it presumably feeds in yak dung.

For a long time Britain was physically attached to Europe by dry land, although much of low-lying Doggerland was marshy, and the Fleuve Manche would have been a major barrier. Nevertheless, plants and animals were able to migrate north and west into Britain as the ice blanket retreated northwards. Hammond (1974) suggested that perhaps half of Britain's modern-day beetle fauna arrived during this time. With the removal northwards of the harsh ice cap, the period known as the Younger Dryas, about 13,000–11,000 years ago, was one of patchy open scrub woodland. As the thaw line moved north, the tundra receded with it, and woodland cover appeared – first the pioneer species birch, aspen, hazel and pine. The birch bark beetle (*Scolytus ratzeburgi*) seems to have been one of the first to arrive (Whitehouse, 2006). Slower-spreading trees followed – small-leaved lime (*Tilia cordata*), elms and oaks – and deep mixed deciduous forest eventually developed. This is backed up in deep-drilled peat core samples by pollen analysis. As open-country plants colonised, so too did plant-feeders like weevils and leaf beetles. True warmth-loving beetles now started to spread north from refugia in southern Europe, with species like the heath and moorland ground beetle *Carabus arvensis* and green tiger beetle (*Cicindela campestris*) appearing at this time.

Although an infinity of variations is possible, there are three main themes of beetle distribution in the British Isles today, and they reflect the species' origins during and after the last ice age, and how they have been able to invade as ice retreated, temperature increased and modern temperate habitats developed. A few, like those ground beetles discussed by Lindroth (1935), or Caledonian pine forest specialists like the timberman longhorn (*Acanthocinus aedilis*), have their home range in Scotland or northern England. These are presumed to be relict tundra species or early scrubland Dryas colonisers, following the cool zone as it moved north, but disappearing from the warmer zone as it developed in the south of the country.[22]

A small number of British beetles have a skewed south-western distribution, hugging the Atlantic seaboard – species like the large, striking black-and-yellow longhorn *Leptura aurulenta*, which reaches from Gower around the West Country and just to Arundel in West Sussex, but which I never saw in East Sussex when I lived there. The golden leaf beetle *Chrysolina banksi* occurs infrequently in the Thames Estuary, but its main populations are west, from the Isle of Man to West Sussex, and around much of coastal Ireland. These types of western distribution are thought to be the consequence of an insect colonising from the south-west, from the previously sheltered, moist, mild Celticland between Britain and France, to Brittany and the Bay of Biscay, a region that has since sunk beneath the English Channel, Irish Sea and advancing Atlantic Ocean.

A large number of species have their heartland in the south-east of England. These make up the majority of the colonising invaders from mainland Europe, right up to the point where the English Channel finally closed off the land bridge. They have marched north and west until some geological or climatic limit has been reached. Some, like the tiny weevil *Kalcapion semivitattum*, have barely made any headway beyond Kent, although its foodplant, annual mercury (*Mercurialis annua*), is common much further across England. It seems to have been spreading in London in the last few years, but this may just be the result of keen coleopterists examining abandoned front gardens in the capital. The stag beetle (*Lucanus cervus*) occurs up to a line from Norwich to Cheltenham, although there are scattered outlying records into Wales and northern England. The metallic green flower beetle *Oedemera nobilis* occurs commonly across England and Wales to approximately the Liverpool–Nottingham–Wash line. The common nettle weevil (*Phyllobius pomaceus*) has made it all the way to the Scottish borders. Meanwhile, *Pterostichus madidus*, the black-clock ground beetle, one of Britain's commonest garden insects, is nearly ubiquitous, occurring right up to northern Scotland and seemingly absent only from the most uninviting of the highland peaks.

Looking at a range of plants and animals, but sadly not beetles, Hewitt (1999) examined molecular markers to determine where British colonies originated as

---

22 Care needs to be used when interpreting distributions. When the rove beetle *Quedius lyszkowskii* (Lott, 2010) was described, new to science from a few localities in Scotland and Ireland, it was thought a putative glacial retreat endemic. But its habitat was under bark in rather drab parks and woodlands, not up mountains. One suggestion was that this was an exotic species (North America was mooted) introduced here, but as yet unknown in its native range; if so, it had been established in Ireland since at least 1905, and was quite widespread there. It later turned up in Norfolk (Sage, 2014), implying that it was, indeed, a 'recent' migrant.

they followed the ice retreat. The rationale is that large, intermingling, genetically diverse, southern European populations became genetically less diverse as colonisation by a limited number of adventurous individuals proceeded north. The meadow grasshopper, *Chorthippus parallelus* (which has flightless females), seems to have arrived from the south-east, from glacial refugia in the Balkans, but our race of the European hedgehog (*Erinaceus europaeus europaeus*) moved north from Iberia. Hewitt's estimate was that the grasshopper must have expanded its range by about 300 metres per year to get here from southern Europe before the English Channel cut off the land bridge. Pollen core analysis also suggests that plants moved north at about 50–500 metres per year.

By about 6,000 years ago, Britain was mostly covered with climax woodland, the wild wood or Urwald, dominated by small-leaved lime. There would have been areas of open country, but these would have been extremely limited, probably mostly coastal. At this time, saproxylic beetles were a major portion of the fauna (20–30 per cent of the terrestrial species, according to Kenward & Whitehouse, 2010), with many beetles associated with shade-tolerant trees (elm, lime), indicating a dense woodland canopy. There was also a pronounced swamp fauna, especially in river floodplains; wetland carabids like *Chlaenius sulcicollis* and *Oodes gracilis* used to occur. Two coincidental effects now start to impinge on what had previously been a free exchange of colonists with the rest of Europe. Meltwater from the Earth's polar regions finally raised sea-levels to the point where the British Isles became islands again,[23] and humans started to clear the wild wood. Nothing would be the same again.

Subfossil beetle remains from 4,000–2,000 years ago reveal that they had already started to feel the effects of deforestation, as species of open ground began to diversify, especially those associated with herbaceous foodplants, meadows and disturbed ground (Whitehouse, 2006). Dung beetles started to proliferate, initially because of grazing in woodland, but gradually as pastures were created from cleared forest (Robinson, 2000). For the next several thousand years, woodland in Britain has decreased, to the point where all we have left now are small relict pockets, like those typified when calculating the saproxylic beetle diversity indices discussed in Chapter 6. Not surprisingly, the drastic destruction of Urwald habitats has had a devastating effect on Britain's beetle fauna.

From almost the point when humans arrived in Britain, one history of our beetle fauna has been a history of woodland beetle extinctions. These

---

23 Ireland had already become separated from Britain by rising sea-levels, probably more than 10,000 years ago, which explains why the Irish woodland beetle fauna is especially depauperate – about 650 British woodland species do not occur in Ireland (Hammond, 1974). Contrary to popular myth, they were not driven out by St Francis.

mirror the similar extinctions of larger forest animals like wolves, bears and wild boar. Two of the most significant beetle losses were *Rhysodes sulcatus* and *Prostomis mandibularis*, representatives of two wildwood families (Rhysodidae and Prostomidae, respectively), which in Britain are known only from subfossil remains preserved in peat and soil from more than 3,000 years ago. They are still declining in Europe as the few remaining scraps of Urwald come under increasing pressure from agriculture and urbanisation. Forest clearance continues in Britain; tree-felling may occur on a small scale, but it is cumulative; few fields are ever deliberately returned to woodland, and it takes many hundreds of years for complex woodland communities to re-establish. Major changes occurred during the medieval period, particularly in lowland England, when naval

FIG 140. The conflict between woodland beetle and human woodland clearance for fields is still with us. These 10 or so dead stag beetles (*Lucanus cervus*) were mangled by a tractor-drawn mower cutting the grass of the playing field of Catford's Forster Memorial Park on the morning of 17 June 1999, after they had emerged from pupal cells in the ancient woodland remnants for which the park is known.

aspirations meant pillaging the countryside for trees used for ships' timbers. By the seventeenth century, only about 16 per cent of England was woodland. The pace of change increased with the agricultural 'improvements' of the eighteenth century, the Industrial Revolution of the nineteenth and mechanisation in the twentieth. At its nadir in 1950, only 5 per cent of Britain was woodland (Beirne, 1952), although this has since reversed slightly. Many currently rare ancient woodland beetles are hanging on in Britain in tiny fragments. Mythical creatures like the shining blue stag beetle *Platycerus caraboides*, the slim wasp longhorn *Strangalia attenuata*, the great capricorn longhorn (*Cerambyx cerdo*) and the 'crucifix' click beetle *Selatosomus cruciatus* were all reputedly British, according to the old monographs, but were probably extinct here before 1800, or shortly thereafter.

Extinctions are sometimes difficult to monitor, and the odd surprise is occasionally thrown up. The equally mythical, yet distinctively pretty click beetle *Lacon querceus* was recorded by Stephens (1830) from a single specimen found in Windsor Forest, Berkshire, by J.H. Griesbach. It was subsequently dismissed as a highly unlikely native species, and ignored until its rediscovery in the forest by A.A. Allen in 1936. It remains confined to this one small area of ancient woodland, with extinction constantly threatening it. On the other hand, ever since the first tree was felled using a stone axe, beetles of open country, steppe, grazing meadow or cultured land have been encouraged in Britain, and despite the English Channel and North Sea, they have continued to arrive here in droves.

ISLAND BIOGEOGRAPHY, BUT NOW WITH MODERN TRANSPORT LINKS

The basic tenets of island biogeography suppose that the number of species occurring on any given island is a balance between those becoming extinct and those newly arriving. The rate at which they arrive is high if the island is close to a mainland, and extinctions are likely to be low if the island is large and offers a rich tapestry of available stable habitats. The beetle fauna of the British Isles is one of the best studied in the world, and although extinctions are tricky to follow, it is possible to gauge immigration rates.

Table 1 gives the number of beetle species recorded in the British Isles during the last 300 years. It shows two distinct phases: an initial hesitation; then a sudden surge, grappling with the sheer number of beetle species here; and then the gradual accumulation of more species as time goes on. A simple calculation shows that in the last 40 years, since Pope (1977), 389 species have been added to the British list – that's nearly 10 species a year.

TABLE 1. Number of recorded British beetle species through history.

Date	Authors	Number	Notes
1710	Ray	237	MacKechnie Jarvis (1976)
1756	Linnaeus	410	MacKechnie Jarvis (1976); Linnaeus listed 586 beetles in total, of which 441 have European provenance, and 410 are now known from the British Isles
1802	Marsham	1,307	MacKechnie Jarvis (1976).
1839	Stephens	3,462	MacKechnie Jarvis (1976). Many species subsequently synonymised; 2,800 is probably a truer number
1866	Crotch	3,092	MacKechnie Jarvis (1976)
1874	Cox	3,193	MacKechnie Jarvis (1976)
1893	Sharp & Fowler	3,243	
1909	Sharp	3,300	'About' this number recorded
1930	Hudson Beare	3,566	
1945	Kloet & Hincks	3,690	
1977	Pope	3,729	Also lists an extra 116 casuals or doubtfully established species
1987	Carter & Owen	3,889	
2008	Duff	4,034	
2012a	Duff	4,072	
2017	Duff	4,118	

True, some of these will have been overlooked cryptic species, long present in Britain but only recently described, where species complexes have been split following family or generic revisions. The small but pretty longhorn *Leiopus nebulosus* was recently found to be two sibling species in Scandinavia (Wallin *et al.*, 2009), with *L. linnei* separated on genitalia characters, head shape and subtle differences in the rather variable mottled patterning of pale scales. Examination of historical specimens housed in museums suggests that various Baltic islands support only *L. nebulosus*, and that elsewhere it is more coastal than *L. linnei*. Both species occur in Britain, but no clear habitat or geographic distinction is yet apparent. However, most 'new' British beetles are genuine new arrivals to our shores.

Finding a beetle species new to Britain has always had a certain cachet among British coleopterists, and the entomological literature is littered with formal articles and notices announcing a new find. During the height of coleopterological discovery and dissemination in the mid-nineteenth century, the *Entomologist's Annual* was founded with the expressed purpose of reporting 'notices of the new British insects'. In its first volume (1855), 227 new beetle species are listed, but these were 'new' since Stephens' (1839) manual; even so, over the next 19 volumes for which the journal was in publication, many scores of new species were notified each year.

The *Annual* makes fascinating reading. I pick up the 1870 volume, almost at random, and find that E.C. Rye is enthusiastically adding 88 new species to the British list, including 38 recently described as new to science. These were exciting times. He, himself, had found *Mordellistena brevicauda* 'in profusion, chiefly in the flowers of *Hieracium* [hawkweeds], on the hill-sides at the back of Folkestone, in June last [1869]'. The Mordellidae are warmth-loving, active summer beetles that feed as larvae burrowing in plant stems, or possibly woody roots. They are all southern species, none making it to Scotland that I know; by the looks of them, they are obviously post-glacial invaders from the south-east. Had this beetle just flown over from nearby Calais to establish itself in Kent? It remains a rare insect today, with scattered records mostly from the chalk in south-east England. I found one on a similar June day, 105 years later, also in a hawkweed flower, along the coast a short way, at Beachy Head in Sussex.

Sometimes a new species seemed just to be another tick on the list. When I first picked up a copy of Spry & Shuckard's (1840) *The British Coleoptera Delineated*, I was bemused to see that one of the specimens delineated on supplement plate 1 as a neat but simple black-and-white engraving was a narrow cylindrical ground beetle with only a few of its broad, toothed fossorial legs still attached. It was identified as *Oxystomus anglicanus*, 'found at Peckham'. I later moved to Peckham, in south-east London, and decided to dig a bit further. It was Stephens (1828) who had started it, and his first volume contains an illustration, nicely hand-coloured, of the same creature (with the same missing legs), although this time named *Oxygnathus anglicanus*. I suppose Stephens just wanted to name the beetle, new to science, and get its description into print, even though he knew it was 'of a form peculiar to South America'. He confesses it is not really a native British species and that it had been found in the fields east of Peckham about five years earlier, where an annual fair was held and booths were set up for the 'sale of merchandize, the exhibition of animals, and so forth, I presume that it may have been accidentally dropped therefrom, and has thus found its way into our fauna'. He was really stretching a point talking about it as being part of 'our' fauna, but this was a time when species were considered fixed and unchanging, existing in a world that was ordered and regimented according to the grand schemes of the Creator. Notions of evolution, extinction, glacial cycles, island biogeography, invasion and the movement of tectonic plates were still some way off.

Finding a new species along the south coast of England immediately suggests that it is a genuine migrant, making landfall after a sea crossing of at least the 33.3 kilometres of the Dover Strait. When the curiously rotund ground beetle *Omophron limbatum* was found in the flooded gravel pits of Rye Harbour, Sussex (Farrow & Lewis, 1971), it was assumed that it had just hopped over the Channel from France. That a new mud beetle, *Heterocerus hispidulus*, was also found here at the same time (Allen, 1970) seemed to indicate that this was an optimum place for migrating beetles from France to settle. This has been confirmed of late. Several other new incomers have been found on this same stretch of coast between Winchelsea to New Romney, where the spit of Dungeness sticks out into the sea, including: *Bracteon argenteolum* (Mendel, 1991), *Nebrioporus canaliculatus* (Carr, 2000), *Bembidion coeruleum* (Telfer, 2001) and *Carpelimus nitidus* (Telfer, 2015). Similar assumptions follow for the arrival of the pine-twig-boring weevil *Magdalis memnonia* on the cliff tops near Eastbourne (Allen, 1972), the possibly bird-nest staphylinid *Astrapaeus ulmi* on Newhaven undercliff (Hance, 2007), the flower beetle *Olibrus norvegicus* on the dunes of Sandwich Bay (Telfer, 2013) and the ladybird *Scymnus interruptus* near the coast at Ramsgate (Parry, 2012).

Not all hops are short, though. A wind-funnel effect in the Thames Estuary is credited with bringing in migratory birds from Scandinavia and the Low Countries to the mudflats of Essex and Kent, and a number of new beetle sightings seem directly connected to this same phenomenon. When the small grey anthribid *Bruchella rufipes* first appeared on its regular foodplant, wild mignonette (*Reseda lutea*), it was along the north Kent and south Essex coasts where it gained a foothold (Hodge & Jones, 1995). The delicate pink weevil *Lixus scabricollis* first turned up on sea beet (*Beta maritima*) along the sea wall on the Isle of Grain. Travelling further upriver, two jewel beetles, *Agrilus cuprescens* and *A. cyanescens*, arrived at Beckton, east London (Hodge, 2010), along with a new weevil, *Perapion brevirostre* (Hodge, 2011). The lovely red-and-black longhorn *Stictoleptura cordigera* created quite a stir when it appeared on Hackney Marshes near the London Olympic Park (Richardson, 2014).

Further round the coast at Southwold, East Anglia, the spotted willow leaf beetle (*Chrysomela vigintipunctata*) was recently found in profusion in reed litter and other detritus washed up onto the beach (Lane & Mendel, 2016). Many of the large and pretty beetles were alive, but a local breeding source has not yet been found. Was this the result of flying hordes, making landfall from the Continent?

Or was there an earlier colonisation event in Suffolk from which they had been washed down? In east Norfolk, a moth trap run near the dune system at Horsey Corner attracted a single specimen of the ground beetle *Amara majuscula*, and although no further specimens were found there, the beetle was later attracted to light traps at West Winch and Dersingham, both in west Norfolk, and another was produced in Worcestershire (Hodge *et al.*, 2016). This beetle appears to have had a major immigration into eastern England that year, but whether it has become established remains to be determined. Meanwhile, round in the other direction, west, many hundreds of the attractive white-flecked black chafer *Oxythyrea funesta* (Fig. 141) were found, mostly on oxeye daisies (*Leucanthemum vulgare*), at Horsea Island, near Portsmouth (Harrison, 2016). This large and distinctive beetle has a long history of turning up as a single vagrant, and (more importantly) being found and identified (summarised by Barclay & Notton, 2015); it was only a matter of time before it became firmly established.

That all these species flew across to Britain, or were blown, seems a reasonable judgement. Movements of insects like butterflies, moths and dragonflies are well known, and reports of their long-distance dispersal are as apparently commonplace as migratory bird reports. But plenty of new beetle species have arrived from much further afield, and their entry into Britain must have been anything but under their own steam.

When Charles Waterhouse was sent some large (7–10 mm), wrinkled tubercular weevils that were 'injurious to ferns in greenhouses' in the grounds of the Dublin Science and Art Museum in 1903, he described them, new to science, as *Syagrius intrudens*, presumably because he knew they were intruders, for this was a well-known Australian genus – although this particular species was as yet unknown in the Antipodes. It remained singularly Irish until 1932, when a colony living out of doors was found in the ornamental gardens of Sheffield Park, near Horsham, West Sussex. It has subsequently been recorded in a few other places on the coast of Devon and Cornwall, and on Guernsey. There is only one way this insect could have got to Britain – in the soil or exotic plants shipped from its homeland. It was not alone. There is now a whole slew of Australasian beetles living wild and free in Britain, including *Saprosites mendax* (Scarabaeidae), Arundel Park (Tottenham, 1930); *Cartodere bifasciata* (Latridiidae), now widespread (Allen, 1951); and *Cis bilamellatus* (Ciidae), widespread (Wood, 1884). The narrow New Zealand weevil *Euophryum confine* (Fig. 142) is now a major domestic woodworm pest in Britain, where damp has introduced fungal decay into floorboards and skirting, but is also quite happy in rotten logs out of doors. Having probably arrived in wooden shipping barrels in the late nineteenth or early twentieth centuries, it was only recognised here much later (Buck, 1948), by which time it had spread widely.

FIG 141. The glossy, white-flecked black chafer *Oxythyrea funesta*, once a rare vagrant in Britain, now seems to be firmly established. (Penny Metal)

FIG 142. The small but distinctively cylindrical form of the woodworm weevil *Euophryum confine.*

Exotics continue to appear, and there is no limit to the distance they travel. The tiny woodworm *Mirosternomorphus heali* was described new to science from Kent (Bercedo & Arnáiz, 2010), but belongs to a genus known only from New Zealand and Norfolk Island. A stunning dung-inhabiting black-and-red rove beetle, *Philonthus spinipes* (Allen & Owen, 1997), is probably Japanese; first found in Dorset, it is now frequent across England, north to Nottinghamshire. A tiny reddish-brown weevil, *Stenopelmus rufinasus*, is from North America (Janson, 1921); it feeds on that bothersome smothering pond plant water fern (*Azolla filiculoides*), itself a native of the New World. Even the garden ponds of south London become matted with this clogging nuisance, and although I have never found the weevil by pond-dipping, I sometimes see it on the sheet if I put a moth light outside my back door. *Epuraea ocularis* (Nitidulidae) is widespread in the Old World tropics and Pacific islands (Booth & Galsworthy, 2015), although it has recently been found in Europe and North America; it's obviously on the move. The mottled brown weevil *Simo hirticornis* is just the latest of several exotic plant-feeding beetles to be found at Brooklands motor racing circuit (Mendel & Morris, 2014), where extensive planting of mature trees and shrubs with significant rootballs has recently taken place.

Mass movement of plants and soils by the horticultural trade is now a major conduit for transporting beetles around the world. In North America, Australia and New Zealand, biosecurity is ferociously enforced because of the ecological chaos that can sometimes ensue. In small temperate, damp Britain, we can, perhaps, afford to be a bit relaxed, but we should not be wholly complacent. It is probably due more to luck than effort that Britain remains free of Colorado beetles (*Leptinotarsa decemlineata*) and western corn rootworms (*Diabrotica virgifera*), but who knows where the next malignant agricultural plague will come from.

In the last few years there has been a spate of new discoveries of beetle species attached to exotic garden plants, and the implication must always be that they have arrived through the international plant trade. The dark blue-and-red flea beetle *Luperomorpha xanthodera* has definitely arrived from China with horticultural produce. It is probably quite widespread and has been found in numerous garden centres on a variety of rose cultivars (Johnson & Booth, 2004). A bright, metallic green weevil, *Pachyrhinus lethierryi*, feeds on Leyland cypress (*Cuprocyparis leylandii*) and Lawson cypress (*Chamaecyparis lawsoniana*) trees, and after it was first found in Hertfordshire (Plant *et al.*, 2006) it soon spread across the gardens of England where these prominent boundary trees are planted; by 2016, it had reached Yorkshire. Earlier, the handsome and distinctive black-and-orange longhorn *Semanotus russicus* had been found emerging from Lawson cypress logs in a garden in Sunninghill, Berkshire (Mendel & Barclay, 2008). When the handsome striped and mottled weevil *Charagmus gressorius* turned up at Dawlish Warren, Devon (Cunningham, 2012), it might have been in the middle of a nature reserve, but it was found on tree lupin (*Lupinus arboreus*), originally a garden plant, now escaped and common around the coasts of Britain. The red-and-black bean weevil *Bruchidius siliquastri* occurs in the seedpods of the Judas tree (*Cercis siliquastrum*); found recently in London, it probably originates in China (Barclay, 2014). A single specimen of the striking domed Tasmanian leaf beetle *Paropsistera selmani* was found on rose mallow (*Hibiscus syriacus*) in west London, but this beetle is already well established in Ireland, where it feeds on 14 species of *Eucalyptus* tree (Sage & French, 2015). The New Zealand fig longhorn (*Xylotoles griseus*) feeds, not surprisingly, in the wood of fig trees (*Ficus carica*), and when it was found breeding in Devon (Walters *et al.*, 2016), this was the first time it had been recorded in Europe.

More are set to come, and every year there are reports of odd specimens of this or that being found in garden centres, or intercepted at ports. Most of these will be only casual records, unable to establish themselves in a different climate. But just like Waterhouse's fern-devouring *Syagrius*, some do survive, if only in greenhouses. The heated buildings at Kew Gardens were always producing peculiar animals, although the only post-glacial specimen of the dung beetle *Onthophagus furcatus* in goose droppings there is a bit of a conundrum. The biosphere domes of the Eden Project in Cornwall have imported huge volumes of plants and soil, and several saproxylic, leaf-litter- or soil-dwelling weevils have been discovered breeding there, including the tropical *Pentarthrum elumbe* (Turner, 2011), the south Asian *Conarthrus praeustus* (Allen & Turner, 2012) and the

Pacific *Stenotrupis marshalli* (Turner, 2014). There are parallels here with the many cosmopolitan domestic beetles found invading our heated homes and food stores.

When my brother-in-law Mark Leppard found a carpet of the cellar beetle (*Blaps mucronata*, Tenebrionidae) – thousands of them by all accounts – infesting the disused Walter Parsons Corn Stores in Reading, destined for redevelopment as a pub, you could hear the rustle of their waddling through the littered remains of the grain, rubble and each other's dead bodies. Each day when the builders opened the door to start work, the floor would be covered with them, but they were gone in an instant, scuttling off into dark, broken corners and through holes in the woodwork. This was probably nearer to the beetle's primary habitat when it first spread through northern Europe and finally arrived in Britain, although by the end of the twentieth century, grain stores had better hygiene and the beetle had been reduced to eking out a meagre living on spilled food scraps that dropped through the floorboards of a few run-down domestic houses. Before humans gave it shelter, though, it was likely originally a general scavenger from Mediterranean southern Europe or North Africa, feeding on whatever dry plant material – dead leaves, broken stems, spilled seeds – it could find; many other large, flightless tenebrionids are adapted to the hot, dry conditions there. *Blaps*, a clunky clockwork, waddling beetle, was well known to the ancients (Beavis, 1988). It, along with the grain weevil (*Sitophilus granarius*), several larder beetles (*Dermestes*), mealworms (*Tenebrio molitor*) and the specialist *Blaps* predator ground beetle *Sphodrus leucophthalmus*, are known only from inside buildings in Britain, and according to archaeological evidence started to arrive with the imperial aspirations and trade expertise of the Romans (Smith & Kenward, 2011). Sadly, today's houses are not as welcoming to such creatures. *Blaps* has undergone a significant decline, what with house redevelopments, underfloor chemical sprays against woodworm, and fitted carpets, vinyl or linoleum stopping food morsels slipping through the cracks in the boards. Long gone are cellars as food-storage areas; instead, they now house unpalatable bicycles, tools, spare furniture or boxes of discarded toys.

This has not stopped new introductions though. International trade in food and commodities continues unabated, and some warehouses around the world still show rather lax biosecurity. The khapra beetle (*Trogoderma granarium*) probably arrived in various foodstuffs, including rice, wheat and nuts, sometime around the end of the nineteenth century; it has since been joined by four more in the genus, most recently *T. variabile* (in 1978) and *T. angustum* (in 1999). The domestic pest species of choice in my house is currently *Attagenus smirnovi*, which I regularly find skittering about on the kitchen windowsills or trapped in the bath. Originally from Africa, it was found in Russia in 1961, was finally identified and described new to science in 1973, and arrived in Britain a few years later. When I first moved to London in 1980, the centre of its British distribution seemed to be the Belgrave Square headquarters of the Nature Conservancy Council (a precursor to Natural England). Several friends worked there, so an indoor field trip was arranged to hunt for it – with great success.

CASUAL APPEARANCES VERSUS TRUE COLONISTS

What constitutes a 'new' species, to be reasonably added to the British faunal list, has changed over time. When John Dunsmore discovered two ground beetles under rocks on the shoreline near Gourock on the River Clyde, they appeared to be a species new to science, and the Reverend H.S. Gorham (1901) duly named them *Stenolophus plagiatus*. No more were ever found and the event remained a bit of a mystery. Eventually, Lindroth (1972) realised that it was actually a North American species, *Stenolophus comma*, presumed to have been brought over in ballast in the hold of a ship, and it was duly consigned to the 'non-established introductions' section of the checklist.

FIG 143. This South African cetoniine chafer, *Melinesthes algoensis*, was found alive in a box of flowers delivered to a Nottingham florist. It lived for several months, feeding on cut grapes and other soft fruit, but could never have established a viable colony in Britain. For one thing it was a lone male.

It's all about whether a species becomes established or not. There was no chance that Stephens' dead and broken South American *Oxygnathus anglicanus* could have established an outpost on Peckham Rye, but finding two *Stenolophus plagiatus* might suggest a viable colony. However, as most field entomologists will attest, it is very frequent to find only one specimen of a rare species, leaving open the question of whether the thing is breeding or is just a casual vagrant. I was thrilled to find several specimens of the chunky thistle weevil *Larinus planus* on the Thames Estuary at Tripcock Ness on 16 June 2008. I had rarely seen this scarce beetle before, and noticing that one was much larger and broader than the other, I fancied that I might have found male and female. It was only when poring over the microscope a few weeks later that I realised they were two different species (Jones, 2009). *Larinus turbinatus* is widespread on mainland Europe, and the brownfield site on the wind-funnel Thames would seem a perfect place for it to establish in Britain, but on return visits to the site I could find no more specimens. Nor could anyone else. It remains an enigma, like so many other beetles found here just the once, on that list of non-established introductions.[24] On the other hand, some invasions are obvious – indeed, they are predicted.

A couple of years earlier, on 5 June 2006, I was walking around the gardens of the local church, St Clement with St Peter, in East Dulwich, while my young daughters had ballet lessons in the church hall. The day was sunny and warm, and several hollyhock plants (*Alcea rosea*) were crawling with apionid weevils. Some of these were the common gunmetal-grey *Aspidapion radiolus*, but on one developing flower bud was a mating pair of a species I immediately knew to be different. The female had an unfeasibly long rostrum. Although I was carrying my 15-month-old toddler son at the time and had no collecting paraphernalia about my person, I was able to knock them into the palm of my hand and transfer them into the small Tupperware container that had previously held his snack of breadsticks and raisins. Later, the specimens were easily confirmed as *Rhopalapion longirostre*, a species that had been spreading across Europe for the last 20 years, and which had long been eagerly anticipated by British coleopterists (Jones, 2006).

It wasn't long before I found it on every hollyhock in every garden in the street, and it is now a regular in this part of south-east London, from Forest Hill, Nunhead, Peckham and Dulwich. It feeds in the seeds rather than making raggedy holes in flowers or foliage, and has its stupendously long rostrum to drill egg-laying holes deep into the large flower heads. Seed movement may

---

24 Stop Press. A second specimen of *Larinus turbinatus* was photographed by Harry Rutherford at Rainham Marshes, on the other side of the River Thames, 25 August 2017. So who knows…?

have caused or exacerbated its spread across Europe. Technically, it had already
reached Britain in 1987, when Jim Jobe brought some hollyhock seed capsules
back from Yugoslavia and later found the beetles crawling over clover plants
in his Knaresborough garden and another in his garage. It was not until many
years later that they were identified as *Rhopalapion longirostre*, but they had never
become established in Yorkshire; there were no local hollyhocks to support
them – the reason Jim brought back the seedheads in the first place (Jobe, 2007).
Since it landed in East Dulwich, other records have been few and far between –
Soho, Kensington, Buckinghamshire, Cambridgeshire – but this is quite typical,
a slow start, the lull before the storm. Soon, I'm sure it will be all over the place.
All it needs is for people to look for it; unfortunately, it is a rather small, dull
grey beetle, and it doesn't seem to do too much harm to the hollyhocks.

If *Rhopalapion* were an obvious pest, things would be different. Take the lily
beetle (*Lilioceris lilii*), for instance. We now know a great deal about this species'
spread, not because coleopterists have found and recorded it, but because it has
had gardeners up in arms as it has shredded their prize blooms. This startlingly
crimson-and-black insect had just begun to make an appearance in Britain at the
end of the nineteenth century. Fowler (1890) recorded it from a few smallholdings
and market gardens in south London and Swansea, but it remained very rare,
almost unheard of; establishment seems not quite to have occurred. However,
in 1939, it appeared in several gardens in the Chobham area of Surrey. This time,
it found the right conditions, or a slightly hardier strain was introduced, and it
started to spread inexorably. Halstead (1989) charted its progress through almost
every country in south-east England, although most records occurred within a
40-kilometre radius of Chobham. By the end of the millennium, it had been found

FIG 144. The striking
black-and-red form of
the lily beetle (*Lilioceris
lilii*) made it an easy
species to monitor as it
spread across southern
England during the
latter half of the
twentieth century.

throughout southern England, north to Cheshire and Lincolnshire (Cox, 2001b), and by 2006 had made good headway into Wales, Scotland and Ireland (Salisbury, 2008). It is still going and now occurs throughout almost all of lowland Britain. The Royal Horticultural Society maintains a recording scheme to further monitor its progress, and through which members of the gardening public can angrily submit sightings. After a slow start, it has spread by about 10 kilometres a year.

Hammond (1974) compiled expansion maps for two other recent arrivals. The Australian *Cartodere* (formerly known as *Aridius*) *bifasciata* arrived in Surrey around 1950, and by 1973 had reached Newcastle, Liverpool and Penzance; that's around 15 kilometres per year, not bad for a 2-mm beetle that breeds in piles of mouldy grass cuttings. This small and obscure beetle belongs to a group of potentially confusing species in the family Latridiidae, many of which are difficult to identify. Thankfully, it is one of the most distinctive – testaceous orange-yellow, flecked with a black bar and dots across the elytra – and once its occurrence in Britain was noted (Allen, 1951), it was immediately recognisable to other coleopterists. Its very generalist life history of breeding in accumulations of decaying plant matter also means that it could find suitable habitat almost anywhere in the country, including garden compost heaps. The latest online National Biodiversity Network Atlas distribution map shows it from almost the whole of England and Wales, and various parts of eastern Scotland; in south-east England it is now almost ubiquitous.

Another immigrant, the large and pretty ladybird *Harmonia quadripunctata*, was closely monitored after it was first found in East Anglia in 1937, and by 1973 it had radiated to Cheltenham and Pontefract – say, 7 kilometres per year. Like many ladybirds, *H. quadripunctata* is an aphid predator, but it specialises on those attacking pine trees. Although notionally a common enough group of species, pines in southern England are especially concentrated on the band of nutrient-poor sandy soil that extends across the lowland heaths from the East Anglian Breckland, through London, Surrey, West Sussex, Hampshire and Dorset, and in 60 years the ladybird had more or less completely filled this geographic zone (Majerus *et al.*, 1997). There are signs that it is continuing to spread north, into the Midlands, Wales and the West Country (Roy *et al.*, 2011), but I never saw it when I lived in East Sussex (it arrived there in about 2000, but remains sporadic), and it is still decidedly uncommon in relatively pine-poor East Dulwich.

The appearance of another ladybird, the harlequin (*Harmonia axyridis*; Fig. 145), in 2003 was a completely different story, and stands as an important model for arrival and invasion by an alien creature spreading into the British Isles. A native of central and east Asia, this large, brightly coloured beetle had been used around the world for many years as a biocontrol agent against alien invasive aphids decimating various crop plants. It had been shipped in industrial quantities and deliberately released into North America from 1916 and much of Europe since the 1980s. It was undoubtedly successful in the growers' battles against plant lice, but had the insidious side effect of outcompeting the native ladybirds, eating them and other non-target insects. In North America, it and the European seven-spot, *Coccinella septempunctata* (released since the 1950s), had previously been blamed for drastic and dramatic declines in several native ladybirds, including *Brachiacantha ursina*, *Cycloneda munda* and *Chilocorus stigma* (Colunga-Garcia & Gage, 1998; Harmon *et al.*, 2007). When the harlequin appeared in the garden of the White Lion pub in Sible Hedingham, Essex, on Sunday, 19 September 2004, it immediately set alarm bells ringing.

In what was, at the time, a bold move, Cambridge ladybird researcher and geneticist Mike Majerus issued a strongly worded press release, proclaiming, 'The ladybird has landed… but not just any ladybird, this is *Harmonia axyridis*, the most invasive ladybird on Earth' (Majerus *et al.*, 2016). This had the effect of getting the media's attention, and galvanised an army of ladybird recorders to

FIG 145. Harlequin ladybirds (*Harmonia axyridis*), showing the variety of colour patterns. Many of these have chosen an awkward place to hibernate, among the webs of *Amaurobius* spiders under loose bark. Despite high mortality, this alien invasive species is here to stay, and is still spreading.

FIG 146. The silk mass under the harlequin ladybird (*Harmonia axyridis*) is a cocoon where a maggot of the internal parasitoid *Dinocampus coccinellae* has emerged to pupate. After the harlequin arrived in Britain in 2003, it took about 10 years before this common and widespread ladybird parasitoid started to target the beetle. (Penny Metal)

map and monitor the species' progress as it moved into Britain over the next 13 years. The Cambridge Ladybird Survey had already been launched in 1984, and had been assiduously collecting data ever since. In 1994 it morphed into the UK Ladybird Survey and published distribution maps (Majerus *et al.*, 1997), but kept recording on tick-over. Now it would have to step up a gear.

Because the harlequin is such a large and obvious beetle, and because it has a tendency to come indoors to hibernate, it soon brought itself to the attention of a whole host of previously uninterested people, now curious, or worried, to know what was this new beast invading their gardens, their sheds, their offices and their homes. A veritable blizzard of records ensued. The photo-identification website *iSpot* was inundated – they were turning up everywhere. The spread of this alien ladybird is now probably the best visualised invasion we have in Britain to date. At the end of 2004, it was scattered over East Anglia and Kent, by 2006 it had taken over the Chilterns, and by 2008 it was solid to Sheffield and Leeds. To date, its explosive expansion is knocking at the doors of the Lake District, Newcastle and most of Wales; there are also a large number of scattered outlying colonies in Ireland, Scotland, Shetland, Orkney and the Scillies. Even at a conservative estimate, its spread was something like 50 kilometres per year.

Like so many species that have invaded from the south-east, the harlequin may nearly have reached the geographic or climatic limit of its potential range in the British Isles. It will probably never fully colonise the uplands of Scotland or Wales, but its infill of lowland and coastal localities may yet creep on for a few years more. Monitoring the arrival and spread of the harlequin is not just an academic exercise in range expansion of a new species; there is now a real legacy of ecological disruption to consider. Majerus's press release and follow-up scientific paper (Majerus *et al.*, 2006) about the ladybird's likely detrimental effect on our native species proved to be unnervingly prescient. As the harlequin spread, so native ladybirds seemed to shrink away in its wake. At first these were anecdotal reports; everyone was seeing harlequins but no one could remember when they last saw a two-spot (*Adalia bipunctata*). However, because of the power of the ladybird survey's sheer numbers of records on its database, it wasn't long before genuine statistical analyses started to show clear declines in many of our native species (Brown *et al.*, 2011; Roy *et al.*, 2012).

Adult harlequins show an astonishing range of different colour forms and spot patterns, and these continued to fascinate, and confuse. Harlequin larvae are also much bigger and brighter (and therefore more obvious) than those of our native species, and these were much more frequently seen trundling over the herbage and up walls and fences ready to pupate. Nobody could fail to notice these new beetles, but along with the harlequin invasion came a strange disquiet. Where previously (see p. 286) pretty, spotty, friendly ladybirds had been viewed with mild curiosity and horticultural respect, now there was an alien menace among them. People were asking how to tell the difference between a good ladybird and a bad one. Should they squash a harlequin when they came across it? To the dispassionate biologist, such distinctions are a nonsense. And anyway, the harlequin was here to stay and no amount of control or exterminating could get rid of it.

All we can do, now, is watch, and see how the rest of the fauna adapts. And from this we might imagine how every other beetle, and indeed every organism in Britain, has had to adapt as each new species invasion has taken place, ever since the ice started to retreat 13 millennia ago.

CLIMATE CHANGE – NOWHERE ELSE TO GO

Climate change is nothing new. When the Weichselian ice sheet covering northern Europe retreated, it was just the last of a series of comings and goings on the ice front going back millions of years. What is different now is the pace of change. Beetles have been useful tools for demonstrating the slow retreat of ice and the colonisation of the exposed land during prehistoric times. In addition to the events in Britain and Europe studied by Lindroth (1935) and Kenward & Whitehouse (2010), a similar transition occurred in North America.

Ashworth (1996) lists several carabids, including the now typical arctic species *Diacheila polita, Pterostichus caribou, P. vermiculosus* and *P. sublaevis,* which between 21,500 and 14,500 years ago occurred in Iowa and Illinois, but are now restricted to Alaska and Yukon following the retreat of the North American Laurentide ice sheet.

Looking at ancient subfossil remains from the relaxed position of hindsight, and demonstrating species movements and range expansions over periods of tens of thousands of years long ago, is all very well, but modern climate changes appear to be happening over centuries, if not decades, and it is sometimes difficult to differentiate between genuine climatic influences and the natural inclination for adventive species to take advantage of temporary, short-term or even seasonal weather patterns. That the large black-and-white chafer *Oxythyrea funesta* was found in profusion near Portsmouth is not, in itself, evidence that the climate of southern Hampshire is warmed to Mediterranean standards, despite the long-standing claims of tourist brochures advertising the English Riviera. It just means that this distinctive and powerfully flying insect has managed to start a successful colony here. Who knows how long the colony will persist? It might not last the year, and fail at the first frost. Only time will tell whether this type of sudden invasion is really connected to climate change. Frustratingly, we may have to wait for the hindsight available a century from now.

However, beetles are well placed to offer insights into real climate anomalies, and there have been a spate of articles mapping, modelling or implying climate change. For example, Hengeveld (1985) suggested that between 1890 and 1975, shifts in the geographic distribution of 365 species of carabid ground beetles in the Netherlands were attributable to the greater frequency of warmer and wetter summers that the country had recently been enjoying. Similar warming may also have affected the proportions of melanic (dark) two-spot ladybirds in the country. In 1980, an east–west transect from Ouddorp on the Netherland coast to Eindhoven, 120 kilometres inland, showed a steep morphological cline, with much higher proportions of melanic specimens (more than 50 per cent) away from the coastline, compared to 20 per cent near the sea. During successive surveys in 1991, 1995 and 1998, the proportions of melanics gradually declined, and by 2004 proportions were almost uniform (about 20 per cent) right along the transect line. Melanic two-spots, it was suggested, are cool-adapted because they are better able to absorb heat radiated from the sun, even in cold winds. However, as the Dutch climate has changed, generally warmer weather means that radiation has become less significant in heating up the beetles to get active in the morning, and the normal spotty red form is much more at home in the now balmy hinterland (Brakefield & de Jong, 2011). Britain's notoriously fickle weather, on the edge of a major ocean and major continental landmass, makes models of climate change very difficult to assess, but these studies will surely follow, and beetles will be lead organisms in this work.

In North America, there is great concern about how changing climate might affect crop pests, especially in the forestry industry. To some extent insect pests of lowland temperate and subtropical areas are already known and understood, and can be controlled, but softwood forestry is a northern temperate or upland enterprise. Pines grow well in this relatively hard habitat, partly because it is cool and they avoid the ravages of many wood-boring species. In British Columbia, the mountain bark beetle (*Dendroctonus ponderosae*) is a significant pest of pine trees and sometimes explodes into destructive plagues, killing the trees across large areas. Its Canadian distribution is not limited by the availability of its host trees, but by the climate, which eventually reaches a cold point unfavourable for brood development. The beetle's range expansion since 1970 has been linked to historic weather records (Carroll *et al.*, 2004), which show that formerly climatically unsuitable regions have now been colonised. If average global temperatures rise by the predicted conservative 2.5 °C, it is also predicted that the beetle will shift its range by seven degrees of latitude north – that's something in the order of 775 kilometres. Similar worries have been expressed for other notorious forestry pests, including *D. frontalis* in North America (Williams & Liebhold, 2002) and *Ips typographus* in Europe (Seidl *et al.*, 2008).

Climate change doesn't just allow warmth-loving species to expand north; it also drives cold-adapted species even further north (obviously the reverse, south, in the southern hemisphere). But in the British Isles, we haven't got an even further north – we soon run out of land beyond Scotland, Shetland and the Orkneys, leaving only the wild expanse of the inhospitable North Sea. The mountain ground beetle *Curtonotus alpinus* is already stretched in Britain, limited to a few relict sites in the Cairngorms. It has only 150 kilometres of northward land to go.

A similar geographic limit applies to cold-adapted species living high up cool mountainsides. If temperatures increase to the point where these montane specialists are uncomfortable, or pressed by lowlanders coming into the now warming habitat, their only option is to migrate higher up the mountain. But the mountain will soon run out of height. A neat example of this was mapped on the Ecuadorian volcano Mt Pichincha (Moret *et al.*, 2016). Precise altitudinal measurements during an expedition in 1880 (quite unusual foresight for the time) meant that comparisons could be made with pitfall trapping transects in 1985/86 and 2013/15. Twelve species of carabid from this well-known high tundra habitat (called páramo) form a community adapted to the sparse vegetation and rocky, sandy soil, but they all shifted their ranges up the mountain by at least 100 metres between 1986 and 2015. One of them, *Dyscolus diopsis*, a flightless páramo specialist, has disappeared from the lowest 1880 site at 4,300 metres, and now occurs only above 4,600 metres. A shift of 300 metres may not seem very much, but on a steep mountainside this means that its 1880 range of 12 square kilometres above the 4,270-metre contour line is now reduced by 90 per cent to less than 1 square kilometres above the 4,560-metre contour. Many peaks hereabouts support microendemic species. *Dyscolus diopsis* occurs in island-like populations on the summits of 11 other mountains in the Andes of northern and central Ecuador. It is in no immediate danger of total extinction, but at current rates of páramo upslope migration, it could soon disappear from four mountains that have peaks lower than 5,000 metres.

Climate change is now a buzzword for environmental upheaval, ecological disruption and wildlife extinctions, but also (admits the terrible cynic in me) for research grant applications. Climate change will undoubtedly be a driving force for beetle study in the next few decades at least, partly because beetles can be recorded, monitored and measured so well. There is a long and noble tradition of wildlife study in the British Isles, and almost everything we know about the ecology, life histories, distribution, abundance, range and changes in fortunes of British beetles we owe to the dedication and perseverance of field coleopterists. In Britain, beetles are second only to the Lepidoptera in the popularity they enjoy through the study of both professional and amateur entomologists. As and when climate change starts to affect the British fauna (and the environment as a whole), it will be through the patterns revealed by hands-on beetle study carried out in this tradition that we can begin to acknowledge and understand it.

CHAPTER 11

A History of British Coleopterists

FIRST FOUNDERS OF ENTOMOLOGY

There is no date at which the study of British beetles began. Beetles are an obvious part of the environment – familiar on flowers, in water, running about on the ground or trundling through the cellar. There are not many, but a smattering of common names for the largest and most frequently encountered species shows an awareness, if not quite a warmth, for them. Most common names were coined for creatures that came to human attention because they were striking and/or beautiful, dangerous or frightening, or ate crops, infested food stores or harmed our homes. There are ladybirds, dumbledors, cockchafers, glow-worm, stag beetle, devil's coach-horse, oil beetles, deathwatch and weevils.[25] What sets apart entomologists, and within them coleopterists, is the curiosity to delve beyond just looking and seeing, and make an active effort to observe and record. Until the advent of the Internet, this meant publishing in printed form – in books and journals.

There are several general histories of entomology, but I'd recommend the many thorough essays in the volume by Smith *et al.* (1973), which cover all eras from ancient Egypt and the Far East, through sericulture and beekeeping, to modern scientific disciplines. In that volume, the papers by Tuxen (1973) on the period 1700–1815, and Lindroth (1973) on 1800–59, are particularly relevant to the

---

25 Clock is an old word for beetle (especially large black ones like dors), which appears to have fallen out of favour during the late nineteenth century, although it has recently been resurrected in the coining of the modern English name black-clock ground beetle for the common carabid *Pterostichus madidus*.

scholar of coleopterological history. And although it is somewhat disconnected to my thread here, since it deals mostly with relatively recent North American economic entomology, there is also the huge tome by Leland Ossian Howard (1930), with one of my favourite book titles: *A History of Applied Entomology (Somewhat Anecdotal)*. As far as I'm concerned, histories need more anecdotes.

Inevitably, a history of British coleopterists needs to start with Thomas Mouffett (also variously spelled Moufet, Moffett, Muffet, etc., 1553–1604), who is credited with publishing the first book in English on insects. His *Insectorum Sive Minimorum Animalium Theatrum* first appeared in Latin in 1634, and was actually mostly written by the Swiss naturalist Conrad Gessner (1516–65), and the English physician and zoologist Edward Wotton (1492–1555), and incorporated much information from an unpublished manuscript by English physician Thomas Penny (1532–89). Such were the machinations of seventeenth-century publishing that these posthumous composite works were common, and plagiarism no different from homage. It is possible, however, to separate Mouffet's prose, delightfully summed up by Raven (1947): 'He has a touch of fancy, a flair for a telling phrase, a literary gift which is wholly unlike the dry scientific precision of Penny's style'. Take note, modern authors, I say. The English edition was finally produced in 1658, and although 90 of its 326 pages covered beetles, they did not make it into the book's full title – *The Theatre of Insects: or Lesser Living Creatures, as Bees, Flies, Caterpillars, Spiders, Worms etc., a Most Elaborate Work*. Mouffett used a broad classification of insects based on wing and leg morphology, which was originally proposed by Ulisse Aldrovandi (1522–1605) in his *De Animalibus Insectis Libri Septem, cum Singulorum Iconibus ad Viuum Expressis*, published in Bologna in 1602. This was a time before the modern system of binomial scientific nomenclature, when plants and animals laboured under cumbersome wordy names, thus *Attagenus pellio* was 'the black beetle with a white spot on each of the cases of the wings'. Many of Mouffett's species are only vaguely identifiable from his text, but a few are helped by the sometimes crude woodcut illustrations, including the huge chafer *Polyphylla fullo*, wasp beetle (*Clytus arietis*) and minotaur beetle (*Typhaeus typhoeus*).

John Ray (1627–1705) was primarily a botanist, but produced his *Methodus Insectorum* in 1705 and *Historia Insectorum* in 1710. He worked closely with Francis Willughby (1635–72), now rather relegated to a footnote in history, but who may have been an equal force in the partnership (Ogilvie, 2012). Ray's general classification is more sophisticated than Mouffett's, and he follows Swammerdam's (1669) divisions based on partial or complete metamorphosis, a phylogenic basic still implemented today. His beetle section (about 240 species) starts with the 'common dor or clock' (*Geotrupes*) and the 'stag-fly' (*Lucanus*

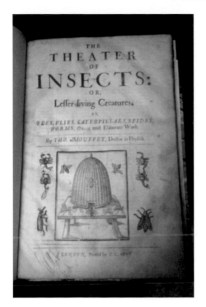

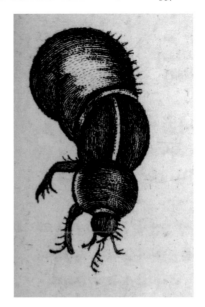

ABOVE LEFT: **FIG 147.** Mouffet's *The Theatre of Insects* (1658), the first book on insects in English. Sadly, the last time I saw one for sale (January 2014), I didn't have the £1,850 on me.

ABOVE RIGHT: **FIG 148.** Mouffet's text and pictures were widely used elsewhere. This woodcut from Brookes (1763) is a clear copy of the 'scarnbug' dung-roller scarab from *The Theatre of Insects*. 'It is a very strong insect, for if one of them be put under a brass candlestick it will cause it to move backwards and forwards as it were by an invisible hand, to the no small admiration of those who are not accustomed to the sight.'

cervus), which he found at Bryanston in Dorset (there is a modern record from Durweston, nearby). He includes 13 ladybirds (which he also called 'ladycows'), the 'gloe-worm', several water beetles and longhorns ('capricorns'), and he comments on the peculiar smell of the 'goat-chafer' (musk beetle, *Aromia moschata*). He describes the common cockchafer (*Melolontha melolontha*) as the 'common tree beetle' or 'blind beetle', and suggests that this is the source of the expression 'as blind as a beetle' for its apparently unseeing, bumbling flight, bashing into things and people. He concludes with the devil's coach-horse, his 'Staphylinus major totus nigra', an apt descriptor for the large, totally black staphylinid. One new thing Ray tries is to classify beetles by breaking them down into 14 groups of similar species (anticipating the suborders of later workers), and this he does using the structure of the antennae. This was Ray's genius, even though it was swept away and largely forgotten come Linnaeus.

EUROPEAN SYSTEMATICS LEADS THE WAY

Leaving Britain's shores for a moment, the whole of natural history was revolutionised by the naming standards introduced by the Swede Carolus Linnaeus (Carl von Linné, 1707–78), with his invention of the system of binomial names (genus and species) we know today. Although this began with the initial publication of his *Systema Naturae* in 1735, it is the 10th edition of 1758 that is now used as the standard baseline – the zero hour from which all subsequent names are taken. He included 586 beetle species in this edition, of which 441 occurred in Europe and 410 in Britain (MacKechnie Jarvis, 1976). Like Ray, Linnaeus was a botanist at heart and his main contribution was in plant naming, but his binomial transformation of the biological nomenclature landscape was the catalyst to a revolution in finding, naming and cataloguing all living organisms. His pupil, the Dane Johan Christian Fabricius (1745–1808), took the entomological baton and ran with it. His many books covered all insects – *Systema Entomologiae* in 1775, *Genera Insectorum* in 1776 and *Entomologia Systematica* volumes 1–4 and *Supplementum* in 1792–99 were the most significant. His summary of the Coleoptera, *Systema Eleutheratorum* (two volumes), was published in 1801 and should really be considered the first beetle book ever written.[26] Almost all of his generic names (and species, come to that) are still extant and immediately recognisable, and his breakdown of the order into manageable units was based, like that of John Ray, on antennal form.

Over the next century, various other great European systematisers, compilers and organisers offered bigger and better systems for interpreting and arranging the Coleoptera, and their names are familiar to anyone flicking through a checklist of British beetles. French entomologist and pharmacist Étienne Louis Geoffroy (1725–1810) is best known for his *Histoire Abrégée des Insectes Qui se Trouvent aux Environs de Paris* (1762); he was the first to use tarsal segment numbers to divide the Coleoptera. Giovanni Antonio Scopoli (1723–88) was born in Italy, but lived most of his life in the Austrian Empire; he wrote *Entomologia Carniolica* (1763), which covered 1,153 insect species (329 beetles) from Carniolia, in what is now part of Slovenia. The red soldier beetle, *Rhagonycha fulva*, is one of his species names. German priest Johann Friedrich Wilhelm Herbst (1743–1807) produced his *Natursystem Aller Bekanten In- und Auslanändischen Insekten* (10

---

26 Fabricius's coining of the name Eleutherata instead of the Coleoptera comes from his classification of insects based on their mouth structure. The Mandibulata (biting-jawed insects) were subdivided into those with the maxillae connected to the labium (Neuroptera, Hymenoptera, etc.), those with the maxillae covered by a broad plate (Orthoptera), and those with the maxillae free and uncovered (Eleutherata = Coleoptera). Ελευθερος, or *eleutheros*, is Greek for 'free'.

volumes, 1785–1806), which contained one of the first real attempts at a complete beetle catalogue. The large black-and-purple *Carabus problematicus* was a Herbst split from the widespread, common and very familiar violet ground beetle, *C. violaceus*, a mixed series of which was presumably giving him problems. Gustaf von Paykull (1757–1826) produced his *Fauna Suecica* in three volumes (1798–1800), all of which were dedicated to the Coleoptera (although he still called them Eleutherata, like Fabricius). He also produced the *Monographia Histeroidum* (1811), one of the first publications to deal with a single beetle family, complete with 13 exquisite engraved plates of these attractive stocky creatures. As a teenager, I was very pleased to come across the short, squat, tessellated dung beetle, *Chilothorax* (then *Aphodius*) *paykulli* in horse dung on some of the more sunny slopes of the South Downs – quite a rarity.

The great French zoologist Pierre André Latreille (1762–1833) took on the mantle of general insect systematiser, and he constructed a classification based not just on wings and legs (as per Linnaeus) or mouthparts (as did Fabricius), but on characters from the entire organism's structure. He had a great influence on entomologists to follow, and if he were alive today he would be completely familiar with the modern phylogenetic arrangements of insect orders, although he called them classes. He was the first to limit the term 'insecta' to the hexapod arthropods, and he also invented the concepts of family and tribe as intermediate categories above genus. He nearly didn't make it though. As a priest at the start of the French Revolution, he was required to sign an oath of allegiance, but failed to do so and was imprisoned in November 1793. When the prison doctor visited, he found Latreille in the dungeon examining a beetle, the rather scarce carrion- or bone-feeder *Necrobia ruficollis* (described new to science by Fabricius a few years earlier, in 1775) and was impressed by his erudition and knowledge. The doctor sent the specimen to a local 15-year-old naturalist, Jean Baptiste Bory de Saint-Vincent (1778–1846), who knew of Latreille's work and succeeded in getting him freed. This is exactly the sort of tabloid-style story later dismissed as fanciful, or at best apocryphal, but it appears to be genuine and in later life Latreille often spoke about his adventure. Bory de Saint-Vincent went on to become a well-known French traveller and naturalist. Latreille went on to become recognised as the greatest entomologist of his age and published a large body of work across all insect orders.

Latreille's most important beetle book, *Histoire Naturelle et Iconographie des Insectes Coléoptères d'Europe* (1822–24), was written jointly with Pierre François Marie Auguste Dejean (1780–1845), who was about the first significant entomologist to specialise almost solely on the Coleoptera. When Dejean died, he had accumulated the greatest collection of Coleoptera in the world at the time, and his catalogue stretched to more than 22,000 species. Dejean's personal commitment to beetles is exemplified by a well-known entomological anecdote about his behaviour on the field of battle. As an aide-de-camp to Napoleon, he was a general at the Battle of Alcañiz (1809) during the Peninsular War and was about to sound the cavalry charge when he spotted a new beetle. Dismounting, he caught it and pinned it into his helmet, which was specially cork-lined for just such an eventuality. During the battle (which he won), his helmet was destroyed, but he retrieved the cork piece and pinned insect. There is some doubt in my mind about the story, since history now tells us that the French lost the Battle of Alcañiz (now in the province of Teruel in western Spain), but in the same year won the Battle of Alcántara (now in Cácares on the Portugese border). Dejean named his beetle *Cebrio ustulatus*, which is a *nomen nudum* anyway, a non-name – there is no published description, and the name appears only in a bald directory in his catalogues (effectively checklists of names to show how to order species in the collection). There is perhaps no way of knowing which *Cebrio* species it really was, or where it was found.

Beetles occupied the forefront of European entomology at this time, sometimes if only by default. Many grand, comprehensive national insect faunas were planned, but in the end managed to cover just one order, usually the Coleoptera. Apart from Paykull's *Fauna Suecica*, there was *Entomologie, ou Histoire Naturelle des Insectes* (Olivier, 1789–1808), *Fauna Austriae* (Duftschmidt, 1804–25), *Deutschlands Insecten* (Sturm, 1805–57), *Insecta Fennica* (Sahlberg, 1817–34) and *Histoire Naturelle des Insects* (Lacordaire, 1854–76). It's tempting to suggest that these enthusiastic and driven entomologists started out on a quest to publish on all orders, beginning with their favourites (beetles, obviously), but then they realised that they had bitten off more than they could chew, found that they had exhausted themselves, and couldn't summon the energy to do all the other groups after all.

EARLY BRITISH BEETLERS[27]

Meanwhile, back in Britain, several notable and important works were also being published. These, however, were not full treatises on beetle (or indeed insect) classification, nor were they attempts to document the entire national fauna, but

---

27 A huge wealth of information on almost any Brit who ever recorded a beetle anywhere is available under the 'Biographical Dictionary' tab of *The Coleopterist* website, compiled by Michael Darby between 1981 and 2009. I have raided this resource remorselessly.

were what Lindroth (1973) aptly describes as 'collector's descriptions'. Issued in fascicles, a few leaves of text and a coloured plate or two at a time, they randomly skipped from one insect order to another from month to month. They eventually formed impressive volumes, but did not achieve comprehensive coverage. Among these are *Illustrations of Natural History* (Drury, 1770–82), *The Natural History of British Insects* (Donovan, 1792–1813) and *British Entomology* (Curtis, 1823–40).

Today, these exquisite books carry the price tags of expensive antiques, well beyond the means of us lowly field coleopterists. I have a few random segments bound together in rather higgledy-piggledy volumes, from my father's library, which he picked up as bits and pieces over the years, and although they are fascinating and wonderful to pore over, they are of limited modern scientific worth. Donovan's often life-sized pictures are either tiny or a bit garish, and his text little more than might be expected on the back of collectors' cigarette cards issued a century and a half later. A whole plate is dedicated to two views of the

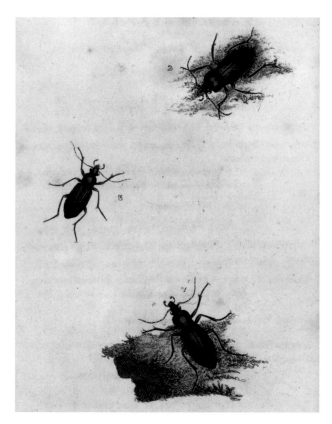

FIG 149. One the better plates from Donovan (1792–1813), showing: 1, *Carabus violaceus*, 'margin of the thorax and wing cases glossy violet'; 2, *C. nemoralis* (he called it *gemmatus*), 'three rows of indented double spots, bronzed'; and 3, *C. granulatus*, 'three rows of elevated, or convexed-oblong spots, with an intermediate elevated line'.

bee chafer, *Trichius fasciatus*, but apart from a few lines of bland description, some in Latin, all he offers is: 'In Germany this insect is not uncommon: we believe it is very rare in this country. Found generally on umbelliferous plants.' Not really very helpful. On other occasions he does go into more detail. For the oil beetle that he calls *Meloe tecta*, he carefully describes the pronounced kink in the antenna, and the proportionally longer elytra, which he compares against the well-known *M. proscarabaeus*. However, it is clear he is actually describing the male *M. proscarabaeus*, and comparing it to the larger, more bloated, more straight-horned female. He found it in Epping Forest, a highly unlikely site nowadays for this declined and now mainly coastal species.

The hand-coloured engravings by Curtis (1823–40) mark a zenith of biological illustration and are the equal to any found in modern identification guides.

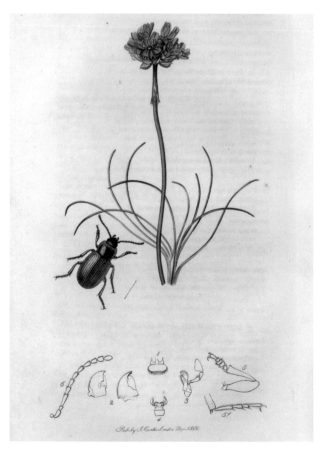

FIG 150. Typical plate by Curtis (1830) showing *Xanthomus* (he called it *Helops*) *pallidus*, described, illustrated and named for the first time. In this case, the plant, thrift (*Armeria maritima*), may well have had specimens of this sand-dune and undercliff scavenger at its roots.

His descriptions are detailed, extensive and clear, and to his credit he often lists other species in a given genus, with short extra descriptions, and notes or comments on how they differ. I would have no difficulty distinguishing his *Cafius fucicola* (his genus and species names still stand) from the more common *C. xantholoma*, which occurs under rotten seaweed around the foreshores of Britain. He found it on the Isle of Wight, and I shall keep a careful lookout on my next visit there, 189 years after its first description. His offerings are not without incongruity, however. Each plate shows a whole insect, and various diagrammatic enlargements of legs, antennae, mouthparts or other useful features, all against the backdrop of a cut plant sprig or flower. Sometimes this might be a foodplant, but in the case of (for example) the large, rare dung beetle *Onthophagus taurus*, he decided to include an inflorescence of sneezewort (*Achillea ptarmica*); perhaps he thought the sensibilities of his subscribers might be rather offended if he featured a cowpat.

The first 'proper' beetle book was produced by Thomas Marsham (?–1819) in the form of his *Entomologia Britannica* (1802); the same thing had happened as on the Continent though – despite its title, it contained only beetles, and it was also published with a title page announcing it as *Coleoptera Britannica*. Marsham broadly classified 1,307 beetle species following Linnaeus, using the structure of the antennae, and his erudite Latin text is detailed enough that many of his species are still extant, and even among those that are not, almost all are recognisable. I'm particularly pleased to see that he starts with the dung beetles, and that the very first species is the minotaur, then known as *Scarabaeus typhoeus*. A few pages later he is the first to distinguish the three British species of dumbledor, recognising the Linnaean *Geotrupes stercorarius* and newly describing *G. spiniger* and *G. mutator*, both of which he records from *stercore humano* – human dung. This seems like a good lead to rediscovering the now very rare *G. mutator*. His only fault is that he also includes the earwigs, which at that time were still usually considered to be part of the Coleoptera.

Marsham's book directly stimulated the first local list of British beetles, when the Reverend John Burrell, sometime rector of Letteringsett, near Holt, published a list of Norfolk beetles in the transactions of the newly formed Entomological Society of London in 1807 and 1809.[28] As was the convention of the day (and following Marsham), he starts out with the dung beetles, and again the minotaur

<hr>

28 This was not the current Royal Entomological Society, but one of many short-lived, transient societies formed by like-minded, mostly wealthy gentlemen (it was almost always men in those days). The histories of British (mostly London) entomological societies are interesting, although a little outside the scope of this chapter; anyone wanting to learn more can find details supplied by Neave, 1933; Poulton, 1933; Griffin, 1940; Stanley-Smith, 1955; James, 1973; and Gilbert, 2005.

(*Scarabaeus typhoeus*), which he calls the 'bull-comber', a word in need of resurrection by coleopterists I feel. His list includes a good number of species and contains some interesting comments, not least being his note on *Corticaria fumata* (now *Cryptophagus dentatus*): 'This insect frequently drops into the milk pail whilst my servant is milking in the latter end of March and beginning of April. Does it arise out of the fodder or putrescent litter; or does it for any cause live amongst the hair of the animal?' His first suggestion is probably correct, as anyone sieving mouldy hay or putrescent litter could attest.

In British entomology, the first half of the nineteenth century is dominated by four towering figures: Frederick William Hope (1797–1862), John Obadiah Westwood (1805–93), William Kirby (1759–1850) and James Francis Stephens (1792–1852). Hope came from a wealthy family (he married into wealth too), and although he joined the clergy in 1823, he soon retired because of ill health. His main publication was *The Coleopterist's Manual* (three volumes, 1837–40), a catalogue of the world Lamellicornia known at the time, but his greatest contribution to British entomology was the donation of 31 cabinets (961 drawers) of his insect collection, 1,800 books, 249 solanders (store boxes), and roughly 230,000 portrait and topographical engravings, to the University of Oxford in 1849, where it formed the basis of the Hope Department of Zoology (Entomology), now the Hope Entomological Collections. In 1858, he endowed the first chair of zoology at the university. Hope's copy of Marsham (1802) is interleaved and heavily annotated with details of his own and others' captures around Oxford when he was an undergraduate. Against the now extinct blue stag beetle *Platycerus caraboioides*, there reads: 'This species long doubted as a native of this country was taken in July at Oxford not far from Witham cut out of a dry and almost sapless oak, another was taken flying not far from the city in 1820'. Of the approximately 60 papers he published, the majority were on beetles, many describing new species.

John Westwood had a chequered early career. Born in Sheffield into an industrial family (his father was a medallist and die-sinker), he left school aged 14 to become an apprentice engraver, before joining a firm of solicitors. He soon gave up law for entomology, having met Hope in 1824, and eventually became his personal curator on £30 per year for one day a week. When Hope endowed the first chair of zoology at Oxford in 1858, it was Westwood who became the first incumbent. Westwood is rightly regarded as being one of the fathers of British entomology. His *Introduction to the Modern Classification of Insects* (two volumes, 1839, 1840) is still a classic, and thoroughly readable. Westwood was instrumental in setting up the Entomological Society of London in 1833 (now the Royal Entomological Society, having gained a charter from Queen Victoria in 1885), became its secretary, and was made honorary president for life in 1883. He produced a prodigious output of published material in the increasing number of scientific journals of the day, and although much of his interest was focused on the Hymenoptera (especially the Parasitica and the chalcid fig wasps), beetles obviously held a particular place in his affection. Writing in his *Introduction* (1839), he claims the Coleoptera 'must be regarded as occupying the foremost rank amongst insects', although he somewhat lets himself down, in modern eyes, by continuing: 'From our earliest childhood, when at school, we learned the cruel trick of putting a pin through the tail of a cockchafer to see it spin.'

William Kirby was rector of Barham, Suffolk, and was one of many eighteenth- and nineteenth-century clergyman to dedicate their spare time (which, after all, they had plenty of) to the study of natural history. He made a

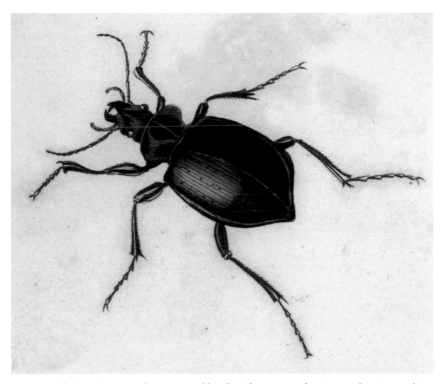

FIG 151. The handsome predatory ground beetle *Calosoma sycophanta* specialises in attacking Lepidoptera larvae, from plate 1 of Kirby & Spence (1815). Quite why Linnaeus decided on the specific name is unclear. A sycophant, now tinged with the notion of obsequiousness, originally meant an informer (a spy maybe, or someone secretive?). Back-translating sycophant to yes-man, one author recently called it the 'agreeable caterpillar hunter'.

name for himself as a hymenopterist with the publication of his monumental *Monographia Apum Angliae* (1802), covering the British bees, but he is now best known for the four-volume *Introduction to Entomology* (1815–26), co-authored by William Spence (1783–1860). One of the first truly popular insect books, it was effectively a how-to all-rounder, and was reissued as an abridged single-volume people's edition in 1856 and subsequently reprinted over several years. Kirby also liked beetles (and Strepsiptera – see p. 324), and a large number of his beetle names (species, genera and families) are still used. He is particularly known for his paper on the weevil genus *Apion* (Kirby, 1808, with a supplement in 1811), which listed 68 species, 31 new to science.

As a clergyman, Kirby wrote several theological books, including *On the Power Wisdom and Goodness of God* (1835), which followed the natural theology of the day

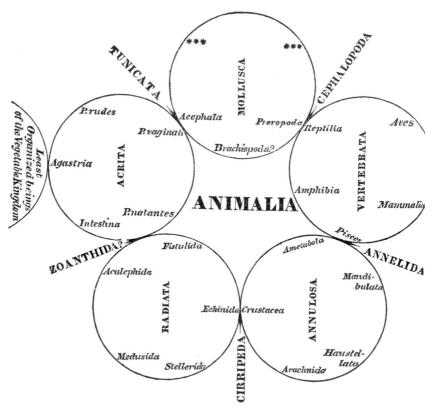

FIG 152. The bizarre quinarian system of zoological organisation contrived by Macleay and supported by Kirby & Spence (1826).

by promulgating all natural history as evidence of a divine architect. In the third volume (1826) of their *Introduction*, Kirby may have overruled Spence when it came to the suggested higher classification of insects, because he firmly supported a strangely artificial arrangement earlier promoted by William Sharp Macleay (1792–1865),[29] called the quinarian system. Organisms, particularly the arthropods, were organised into five groups, subdivided into five more subgroups, and so on, each of the fives forming a circle, organised according to their affinities or resemblances. Kirby may have favoured this cumbersome artificial formula because it suggested the orderly hand of the Creator. Going even further, Kirby wonders: 'The number *five*, which Mr Macleay assumes for one basis of his system as consecrated in nature, seems to me to yield to the number *seven*, which is consecrated both in *Nature* and in *Scripture* [his italics]'.

BEGINNINGS OF MODERN COLEOPTEROLOGY

There can be no doubt that what we understand as British beetle study today has its foundations in J.F. Stephens' *Illustrations of British Entomology*; the first five volumes (1827–34) of the Mandibulata consider the Coleoptera, and they remain as useful and compelling today as when the original fascicles were delivered to the eager subscribers nearly 200 years ago. The son of Captain William James Stephens, James Francis was introduced to the Admiralty Office by his uncle Admiral Stephens in 1807, but his entomological interests were more pressing. In 1818, he was requested by the trustees to join the British Museum and went on secondment to order the insect collections there. When he returned to the Admiralty in 1845, he apparently encountered some animosity from his superiors – maybe they took a dim view of his having had 27 years off.

In 1832, Stephens was involved in a famous plagiarism spat with James Rennie (1787–1867), whom he accused of infringing his copyright by copying text from the *Illustrations* for his *Conspectus of British Butterflies and Moths*, published earlier that year. MacKechnie Jarvis (1976) succinctly suggested that this was a time when the courts thought little of natural history writers and Stephens lost his case, and a lot of money on legal fees. But such was the concern and outrage from his fellow entomologists, most of whom sympathised with and supported

---

29 Incidentally, Macleay was one of the first zoologists to recognise the important difference between morphological similarity due to evolutionary relationship (homology; he called it affinity) and similarity due to function (analogy).

Stephens, that they publicly subscribed financial aid to cover all his costs.[30] When John Curtis published a snide comment about the affair in 1833 (attached to plate 461 of his *British Entomology*), he was roundly condemned by a long and strongly worded put-down in the *Entomological Magazine* from its editor Edward Newman, whose masterly rhetoric concludes: 'We fear that Mr Curtis will find he had better, far better have committed the whole copy of the tainted number to the flames, than have ventured to risk it on the excited wave of public opinion.' Well said, sir.

The *Illustrations* stands as a landmark publication. Its layout is clear and proper, its orderliness is consummate, its tone is engaging and informative, and its content is rich and inclusive. You can tell that I really like this book. There is no doubting Stephens' ability to describe species and order them well, as he does also with genera, families and suborders. His prose is at once scientific and charming, as exemplified in his introduction to the Heteromera (volume 5, 1832):

> *This section of the Coleoptera is decidedly an artificial one, but, from the poverty of this country with regard to the groups, it is utterly impossible to arrange them agreeably to their affinities, the chasms being so numerous: the greater portion of the species are of dull and sombre colours, those of the first three families [Tenebrionidae, Blapsidae, Helopidae] usually black or obscure shades of brown, rarely with metallic tinges; but amongst some of the typical families are several insects of gay and lively hues, frequently brilliantly metallic.*

Yes, yes. Completely right; I couldn't have put it better myself.

There are a few quibbles though. Stephens described many species new to science, increasing Marsham's 1,307 beetles to 3,462 by 1839, when he published his one-volume *Manual of British Beetles*, an annotated catalogue of the entire British fauna. There was quite a lot of duplication, as Stephens appears to have been rather zealous in allocating new species names to only slight variations, and these sometimes obscure differences proved too subtle for others to validate. This, despite an apparent aversion to using a microscope – he seems to have taken the view that what could not be seen through a hand lens was of little practical use.

---

30 I'm tempted to support Stephens, too. Rennie follows the species order of the *Illustrations* to the letter, and although he paraphrases, he has effectively produced a short précis of it for his own profit. My copy of Rennie's small book, from the Epping Forest and County of Essex Naturalists' Field Club Library, has a contemporary handwritten note above the library regulations sticker: 'The sale of this book was stopped by injunction of the Court of Chancery, being to a great extent a piracy of a similar work published by Brown [actually Baldwin & Cradock].'

FIG 153. Superb plate 1 from Stephens (1827), showing: 1, *Cicindela hybrida*; 2, *Drypta dentata*; 3, *Polistichus connexus*; 4, *Calodromius spilotus*; 5, *Philorhizus melanocephalus*; and 6, *Lebia cruxminor*.

The true number of his species is now known to be closer to 2,800. One of his main critics was George Robert Waterhouse (1810–88), whose *Catalogue of British Coleoptera* (1858) eventually helped sort out much of the Stephens synonymy.

Then there was the continuing cavalier attitude towards data. The idea that entomologists should accurately record where species were found was still secondary to the primary possession of a fine series of specimens in the collection. Stephens (1828) knew of only three reputed British specimens of the streaked bombardier, *Brachinus sclopeta*, but recorded a locality for only one: 'captured by Dr. Leach in Devonshire, two years since'. He did not know the locality of a specimen in the Hope collection and in an almost comical example of the parlous state of nineteenth-century labelling did not even know of the origin of a specimen in his own collection; all he could report was 'I rather suspect that mine came from Hastings'.

Stephens regularly mentions his correspondents and colleagues – people who have sent him specimens or information; thankfully, they were a bit better at letting him know the localities. These included: William Elford Leach (1790/1–1836), James Charles Dale (1791–1872), Leonard Jenyns (1800–93), William Chapman Hewitson (1806–78), Henry Doubleday (1808–75), Charles Cardale Babington (1808–95) and Andrew Hughes Matthews (dates unknown), among many others. A regular was Lewis Weston Dillwyn (1778–1855), collecting in Wales, whose Swansea beetle list was privately published in 1829 (the second British local list after Burrell in Norfolk). He ran a pottery at Swansea, was a magistrate and high sheriff of Glamorganshire, then Member of Parliament for the county (1832–37), and alderman and mayor of Swansea (1835–40). In his preface, he self-deprecatingly claims that entomology was never a serious study, just an amusement, but his list is rich and thorough, and he makes many astute observations. He sent many specimens to Leach to name, and the reply came back that 49 had never been described before. Some of these, and a few of Dillwyn's generic names, survive today, but sadly both *Otiorhynchus dillwyni* and *Onthophagus dillwyni*, which Stephens named in his honour, do not.

In later life, Charles Darwin often alluded to his great affection for beetles: 'I feel like an old war-horse at the sound of a trumpet when I read about the capture of rare beetles – is this not the magnanimous simile for a decayed entomologist?' (letter to John Lubbock, September 1854). As a young man, before he found his international fame, he was one of the keen subscribers to the *Illustrations* and, as did many others, he was quick to write to Stephens to share some of his finds. Stephens duly published several pages of these addenda sent in by correspondents in an appendix at the end of each subsequent volume, and about 30 Darwin records were eventually listed. Technically, Darwin's first record is now a non-species (*Ocys tempestivus*, which has since been synonymised as just a slightly larger and brighter form of the very common *O. harpaloides*). But *Blethisa multipunctata* is a beautifully sculpted metallic ground beetle from riversides and marshes. Studying at Cambridge University, Darwin was well placed to find it in the fens, still a major stronghold. I've never found it. If I did, I'd certainly let someone know and hope to see my record in print, with my name beside it. Darwin expresses this sentiment well: 'No poet ever felt more delighted at seeing his first poem published than I did at seeing in Stephens' *Illustrations of British Insects*, the magic words, "captured by C. Darwin Esq."'

One of Stephens' other claims to fame is his advocacy for the killing bottle. He declaims his abhorrence for the previous practice of directly pinning still-living

---

Charles Darwin's biggest beetling mistake

The famous story about Charles Darwin collecting when an undergraduate at Cambridge, around 1828, has sometimes grown in the telling. In his autobiography, Darwin (1887) recounts how he was tearing off some old bark from a dead tree and saw two rare beetles, which he picked up, one in each hand. He then saw a third that he could not bear to lose. Oddly, his reason seems to have departed him at this point, because – and I still boggle to read his account – he popped the one in his right hand into his mouth. Darwin then feigns surprise: 'Alas! It ejected some intensely acrid fluid, which burnt my tongue so that I was forced to spit the beetle out, which I lost, as was the third one.' I really cannot imagine any experienced coleopterist (or novice come to that) carrying on in this foolhardy manner.

   We never discover what the beetles were; maybe in his embarrassment Darwin blotted them out from his memory. Some have suggested his description implies a bombardier beetle, *Brachinus crepitans*, but Darwin would have known that particular very distinctive species, and anyway bombardiers are denizens of chalk downs and sea cliffs – dry, crumbling soil, plant roots and loose rubble – not fungoid tree bark in fenny Cambridgeshire. Ejecting the foul contents of the cloaca, combined faeces and urine, is a standard defence for many beetle groups, and would perfectly well have given Darwin his unpleasant mouthful. Then there are the huge range of complex toxins and repellents served up by beetle alchemists, as discussed in Chapter 4.

   Darwin learnt his lesson and was more circumspect in all his subsequent beetle captures. When he visited Van Diemen's Land (Tasmania) in early 1836, he reports finding two species of *Onthophagus*, and two of *Aphodius* in the moist cowpats. He astutely commented (Darwin, 1839) how unusual it was since the native marsupial droppings (hard, dry nuggets) were very different from those of the cattle, which had been introduced with European settlers only 33 years earlier, but he did not attempt to put any of the beetles into his mouth.

---

insects into the collecting box, then letting them expire in 'agonized writhings'. In a short note to the *Entomologist's Magazine* (1834), he promotes the use of crushed cherry laurel leaves (*Prunus laurocerasus*, which releases a cyanide compound) in an airtight box to dispose of the insects humanely, a technique still useful today.

   The *Illustrations* remained a benchmark for the next 70 years, and in its wake the study of beetles spawned several generations of coleopterists and beetle books. This was a great time for scientific journals, and titles like the *Entomological Magazine* (1832–38), succeeded by the *Entomologist* (from 1840),

Transactions of the Entomological Society of London (from 1836), *Journal of Natural History* (from 1841), the *Zoologist* (from 1843), *Entomologist's Annual* (1855–74) and *Entomologist's Monthly Magazine* (from 1864) are full of important sub-monograph papers to short notes on Coleoptera. These old magazines are still a huge source of useful information and fascinating insight.

Now also began a time of increasing specialisation as some workers started to concentrate on particular groups or families. The 1825 monograph on Pselaphidae and Scydmaenidae (now both subsumed into the rove beetle family Staphylinidae) by Henry Denny (1803–71) was well ahead of its time in terms of detail, structure, style and layout; the colour plates are excellent too. A sometime curator of the Leeds Literary and Philosophical Society, Denny later (1842) published a similar superb monograph on lice, the Anoplura. John Frederic Dawson (1802–70) started

FIG 154. Typical plate from Dawson (1854) showing the appearance of identification thumbnails as well as whole-beetle portraits.

a long tradition of writing specialist books on the ground beetles, Carabidae, and his *Geodephaga Britannica* (1854) remains an important and authoritative work. At the other end of the popularity scale, Andrew Matthews (1815–97), son of Stephens' correspondent A.H. Matthews, wrote the most important book (though in Latin) on Britain's smallest beetles, the Ptiliidae, then called Trichopterygidae, in his *Trichopterygia Illustrata et Descripta* (1872). If a monograph could not be published as a free-standing book, it could certainly be published in one of those many competing scientific journals. Thus David Sharp (1840–1922) placed his major work (182 pages) on the large and difficult staphylinid genus *Homalota* (now subdivided into smaller genera) in the *Transactions of the Entomological Society of London* (1869) and his mammoth (1,000-plus pages) revision of the world water beetles into the *Scientific Transactions of the Royal Dublin Society* (1882).

Along the way, there was also time for some less technical consolidating works. Waterhouse's 1858 catalogue, followed by those of Crotch (1863), Rye (1866) and Sharp (1871), allowed coleopterists to catch up on which species were considered valid, and which had been sunk in synonymy. There were also some more introductory works like the 'picture book' of Spry & Shuckard (1840), which nominally offered a line illustration of each genus, but precious little in the way of description or biology.[31] What must count as the first popular guide to beetles (1866) was produced by Edward Caldwell Rye (1832–85), one of the founding editors of the *Entomologist's Monthly Magazine*. Aimed very much at the beginner, it starts with a look at insects in general, considers how to go about finding beetles and then sets off on a whistle-stop tour of the families. It was greatly helped by the 16 elegant hand-coloured engravings. Rye starts his preface with a warning that the book is not a complete dictionary, nor a stark grammar (analogies previously used in other insect books), but is instead a *delectus* – a book of selected passages – a great word that I wish I'd used.

An unusual aside occurred in 1874, when the almost unknown Hon. Herbert Edward Cox (1838?–1914) published his two-volume *Handbook of the Coleoptera*. Virtually nothing is known about this man, except that he moved from London in 1894 to Jamaica, to take a post in local government that earned him his honourable title. He seems to have published nothing else on beetles, although some of his specimens are now in the Hope Entomological Collections at Oxford. His book was the first (in Britain) to use dichotomous keys throughout, and although sometimes unwieldy, this was the first book with which I tried to identify some of my beetles when, as a 16-year-old in 1974, I was bought a copy by

---

31 It served, nevertheless, to provide most of the thumbnail illustrations for the key to families on p. 262–79.

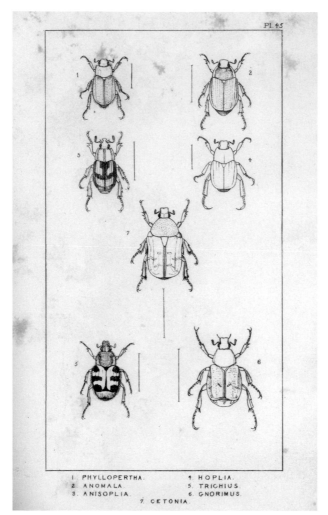

Pl. 45

1. PHYLLOPERTHA.
2. ANOMALA.
3. ANISOPLIA.

4. HOPLIA.
5. TRICHIUS.
6. GNORIMUS.

7. CETONIA.

FIG 155. Typical plate from Spry & Shuckard (1840) showing simple but clear outline diagrams for each genus.

my parents – a reward for passing my O levels (equivalent to GCSEs nowadays). Antiquarians might shudder to hear, but when I received the book, a century old, it was uncut: the printed sheets had been folded and bound into the two volumes, but it had never been opened, read or used. Cox's volumes were interesting, but suffered the age-old problem with conciseness – the couplets were sometimes difficult to appreciate and he had an annoying habit of describing species as 'not uncommon', 'moderately common' and 'tolerably common'. In the end I found it curious, but of limited use.

THIS IS THE MODERN WORLD

I'm going to let my continuing personal bias dictate this final segment of the chapter. As far as I'm concerned, the modern era of British beetle study began in 1887 with the first of five volumes in the magnificent series *The Coleoptera of the British Islands* by the Canon William Weekes Fowler (1849–1923). The full work followed in quick succession (1888, 1889, 1890, 1891), and together with the supplement, co-authored by Horace St John Kelly Donisthorpe (1870–1951), published in 1913, it remains the most complete and authoritative set of books to cover the entire British beetle fauna. Born into a clerical family, Fowler was ordained in 1875 and then spent more than 20 years as headmaster of Lincoln Grammar School, before taking the rectorship of Rotherfield Peppard, Oxfordshire, and then Earley, near Reading, a few years before his death. He published a great many articles on beetles, and was on the editorial board of the *Entomologist's Monthly Magazine* from 1885 for 38 years. He also wrote significant portions of the *Fauna of British India* and *Biologia Centrali-Americana*, prestigious multi-volume works being pumped out under the umbrella of empire. This erudite man was completely immersed in beetles, a doyen, although there is a charming anecdote quoted by MacKechnie Jarvis (1976) about Fowler's relationship with John Arthur Power (1810–86), a top coleopterist of the day. In the early 1880s, Fowler took apartments in the house next to Power's, in Ashburnham Road, Bedford, in which he stayed on his many visits from Lincoln, and he obviously availed himself of Power's great coleopterological knowledge. Later, Power's youngest daughter, Beatrice, would recount: 'Who was the young clergyman we always had in the house? My mother said she thought that he did most of his collecting in my father's cabinets.' Fowler always acknowledged his deep debt to Power, whose name is the first mentioned in the preface of *The Coleoptera*, and it is easy to find reference in his species accounts to localities from 'Dr Power's collection'.

It is with his six beetle volumes that Fowler's name is now synonymous. The 'large paper' edition contains 200 hand-coloured plates, but with a print run of only 250 copies it is scarce and now very expensive; however, the 'small paper' unillustrated volumes are more obtainable. Fowler's strength lies in the detailed descriptions for each species, genus and family, and although there are short dichotomous keys, these suffer the age-old problem with beetle identification in that they are often comparative (larger/smaller, broader/narrower, more or less heavily punctured, etc.) – not much help if you have only a single specimen to examine under the microscope. Reading through his text, though, usually offers clarity and confirmation. Not until the recent offerings by Duff (two, of an

PLATE 99.

t. Morgan, del. et lith.

Vincent Brooks, Day & Son, Imp.

L. Reeve & Cº London.

FIG 156. Typical colour plate from the large paper edition of Fowler (1890). Although the illustrations are pretty and helpful, it was Fowler's meticulous, detailed text and clear prose that made his descriptions and species discussions so useful.

envisaged four volumes, in 2012 and 2016) has any attempt been made to equal Fowler's depth and coverage.

This is not to say that the last 120 years have been empty – far from it. Water beetles, sometimes not taken seriously by students of terrestrial Coleoptera, because they belonged to that other-worldly pond-dipping fauna of caddisflies and leeches, got a huge boost from the activities of William ('Frank') Alexander Francis Balfour-Browne (1874–1967). He produced a large output of material, including three important monograph volumes published by the Ray Society (Balfour-Browne, 1940, 1950, 1958), and his own semi-autobiographical *Water Beetles and Other Things* (1962). Balfour-Browne was encouraged by David Sharp, and would later entertain readers with the famous *Agabus striolatus* anecdote. Apparently, Sharp had been sent a dozen specimens of this very rare East Anglian water beetle, and was softening them up in a glass of water prior to mounting them on card for his collection. He managed the first two while a 'guest' wandered about the room. On looking up from setting the third, he caught sight of his visitor downing the last drop of water from the tumbler, beetles and all. Although Sharp considered giving an emetic to recover the precious beetles, he resisted.

The major twentieth-century British beetle book was produced by Norman Humbert Joy (1874–1953), a medical doctor who had published numerous notes and articles on beetles (particularly those in bird and animal nests) before his *magnum opus* appeared in 1932. His *Practical Handbook of British beetles* was exactly that – practical. A large volume (622 pages) of keys, and a second of 168 black-and-white plates, revolutionised beetle study by making it readily accessible to those who did not care to trudge through Fowler's meticulous but sometimes dense technical prose. Joy's aim was to use easily observable characters, backed up with whole-insect drawings and thumbnail sketches, to produce an identification key for the novice, and this he certainly achieved. The book's importance was so great that it was reissued in facsimile, in a slightly smaller format, in 1976, and a not-quite-supplement of recently added or split species was produced by Hodge & Jones (1995). It is still, very often, the first book I reach for when I start trying to key out a beetle. Joy wrote a small popular guide, *British Beetles, Their Homes and Habitats*, in 1933, and although this was reprinted several times up to 1949, he very soon afterwards appears to have lost interest in the Coleoptera. He gave away his collection in 1933 and wrote a book on birdwatching in 1936.

The other key development in the last half of the twentieth century was the series of identification *Handbooks* produced by the Royal Entomological Society. Each covered a single family, or part, or group of closely related families. Production was piecemeal, depending on who was interested in what groups and had time available to write them. They started off being very brief (that of F.D. Buck on 15 families of the Heteromera, 1954, was only 30 pages in total), although all were illustrated with useful line drawings and thumbnail sketches, and some have now developed into mini-monographs (Luff's 2007 Carabidae is 248 pages), with detailed descriptions, copious biological and life-cycle information, and coloured plates. To date, 29 have been produced, although this includes second-edition revisions of Carabidae and Scarabaeoidea. Popular introductory works (with lots of pictures) for the beginner continued to appear, with those by Bechyne (1956), Linssen (1959), Lyneborg (1977) and Harde (1984) holding special places in my heart.

Despite these, there was a definite lull in British beetle book production during the second half of the twentieth century. Social changes perhaps meant that the leisured classes were less inclined to write natural history books, while the professionals and academics in museums and universities were finding it hard to justify time spent on the 'trivial' science of insect taxonomy. However, this did not mean that British beetle study was in the doldrums. Luckily, important beetle books were emerging out of Europe, with *Die Käfer Mitteleuropas* (Freude *et al.*, 1964–83, supplements and revisions ongoing) taking centre stage, along with several volumes in the *Fauna Entomologica Scandinavica* (written in English) and *Faune de France* series. That British coleopterists were anything but insular is borne out by the fact that with the issuing of each *Käfer* volume there was a spate of publications in the journal literature in response, including new species found, revisions of families and genera, checklists and synonymies. It was a very busy time.

Not everyone has the time and facilities to sit down and write books and monographs, but sharing my and Darwin's delight at seeing our names in print, everyone can take satisfaction from reporting their new observations or unusual finds in the burgeoning entomological press. It is easy, and engrossing, to while away the hours reading through the last 150 years of entomological journals finding curious snippets and hidden records. Seymour Bingham, of Newnham, Gloucestershire, seems to have been a lepidopterist, but as far as I can ascertain published just one single note in the *Entomologist's Weekly Intelligencer* (price one penny) on beetles. He announced, on 1 May 1858, that he had found the scarce fungus beetle *Biphyllus lunatus* 'in its old haunt', and offered to mail live specimens to anyone who cared to send him a stamped envelope. He is typical of the many thousands of names that crop up, once or often, and through which the history of Coleoptera (and history in general) is intertwined. Whether or not it was connected, a few weeks later, on 29 May, Newark entomologist George Gascoyne was bemoaning the postage costs of sending specimens and foodplants through the post at the exorbitant letter-rate charge of two pence per ounce, whereas solicitors could send parchments, merchants could post documents, and printers books at the more reasonable penny for four ounces, thus 'trade is encouraged'. He suggested a campaign to write to Rowland Hill, then secretary to the Post Office, to set the situation right. These sociological nuggets are equally informative, bringing colour and depth to the lives of often long-dead entomologists.

From the early 1930s, one coleopterist's name repeats more than any other; among the most important British entomologists of the twentieth century, Anthony (Tony) Adrian Allen (1913–2010) wrote no book, but will be remembered for the very many hundreds of notes and articles he published. A chemist by training, he seemingly found half the beetles on the British list in his small back garden in Blackheath, and acquired an encyclopaedic knowledge of all things coleopterous. A private and self-effacing individual, working on his own, he nevertheless published copious notes on rare and unusual finds, generic revisions and complex synonymy; he reported endless new species to the British Isles and described several species new to science, and the small and obscure mould beetle *Corticaria alleni* is named in his honour. He was always very friendly and helpful to anyone who contacted him. He kindly read through the entire

FIG 157. Tony Allen (1913–2010), on the left, discusses beetles with David Appleton (1944–2006) in the handsome surroundings of Chelsea Old Town Hall, at the Annual Exhibition of the British Entomological and Natural History Society, 27 October 1984. Meetings such as these mix social event with scientific discourse.

typescript of *New British Beetles* (Hodge & Jones, 1995), and offered me a glass of peach schnapps when I called in to collect it. His beetle collection is now in the Natural History Museum, but other than a brief tribute by Barclay (2010), no obituary appears to have been published – he seems to have outlived all the contemporaries who might normally be called upon to write such a piece. I found him delightfully curious and quaint, possibly how people might remember me in the future. Over his long and active beetling life, he was a bridge between the wealthy collectors of the Victorian era (he knew Donisthorpe well) and the citizen scientists who are now taking the study of Coleoptera forwards into the twenty-first century.

CHAPTER 12

The Future – How to Study Beetles

B Y THE END OF THE TWENTIETH CENTURY, entomology had transformed from the urbane pursuit of the leisured gentleman (yes, it was almost exclusively a male pursuit in the early days), to the rigorous discipline of the modern citizen scientist. I have straddled the end of this metamorphosis. Where previously top scholarly or academic workers published monographic revisions, and the collectors stocked their cabinet drawers with immaculately set museum specimens, there is now a consensual study of wildlife recording, monitoring and conservation – everyone and anyone can contribute.

First, though, there is a tricky issue that sometimes seems contradictory to the newcomer – the insect collection. Why is it that those claiming to support insect conservation are so happy to kill the objects of their study? The answer is relatively straightforward – you do not have to kill insects to study them, but if you do, that study will be greater and deeper and longer lasting.

There are, of course, the well-rehearsed arguments – the sheer unimaginable numbers of insects hidden all around us, their prodigious fecundity, and the certain impossibility of entomologists doing any harm by collecting a few specimens, or perhaps even a lot of specimens. When I spoke to some novice potential entomologists recently, I tried to dumbfound them with a rough calculation I did on the back of a proverbial envelope: more insects were killed in the building of the 2012 London Olympics than have been collected and killed by all the entomologists in the world who have ever lived.

My calculations went something like this. The Olympic Park in east London covers 2.5 square kilometres, or 2.5×10^6 square metres, so with a nominal insect density of 1,000 per square metre, we get a total of 2.5×10^9 (2.5 billion) insects destroyed. This is roughly 80 times the size of the Natural History Museum's insect holdings (more than 30 million specimens). The museum has the largest collection (by a long way) anywhere in the world, so 80 of them ought to be enough to equate to worldwide entomological activity during, say, the last 300 years. I hope I've built enough latitude into the equation; actually, insects are often quoted at densities greater than 10,000 per square metre, and I did not take into account any of the other Olympic venues around the country.

It's a statistic to get people thinking, at least, but there is still some unease at the idea of a 'collection'. For what is a collection, if not a vainglorious display of prize trophies? It's here that I have to start making the other familiar claims about not being able to identify most insects unless a specimen can be minutely examined under the microscope – often by checking individual bristles on individual legs. My stock statistic is that I reckon on something like 1,500 British insects being easily identified from a photograph; but since there are more than 25,000 species here, it means that nearly 95 per cent of British insects need confirmation of a specimen down the stereomicroscope. This proportion goes up for small creatures like beetles, where I'd hazard an easy guess that only 1 per cent of British species can definitely be named from a picture. Even the common seven-spot ladybird (*Coccinella septempunctata*) cannot reliably be separated from the scarce seven-spot (*Coccinella magnifica*) from a photograph. I, for one, would not care to pronounce on any snapshot of a huge dumbledor, given that the distinctions between two of our commonest species, *Geotrupes spiniger* and *G. stercorarius*, rest on the length of the sharp ridges on the hind tibiae, the punctuation under the abdominal segments and the shape of the jaws. Yet these are giant beetles, in the order of an inch long; the task becomes even more extreme for the vast majority of Coleoptera, which are under 5 mm long. So many beetle species also have a variety of different sizes or colour forms, males and females differ, and all have very different larvae too.

The trouble is that the physical presence of the 'collection', with its neatly aligned rows of carefully mounted insects and painstakingly presented labels and headings, looks very much like a hangover from the past – a past where collecting something for the sake of it, to own it, to savour it, and to display it proudly and overtly, was still a perfectly acceptable scientific procedure. For half a century, though, there has been a formal published code for collecting. It appeared at a time when butterfly collecting began to be viewed with mistrust, and egg collecting became the criminal activity of a rogue underclass. Just over 15 years ago, I was pleased to be involved with reviewing it, and if I may make so bold, my greatest contribution was to get its title changed to 'Code for collecting insects' where previously it had been a code for 'insect collecting' (Invertebrate Link, 2002). Here is a subtle, but fundamental difference. One, like stamp collecting

FIG 158. Even a collection of easily identified species has its positive uses. Apart from the difficult smaller species that do require microscopic examination, the familiar bright, spotted forms of ladybirds immediately help get across concepts such as genus, species, rarity and variation to an enthusiastic but non-expert audience.

or egg collecting, is all about making that private, often secretive, vanity-stoking display to show off magnificent captures. The other emphasises the workaday need to maintain an ordered and accessible scientific reference collection to aid and support identification, and to advance the science of entomology further. Sometimes, the distinctions between one species and another are so fine and so subtle that a large number of specimens from various localities, found over many years, often have to be compared side by side to identify them firmly. *This* is why a collection is necessary.

Once a decision is made to study beetles by creating a reference collection, the best possible news is that this is a relatively cheap and easy process. This is one of the reasons that entomology has much longer been an egalitarian pursuit than, say, ornithology. Two hundred years ago, anyone wanting to study birds needed to be aristocracy, landed gentry or perhaps clergy, the reason being that bird study required a gun to down the quarry, and land on which to legitimately down it. Insects, on the other hand, could be studied by anyone who walked along a country lane or across the local common with a home-made insect net and a pocket full of cheap glass vials. This accounts for the huge diversity of individuals

publishing notes on their beetle findings over the years, from the one-off
Seymour Bingham and his *Biphyllus lunatus* swap offering, to the Reverend Canon
W.W. Fowler, learned headmaster and world-class scholar masterminding his
monumental monographs, and eventually to 19-year-old me scratching about
with my home-made insect net at Rye Harbour.

FINDING BEETLES

Hardly any specialist equipment is necessary to collect beetles, and the
techniques have hardly changed in 200 years. The great naturalist Edward
Donovan (1768–1837) was one of the first to publish a how-to guide (1805),
and this has been followed by a spate of publications down the years, mostly
aimed at expat travellers visiting exotic places where exotic creatures were still
waiting to be discovered. These would be shipped back to Britain to be sold
for profit at the popular natural history auctions of the day, or to populate the
burgeoning museums around the country. Taylor (1876) was riding on the back of
'recent' adventures by Darwin, Wallace and Bates in describing how to preserve
everything from fleas to flying fish, but Smart (1940) was addressing a more
modern entomologist sending back scientific samples from the Commonwealth
for research and study. Today, just as for the hard-pressed Victorian voyager,
travelling light is the best option; there is little need for copious equipment and
much can be cobbled together from common domestic items, easily repurposed
for the laboratory. Table 2 gives a list of equipment traditionally associated with
entomology, together with easily sourced alternatives. What follows is my very
personal prescription; a lot more detailed information of the hows and whys of
beetling can be found in the hugely practical *Coleopterist's Handbook* (Cooter &
Barclay, 2006), and in the hands-on introductions to finding and setting insect
specimens provided by White (1983), Wheater & Cook (2003), Gullan & Cranston
(2014) and the like.

In the field, I travel as light as I can, the heaviest things in my rucksack being
my packed lunch and flask of coffee. A light, short-handled, folding insect net
suits my nearly every purpose, and apart from a few small glass tubes, the most
important outdoor implements are a notebook (and pens) and a hand lens. I
mistreat my rather flimsy net horribly, well beyond its intended delicate purpose,
using it as a sweep net, trawling through the rough herbage to knock beetles
off the plants I'm attacking down into the bag. This is a good way to find plant-
feeders – weevils, leaf beetles, longhorns – and just about anything sitting in
flowers, on leaves or in the long grass. Inevitably, the net often gets torn, so I have

FIG 159. The author
showing the correct
use of a sweep net
in a flowery Greek
field (1986). A brightly
coloured shirt is not
obligatory, but it helps.

to repair it with needle and thread. On one nearly disastrous occasion, I realised
I had no needle, not even an odd bodkin, so I used a fresh pine needle instead,
and it worked a treat. I have occasionally snatched my net at a whirligig beetle,
Gyrinus, skittering about on the surface of the water, but proper pond-dipping –
coarse-dredging through the thick water weeds – needs a much stronger, stouter
construction and a heavy-duty canvas-and-net bag. Looking for water beetles is
equal to the fun of finding terrestrial species, but requires a bit more forethought
to take the right equipment at the right time.

There is great delight in the simple art of visual searching, carefully
approaching flowers to see what is feeding, or resting on leaves, sitting on tree
trunks or running about on the ground. No log or stone is passed without rolling
it over; indeed, any item of rubbish from broken planks to dumped mattresses can

TABLE 2. Heath Robinson entomology: cheap home-made alternatives for the citizen scientist.

Traditional equipment	Alternatives
Collecting	
Commercially available tough, reinforced **sweep net**	Home-made net using broom handle, metal tubing and cotton fabric, muslin or net curtain
Commercially available **'insect' net** – folding frame is neat and easy	Home-made net using several twisted wire coat-hangers, webbing and net muslin or curtain. Does not fold for packing into bag, but can be made any size to fit
Commercially available folding **beating tray** – cotton fabric stretched on folding wooden or bamboo fan frame	Pale-coloured umbrella, or just spread a white plastic or cotton sheet on the ground under the tree
Glass collecting **tubes** in a variety of sizes ('bottle' style with reinforced rim/neck are my favourites), glass-topped 'pill boxes', tins for larvae	These are very cheap, but domestic alternatives include small glass/plastic containers such as herb/spice bottles, mint packets, medicine bottles, micro Tupperware, hummus/dip containers, cream-cheese packets, gift boxes, sweet tins, take-away packets – the list goes on
'Pooter' small suction collector – several styles available commercially	Make your own using a small bottle or jam jar and rubber or plastic tubing (used in home-brew set-ups)
Sieve for sifting grass cuttings, straw debris, flood refuse, etc.	Standard garden sieve; stacking sieve nests also useful
Mercury-vapour moth traps – available in many different designs and configurations, including mains- and battery-powered, with or without generator	Mercury-vapour light run on its own on a white ground sheet. Any powerful bulb works to some extent. Lighted windows, porches and street lamps worth checking
Plastic **sheeting**, for sieving, sorting or beating	Plastic sheeting easily available. Also useful for sitting on during damp picnics
Killing	
Ethyl acetate 'killing fluid' – commercially available from naturalists' suppliers, some chemists and chemical companies	'Acetone-free' nail-varnish remover is mostly ethyl acetate – check ingredients list. Tippex thinner (trichlorethane) no longer available
Drop insects into storage **ethanol**	Various alcohol preparations easily available from chemists – see 'Storage' below

Traditional equipment	Alternatives
Killing *continued*	
	Leave insects in tubes overnight in freezer – they need to be defrosted before mounting
	Chopped leaves of cherry laurel (*Prunus laurocerasus*) release cyanide
	Very small beetles dropped into boiling water (note that wetting can damage other insects)
Mounting and labelling	
Bristol **board** (four-sheet) – 'traditional' card for mounting small insects	250 gsm white card from art suppliers is perfectly good
	Try cutting up old Christmas and birthday cards, but use the matt, not the shiny, surface, and use expensive luxury cards, as these are stiffer than cheap flimsy ones
Gum tragacanth preparations, smooth and tacky, are available as specialist Coleoptera mounting glue from commercial suppliers	Wallpaper paste made fairly stiff, PVA at a pinch for beetles with convex undersides, or mixture of the two
Stainless-steel **micro pins** in a variety of thicknesses (0.1–0.5 mm) and lengths (10–35 mm), for direct pinning of insects – available from specialist suppliers	You really need to use micro pins. They are relatively cheap and easy to obtain by mail order. Ordinary pins are much too thick and will eventually corrode over the years
Entomological **mounting pins**, for pinning through the card or mounting strip holding the insect – variety of sizes available	Again, mounting pins are cheap and easy to obtain. Fine dressmaking pins can be used at a pinch, but these seem thick and ungainly next to the real thing
Polyporus **mounting strips**, for holding a micro-pinned insect and acting as a buffer against vibration. Plastazote strips are also used now	Plastazote (expanded polythene) packing can be finely cut, although this is an inexpensive item to buy
Watchmakers' and other **tweezers/forceps**, for picking up insects and holding the pins used for pinning and mounting.	Cheap eyebrow-plucking tweezers come in a variety of shapes and sizes; fine tweezers are available from modelling shops
Cork **setting boards**, for pinning out insect wings, especially for wing-display purposes	Most insects can be roughly pinned using small pieces of plastazote and tracing-paper strips

TABLE 2 *continued*

Traditional equipment	Alternatives
Mounting and labelling *continued*	
Setting needle, for teasing out legs and wings during gluing and carding, or on setting boards. Old books recommend a thin, stiff animal bristle	Micro pin stuck into a matchstick. Use a similar item, the pin tip knocked on a solid surface to curve it, to create a crook-like hook for genitalia extraction
Ready-printed data labels using permanent Indian ink on acid-free board	Computer-generated data labels using inkjet or laser printers. Set to 5- or 6-point type, or simply reduce reproduction ratio when printing
Mapping pen for handwriting or annotation of individual data labels	Narrow-tipped pens, 0.1 mm diameter – easily available from high-street stationers
Storage	
Custom-made insect cabinets, with air-tight, fitted glass lids on cork-lined drawers	Available second-hand, but can still be expensive or in need of renovation
Wooden store boxes, lined with cork or plastazote, in various sizes	Available second-hand, although new ones are not prohibitively expensive. Smaller containers can be crafted from wood, cardboard or plastic boxes by lining with plastazote. Any box that is less than perfectly airtight needs to be checked regularly for pests like museum beetles (*Anthrenus verbasci*), otherwise entire contents can be lost
Storage ethanol, especially for soft-bodied creatures like larvae	An HM Customs and Excise licence is needed to purchase absolute (95–100 per cent) alcohol, but alternatives like rubbing alcohol work, as does concentrated white pickling vinegar. Specimens need to be soaked in water for some time to clean them before mounting. Can make specimens brittle
Identification	
Monographs, identification guides and technical articles scattered through specialised entomological journals	Accurate identification still requires some of these, but there is increasingly a cross-over between 'popular' and 'scientific' guides. Plenty of guides available online

Traditional equipment	Alternatives

Identification *continued*

Comparison of specimens with those in the named collections in museums and learned institutions	Still a useful technique, but you have to know where to go and who to see. You also need to be some way along in identification. Local museums often have British insect collections and are keen to foster links with local residents, but they may not have a specialist coleopterist in residence
Hand lens – x10 multiplication is standard for field use, perhaps x20 in bright light	Still the easiest and cheapest item of specialist equipment to buy. Huge variety available, from hobby stores, and medical or technical suppliers
Stereo (binocular) microscope, magnification range x10 to x30 (up to x100 rarely), with either zoom optics or a rotating object lens set, and good lighting. Cost is anything up to the price of small car	Basic student binocular microscope can be bought new for £90 (x10) or £160 (x10 and x30). A cheap anglepoise desk lamp provides ample light

FIG 160. Correct use of the beating tray, sometimes apparently also used to keep off the rain. No need, though, if you have a pith helmet. From Allaud (1899).

yield treasures resting beneath. Turning over a tractor tyre washed up on the north Kent saltmarshes at Lower Halstow many years ago revealed the scarce carabid *Polistichus connexus*, the first time it had been found in the county for some time. Loose fungoid bark can be prised off with a garden trowel or stout knife; large bracket fungi can be tapped smartly over a beating tray, and crumbling heartwood can be minutely dissected. It's rather gone out of fashion now, but I wonder if there might be some mileage in resurrecting a technique for chasing wood-boring beetles out of their burrows – lighting up a pipe of tobacco and by placing the lips against the wood, blowing smoke down into the tunnels (Ingpen, 1839). Maybe vaping could be tried as a modern alternative? Grass cuttings,

FIG 161. Sometimes beetles occur in mind-numbing profusion. Here, there are an estimated 100 *Meligethes* on this single knapweed flower head.

leaf litter, moss, haystack refuse and flood deposits can be sieved or shaken over a plastic sheet, low-hanging tree branches can be bashed over the beating tray or upturned umbrella, and of course every rancid carcass and pat of animal dung should be examined with a garden trowel or stout knife. Beetles are everywhere.

Picking up beetles is remarkably easy, since they are so tough, and even the smallest can be gently held between forefinger and thumb pads until popped into a glass tube, either to be checked with the hand lens and released, or retained for microscopic examination later. I'm a keen advocate of the wet-finger technique, especially for tiny beetles a millimetre or two long running over a hard, flat surface like a log or fence post. Simply lick the tip of your index finger, dab it down gently onto the domed back of the beetle and lift up; the stickiness of the spittle keeps the insect momentarily adhered to your finger. Place an open tube over the beetle stuck on your finger and give your hand a smart tap to dislodge it into the vial. Easy. I did get into trouble once, though, by writing what I thought was a witty comment about the taste of the finger as the day progresses, especially if dunging is involved. An irate letter to the editor pointed out that this was just the way to get a serious gastrointestinal complaint – so please act responsibly, and wash your hands occasionally.

If you do not want to touch the delicate beetle, suck it up with a pooter. This peculiar piece of equipment, apparently devised and first used by American entomologist Frederick William Poos Jr (1891–1987), comprises a glass container and two flexible rubber tubes, or variants thereof (Leather, 2015). Sucking through one tube lifts insects off the beating tray or ground sheet, or out of the net, and up the other one. A gauze barrier prevents them getting sucked up all the way into your mouth, and instead they are trapped in the central bottle container. Pooters come in several shapes and sizes, but a standard is to have a small glass bottle, say 20 mm in diameter and 100 mm long, which can be removed and stoppered to contain the live insects, while a replacement is added to the suction section to suck up more. It's remarkable how many non-entomologists have heard of this device and are eager to see its use first hand, but it can cause confusion. There was the time John Parry was out beetling, net in one hand, hand lens in the other, while the pooter dangled from his mouth, gripped between his teeth; a passing dog-walker made a sympathetic sigh and uttered: 'You poor man, do you always have to wear that thing?'

A good specialist beetler trick is to splash water from the shallows of a stream or pond up over the edges of the sandy or gravelly banks. As the water drains down into the soil, the semi-subterranean beetles, fearful there is a flood coming, emerge to scurry over the damp ground. Typically, ground beetles like *Bembidion* and *Dyschirius* will appear, along with burrowing rove beetles like *Bledius* and *Carpelimus*, and mud beetles such as *Heterocerus* and *Helophorus*. Picking up beetles that are running around on bare ground is not far from fingertip grubbing in grass roots or loose soil, and the hunched, head-down, hands-and-knees grovelling behaviour of the engrossed coleopterist is quite distinctive. Peeling back low-growing plants overflowing the footpath or spreading on bare soil can reveal soil-dwelling weevils like *Otiorhynchus* and *Cathormiocerus*; these rarely climb, so are unlikely to be found in the sweep net. Gently prising back dead leaves around the roots or teasing out the dense root thatch reveals genuine plant root-feeders, and plenty of ground-dwelling carabids and staphs.

During the winter, 'tussocking' is very productive, where the tight-packed stems of a large grass tussock (tufted hairgrass, *Deschampsia cespitosa*, is the best) can be sawed through just above the soil layer. The old books recommend a serrated knife, but my mum was less than pleased when I borrowed the bread knife for such an adventure; now I sometimes carry a small folding garden saw. The severed stalks sieved over a sheet reveal all manner of small beetle packed tight in hibernation torpor among the dry stems. This is the only way I have ever found the scarce ground beetle *Lebia chlorocephala*.

For the mechanically minded, a mechanised suction sampler opens up a new world of beetles living in the dense short turf. A two-stroke garden blower-vac, with a small net bag securely fixed over the air-intake tube with a Jubilee clip,

FIG 162. A converted two-stroke garden blower-vac can suck up insects (into a well-secured net bag) from the shortest of swards. The high-vis vest was to reassure the occupants of Canary Wharf that the person wandering about on the roof with what looked like a rocket-launcher was somehow official. Key occupants of the *Sedum* matting on this living roof were *Helophorus nubilis* (Helophoridae), and *Tachys parvulus* and *Bembidion lampros* (both Carabidae).

allows insects to be sucked out of even the shortest and densest sward; these can then be emptied out and sorted on a plastic sheet. These machines are, of course, rather unwieldy, so best used for specialist surveys on sheep-cropped downland, sea cliffs (please be careful), playing fields or green roofs, where they can find carabids, staphylinids and byrrhids aplenty. I got excited on the one occasion when the interesting seed-shaped beetle in the sweep net turned out to be an actual beetle (the squat jewel beetle *Trachys scrobiculatus*) rather than the usual seed, but was reliably informed that it was quite frequent to anyone using one of these suction devices. They're quite heavy though, and mine stays mostly in the garden shed.

Ground-dwelling beetles like carabids lend themselves to being trapped in pitfalls – usually a small plastic cup (or something like a yogurt pot) set in the ground, its lip flush with the soil surface, and re-examined every so often to see what has fallen in. A loose cover of broken slate, tile or tree bark prevents the cup from filling with rainwater and washing everything away. Dry traps have to be examined very frequently (every day or so), otherwise predators eat each other or decaying remains turn to mush. For longer-term trap-setting, it is possible to use a water-based preservative like ethylene glycol (antifreeze), although this is poisonous to wild and domestic animals, so should be used only with great caution; alternatives include acetic acid (white preserving vinegar) or brine (standard salt mixture). The handsome carabid *Scybalicus oblongiusculus*, long thought extinct in Britain, was recently rediscovered during a pitfall survey in Essex's Thames corridor (Hammond & Harvey, 2008).

Care should always be taken when setting pitfalls, which are apt to confuse or upset members of the public, whose dogs often dig up the plastic cups. There are other risks, too. On hands and knees on a derelict brownfield site in Rochester Harbour in 2004, I was having trouble making a hole big enough for the cup in the crushed brick-and-concrete rubble, so was bashing away with a large sheath knife, heaving great double-handed strokes at the unyielding substrate. When I glanced up, the man looking down at me, in a smart suit and tie and carrying a clipboard, did not seem fazed by my unorthodox actions. He was not, however, with the local planning office, but an inspector with the Rochester Police homicide department, and his overalled colleagues close by were the fingertip search squad examining the land where a sawn-off shotgun had been found earlier. I agreed to let him know if my digging found anything of significance. In the end, I found no incriminating evidence, but the site did yield *Stenolophus teutonus*, *Ophonus parallelus* and *Dicheirotrichus obsoletus*.

Vane traps work on a similar principle, catching flying beetles that crash into a vertical barrier, instinctively fold their wings, and then fall down through a funnel into a closed tub or bottle suspended underneath. They work well hanging in hollow trees or near deep tree rot-holes, catching often scarce and secretive saproxylic beetles. *Quedius lucidulus* was recently discovered in Britain in a vane trap over a fallen beech in The Mens, in Sussex (Telfer, 2012). Traps, whether pitfall or vane, can be baited with rotten fungus, dung, carrion, cheese, fermenting fruit or whatever.

For serious long-term studies, a Malaise trap produces excellent results. With an open-sided tent-like construction, but with full-height central partition, it works on the principle that flying insects climb upwards when they meet the vertical barrier, so at the apex of the sloping canvas roof a small hole leads into a holding-chamber bottle, which traps the insects until they can be examined. The traps are vulnerable to damage from animals or vandals, and really only work in secluded (or secure) private places. For many years my father had one in his back garden, much to the bemusement of the neighbours, and from the trawl of tiny beetles he mailed me each month, I would regularly pick out the minute

(1.2–1.5 mm) hemispherical ladybird *Clitostethus arcuatus*, the only times I have ever seen it.

Just like moths, many beetles are attracted to artificial light for not really very well understood reasons, and although moth traps of varying complexity and cost are available on the open market (usually with powerful mercury-vapour bulbs), just putting a bright light over a white sheet on the ground is enough to get started. My first dung beetle, *Acrossus* (then *Aphodius*) *rufipes*, arrived at the lit window of the lounge when I was a 10-year-old; it was audibly tapping against the glass as it fought to get to the source of the light. Even if they are not attracted to lights, many beetle species fly at dusk, perhaps a predator-avoidance technique by being active after the birds have gone to roost but before the bats come out; the old books frequently refer to beetles caught by evening sweeping. There is something hugely attractive about walking out and about in the early evening, as the sun's shallow, slanting light takes on a golden glow and the air stills. At this time of day beetles have a characteristic flight – straighter than bees and flies, and more purposeful. Although he was describing gnats, the words of Thomas Hardy in *Tess of the d'Urbervilles* ring true: 'capturing the moment of the setting sun's slanting rays... irradiated as if they bore fire'. I remember the calm, idyllic evening of 7 May 1975, when I strolled in the water meadows at Milton Street, near Alfriston in Sussex, catching the irradiated aerial plankton as it came drifting over: *Anthracus consputus* (Carabidae), *Aderus populneus* (Aderidae), *Epuraea melanocephala* (Nitidulidae), and *Grynobius planus* and *Ptinomorphus imperialis* (both Ptinidae), all scarce or secretive species normally.

MAKING THE COLLECTION

Back at the laboratory bench (or kitchen table as I call it), specimens can be dealt with quickly and easily. Just as with field collecting, there is little need for specialist equipment, and many items can be cobbled together from easily available alternatives. I have a travel laboratory kit that I take on holiday with me in case I need to deal with an unexpected specimen; everything fits into a small tin, 10 cm square and 2 cm deep – card, pins, mounting strips, glue, brushes, tweezers, hand lens, setting needle and scalpel blade (Fig. 163).

Ever since James Stephens promoted the humane use of the killing bottle back in the mid-nineteenth century (Chapter 11), various chemicals have been suggested. The standard is now ethyl acetate, a volatile and strongly aromatic liquid reminiscent of pear drops and available from specialist suppliers – although at least one beetle book was still advocating the use of cyanide until quite recently (White, 1983). A twist of tissue dipped into the ethyl acetate, then dropped into a well-stoppered small glass jar, is all that is required. It has to be glass, which is resistant to the ethyl acetate. Alternatively, a thin layer of plaster of Paris at the bottom of the jar can absorb a few drops of the killing fluid; this prevents tarsi becoming entangled. Drop the beetles in and give them about 10 minutes. Check the ingredients list on the container, but acetone-free nail varnish remover is usually dilute ethyl acetate and works well enough if given extra time to be effective, and Stephens' original suggestion of crushed cherry laurel (*Prunus laurocerasus*) leaves is perfectly acceptable, although it takes an hour or so. The glossy evergreen foliage of cherry laurel, a common garden shrub, gives off the scent of rancid almonds when cut up, shredded and crushed – prussic acid or hydrogen cyanide – although not in concentrations toxic to humans. Another easy alternative, and one recommended by hymenopterists especially, is to put the tubed insects in the freezer overnight. This humanely sends them into sleeping roost mode, then into the oblivion of hibernation torpor, before finally killing them. They need to be defrosted before being mounted, but this takes only a few minutes.

One reason why entomology is such an easy study is that the chitinous exoskeletons of insect specimens can be preserved by the simple expedient of just allowing them to dry out for a few days. For mounting beetles, two possibilities present themselves – direct pinning through the insect, or gluing it onto a small piece of card. The decision to do one or the other is usually based on beetle size. I must admit to being rather casual with my technique,

FIG 163. My travel laboratory kit contains everything I need in a small, easily portable tin: tweezers, pins, scalpel, card, gum, mounting strips, paintbrushes, hand lens and a setting needle made from a micro pin stuck into the end of a matchstick.

FIG 164. Beetles can be kept in the fridge overnight, to deal with in the morning. I was given this timberman (*Acanthocinus aedilis*) by Professor John Owen, who had reared it and shown it at a meeting of the British Entomological and Natural History Society (24 July 1986). A tip – ask permission if you're going to do something like this.

but I tend to pin anything over about 20 mm, card anything under 10 mm, and just take a view on everything in between. Water beetles, where underside characters are used in many keys, might be pinned no matter how small. The industrialisation of pin manufacture revolutionised entomology during the eighteenth century, as these once relatively expensive luxury items became mass-produced, commonplace and cheap. Today, specialist stainless-steel pins of varying diameters (0.1–0.5 mm) are available in different lengths from entomological suppliers. These have replaced the old ferrous, brass, zinc or copper alloy pins, which react with the chemicals in the insect body, forming a corrosive encrustation of verdigris that eventually damages the pin and specimen to the point of breakage. By convention, rather than necessity, the pin should pass vertically through the centre of the beetle's right elytron. Large beetles require large pins, which then pass directly through the data labels, but a smaller specimen can take a fine pin that is then inserted into a mounting strip, itself taking a large handling pin for the labels. The mounting strip was traditionally cut from *Polyporus betulinus*, the stiff white birch polypore fungus, but today this has been almost entirely replaced by expanded polythene plastazote strips. These act as a brace and suspension against jolts whenever the specimen is being moved or handled.

Gluing a beetle onto a piece of card may seem difficult and fiddly for the novice at first, but with a little practice it soon becomes a standard procedure. First, start with a piece of card much larger than the beetle; it can be trimmed

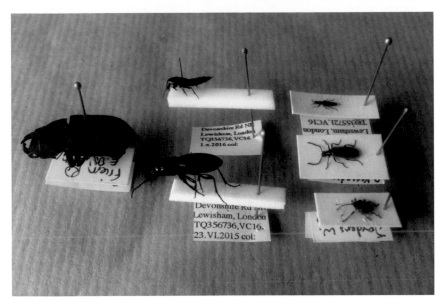

FIG 165. Three mounting styles (left to right): direct pinning through the beetle and data labels with a large pin; micro-pinning the beetle onto a mounting strip with a larger pin through to the data labels; carding the beetle, with pin through card and data labels. None of these is too fiddly; in fact, all these specimens were pinned and carded by complete novices at an open entomology training workshop.

later. Lay the beetle on its back on a piece of tissue, and gently tease out limbs and antennae with a fine, dry watercolour paintbrush. Apply glue – ideally a small amount to avoid miring – to the card, and gently lay the beetle on it. Tackiness should be sufficient to allow all legs, antennae and palps to be gently and neatly spread to aid identification – for this I use a micro pin stuck into a matchstick, which is easier to obtain than the stiff animal bristle suggested in early instructions by Donovan and the like. This limb manipulation can be achieved under low magnification (×5 to ×10) or even with the naked eye. After the glue has dried, the card can be trimmed down, to leave the beetle in the centre of a small, neat rectangle. A large pin through the end of the card then also takes the data labels. Another possibility involves gluing the beetle onto the pointed tip of a small triangular piece of card ('pointing'); this is useful in that it allows easy viewing of underside features. Those specialising in particular beetle families may have a convention they always follow, but no beginner should fret too much about the wrong or right way to mount specimens, or the tidiness with which legs are splayed. Whatever works is fine.

At some point comes the realisation that in many species complexes the external surface characters of the beetles appear uniform, and species can be reliably separated only by examining the dissected genitalia. This is another layer of complexity and, for the expert, often requires another set of kit and chemicals. To start simple, though, the male aedeagus or female spermatheca can usually be hooked out through the tail end of the freshly killed beetle, before gluing it to the card, by simply using a micro pin that has been tapped onto a hard surface to give its tip a microscopic bent crook form. The dissected parts can then be glued onto the same card as the rest of the beetle. With experience, consideration can be given to the orientation of the aedeagus, and whether the separate parts (parameres, phallobase) should be separated. Later, for the trickiest dissections, of the most delicate species, it may be necessary to boil the aedeagus in a tiny crucible containing a solution of potassium hydroxide or other clearing agent to dissolve all the excess soft tissue; this gives an easier view under the microscope. The cleared aedeagus can then be mounted in a blob of resin-like Canada balsam, or similar clear preserving medium, so that its three-dimensional structure can be appreciated from varying angles, even under high power.

Some people go to great lengths to make the set specimen as neat as possible. This can involve dropping the card-mounted specimen into a bowl of warm water a week or two later, dissolving the glue and floating the beetle off its original card, then carefully remounting it on a specially die-punched card rectangle with stylishly bevelled corners. At least one coleopterist used circular punched cards for his prize specimens. Neatness is a just bonus, however, and even the most messily mounted beetle can often be identified with little fuss. True, there is something subtly pleasing about a gracefully carded beetle; it elegantly resembles the aesthetic pictures shown in identification guides, and it is easier to compare two similarly neatly staged beetles when contrasting the sometimes frustratingly comparative distinctions used in keys. As with many things, experience makes it easier, but to get experience you just have to get in there and have a go. The most important thing is that the beetle specimen is preserved, and that it can be closely examined under the microscope, now or centuries from now.

To maintain its scientific importance for centuries, though, a specimen must have data. At least one data label should give, at the absolute minimum, the date and place where the beetle was found. This wasn't always the case. Old collections are often a series of pinned insects, without any accompanying information whatsoever. At least Alfred Russel Wallace labelled every single specimen he sent back from the bountiful tropics; he used the 8 mm circular chads from a hole punch, with an abbreviated code: thus 'Sing' for Singapore, or 'Sar' for Sarawak. That unique specimen of the world's largest water beetle, *Megadytes ducalis*, states simply 'Brazil'. Anyone can do better than that, and a modern data label, hand-written with a fine-tipped pen or printed from the computer in 5- or 6-point type, should offer: locality name; nearest village or town; region or county; Ordnance Survey grid reference to six or eight figures; vice-county; date; and the name of the collector. Additional labels can then record more details, such as what the beetle was doing, how it was found, its foodplant, habitat, height above sea-level, time of day or weather. I have a mangled stag beetle specimen,

FIG 166. The notorious 'woolly bear' larva of the museum beetle (*Anthrenus verbasci*) can wreak havoc in an untended, unexamined insect collection. Its natural reservoir haunt is, as shown here, feeding on the remains of dead insects – prey caught in the spiderwebs under loose bark. The rotund adults are common on flowers, feeding on pollen.

labelled with the address in East Dulwich where my friend found it, the date, and the apology 'Damaged, caught by cat'. My best, though, on a pseudoscorpion *Chthonius ischnocheles*, states: 'Sudbury Town, 18.x.99, in old passport, dumped stolen goods'. Eventually, an identification label should record the species name, who determined it, and when. If it is reidentified many years later, the old determination label is not removed, but another is added. It is not uncommon for museum specimens to accumulate 10 or more data labels, with varying degrees of legibility, over the years.

Insect collections are traditionally housed in glass-topped cork-lined drawers, arrayed in elegant cabinets of varying sizes. Plastazote sheets (expanded polyethylene) have replaced cork, and functional plywood, MDF and soft woods may have surpassed mahogany dovetail jointing in the construction, but these can still be very expensive, as befits any item of precision hand-built furniture. Store boxes are cheaper and easier to obtain, but since they must be air-tight to keep out *Anthrenus* museum beetles, cost should not override quality – many an important collection has been destroyed by voracious pests, leaving nothing but crumbling remains and piles of dust around a forest of naked pins. Until the 1970s, chemical fumigants like naphthalene (mothballs) and paradichlorobenzene were widely used in museums and private collections, but fears over carcinogenicity have led to these being phased out. I find the scent of mothballs strangely pleasant and evocative, but museum workers, by all accounts, soon became accustomed to it and failed to smell it in their day-to-day work, so I can imagine them being concerned about long-term exposure to a more than trivial health risk. The only useful way to combat these collection pests is a rolling regime of regular inspections. If an outbreak of *Anthrenus* is discovered, the affected drawers are put into chest freezers for a month to kill off the beetles, larvae and eggs.

EXAMINING BEETLES

To my immense annoyance, as a tyro, most of the books I tried to use to identify my beetles always started with the same unhelpful advice – the best way to identify any given specimen was to compare it with specimens already named in a reference collection. This caveat always seemed like such a cop-out to me. And a discouragement. What they should have said is that many species *can actually* be identified using the descriptions, diagrams and keys in the books, but that the distinctions between *some other* species pairs, or groups of sister species, are so subtle and difficult that the only way to be sure of their identity is to have examples of each of these species in front of you. Finding a complete reference collection in a museum is certainly one recourse, but another is to accumulate sample specimens gradually from here and there, with the hope that eventually the various different species will be discovered because they appear different from those other specimens already in the collection. When Duff (2014) reported that he had found a weevil new to Britain, *Tanysphyrus ater*, at the edge of a Norfolk pond, it was not because he had worked it through the Continental keys that same evening and found it to be novel, but because he had noticed it stood out as slightly different in the short row of specimens of the common duckweed weevil (*T. lemnae*) when he re-examined them all some two years later. Again, this is why a reference collection is of paramount importance – because it allows identification by comparison and reidentification by recomparison later on. Thus the science of entomology advances.

FIG 167. One of the joys of studying insects is that it does not require complex or expensive equipment. Here, a group of London Wildlife Trust trainees set and mount insects using a selection of low-cost stereoscopes and a minimum of entomological paraphernalia. One of them still managed to find the uncommon flower beetle *Mordellistena neuwaldeggiana* despite our low-tech approach.

My earlier statement that most equipment can be cobbled together from easily available domestic bits and pieces needs some qualification when it comes to identifying beetles. There are two important aspects of identification that may require a capital outlay: a microscope, and identification guides. I spent my first 10 years as a coleopterist making do with a small hand lens, until I bought a very cheap microscope. Even here, though, all budgets can be catered for. At the top end of professional specification, you can buy a spectacular device for the price of a small car. But the increasing popularity of hobby studies means that there are now (2017) a large number of cheap budget scopes available at around the £100–150 price tab. Today, you can buy a new microscope, perfectly adequate for the job, for much less than the price of a new smartphone.

The basic requirement is for a stereomicroscope, not the traditional monocular compound microscope used at school to look at red blood cells or onion skin. Only low magnification is needed – x10 is perfectly good for looking at beetles. This is achieved by having x10 eyepieces (the bits you look through) and x1 objective (the other end, nearest the beetle). Slipping in x15 or x20 eyepieces gives slightly greater magnification. The next level up (still very cheap though) has a swivel-turret with two objectives (either x1 or x3), allowing an easy switch, with a twist of the wrist, so x10 eyepieces now give x10 or x30 views. The lower magnification shows the entire beetle, while the higher is enough to look at antennomeres. Even higher powers can be achieved by slipping in, say, x20 eyepieces to give x20 and x60. Be warned, though, with these low-cost microscopes that the higher the power, the lower the light quality, and the less clear the image reaching the eye. I now aspire to a zoom microscope, which enlarges seamlessly from x4.5 to x45 with the turn of a button. Occasionally, I pop in the x20 eyepieces to check tarsal segments or microsculpture at up to x90.

The cheapest stereoscopes have no built-in lighting, but there is usually the option to pay a bit more and have an integral light bulb just above the specimen. My first scope had nothing, but the most basic of desk lamps was quite adequate to illuminate the object. I often still use an anglepoise to add in extra light or adjust reflections when peering down my mid-range zoom. Top-end professional machines have built-in adjustable lights, superbly clear optics to give higher magnification and greater zoomability, and often a camera, computer or video port. Like so many things, the more you pay, the higher the quality and versatility, but it is worth visiting a showroom or trade fair to get a feel for what costs how much before making that final decision. However much you spend, rest assured that the microscope will last longer than any smartphone. Nearly 35 years on, I still get out my first little scope occasionally, and it works fine.

IDENTIFICATION GUIDES – BOOKS AND MENTORS

The brief introduction to British beetle families in Chapter 7 gives a few sample identification guides for each group. Some are more detailed than others, some are more technical, and some are more up to date. Soon after this book goes to press, there will probably be a few new keys available anyway. For the beginner the most important thing is to find books with good pictures. No birdwatcher ever squints through their binoculars trying to discover if the frolicsome avian in the bush away off yonder is anisodactylic (one toe backwards) or zygodactylic (two toes backwards). They make a rough assumption on whether it might be a wood pigeon or a woodpecker based on its general size and shape, and maybe whether it's cooing. It's the same with beetles. Despite what it says in the keys, you rarely have to count tarsal segments to get to a family; you just need to scour through the pictures until you get close. Picture books come and go. Useful starts can be made in White (1983), Harde (1984), Chinery (1986), Evans (2014), etc. Older books like Bechyne (1956), Linssen (1959), Lyneborg (1977) and Zahradnik (1977) are useful if second-hand copies are available. Eventually, more detailed monographs and papers can be sought out. Some of these are free-standing books of varying costs and complexities, but many of the key works cover just a single genus, sometimes just two or three species, and these are scattered through entomological journals, some mainstream, some very obscure. They can usually be found, but sometimes the hunt for the identification paper is longer and more exhausting than the hunt for the beetle in the first place.

A significant development began in 1980 with the *Coleopterist's Newsletter,* a slightly informal duplicated news-sheet of reports, sightings, keys and articles on British beetles (not to be confused with the journal of the same name published by the Coleopterists Society in the United States). In 1992, it morphed into the *Coleopterist* and is now a full-colour journal published three times a year. It carries detailed papers on all aspects of British Coleoptera, from family or generic revisions, to new species, local lists, unusual observations or the fact that my old friend *Hydnobius perrisi/Sogda suturalis* has recently been found in west Norfolk. Other all-beetle publications include *Latissimus,* the newsletter of the Balfour-Browne Club, dedicated to water beetles (not just British ones) and *Beetle News,* an online open-access newsletter with several very useful illustrated beginners' guides.

Buying books is my number one expense, but I like books and am comfortable with my outlays. An increasing number of specialist guides and keys are also available online (see Appendix); many well-out-of-date (nineteenth- and early-twentieth-century) but still highly useful scientific journals, books and papers are also in the public domain, and as scanned searchable resources they are easily accessible. A few out-of-print Royal Entomological Society handbooks are free to download from their website, including 19 on beetles from the introduction and key to families by Crowson (1956), to the still current and authoritative key to Dermestidae by Peacock (1993). Picture books are also available online, and the famous colour-plate works by Calwer (1876), Fowler (1887–91) and Reitter (1908–16), among others, are now digitised and free to view. In addition, there are immeasurable beetle photographs out there to view and compare. A few current beetle sites are listed in the Appendix.

The Internet really comes into its own, though, when offering up your own pictures of beetles for others to identify. Where previously taking a box of difficult specimens up to the museum reference collection or asking a more experienced coleopterist to cast an eye used to be the norm (it's still a good way to get help), now uploading a picture for others to look at is the way many people get started. Whether subjects are photographed alive in nature or dead down the microscope, the *iSpot* wildlife-identification site, run in conjunction with the Open University, allows pictures to be uploaded for others to view and name. Other slightly more specialised groups also exist (see Appendix), eventually leading to recording schemes for individual beetle families, or private individuals working on particular genera. This process is two-way: the newcomer gets help and insight with naming species, and the namer gets locality information to feed into the database.

Even the most trivial of beetle records (no matter how common and widespread the species) become subsumed into the recording machinery, eventually to be churned out at the other end in the form of atlases, local lists, status reviews or family revisions. These statements of species distribution, range expansion (or decline), population trend and conservation status all add to our understanding of how the Coleoptera are doing in Britain. They range from superb local lists like those for Gloucestershire (Atty, 1983) and Surrey (Denton, 2007), to national recording scheme results like those produced for Carabidae (Luff, 1998) and leaf beetles (Cox, 2007). In the future (possibly even in the very near future), they will also build into our interpretation of the effects of climate change, habitat fragmentation, urbanisation, intensive farming, invasive aliens, pollution, land management and nature conservation.

The harlequin ladybird (*Harmonia axyridis*) remains the best-studied example of an alien organism spreading across the UK; pictures of it dominate the *iSpot* website, and these records formed an important part of the analysis of its move through the country. At the other extreme, when Matt Dodds put up a photograph of a fat beetle grub, cupped in the hand, he set off a cascade of excited

FIG 168. After all that work, I need a sit-down, and where better? My favourite bench, in Crane Park, Twickenham.

action. His initial plea, 'I was hoping it might be a noble chafer but haven't really got a clue', posted on 23 June 2011, was followed two days later by the announcement that Mark Telfer and Martin Harvey had visited the old orchard site and confirmed the presence of living larvae and the fragmented remains of several adults of the extremely rare and very handsome *Gnorimus nobilis* (Harvey *et al.*, 2014).

This is not an isolated occurrence. With so many interested citizen scientists now armed with cameras, computers and the curiosity to look deeper, all manner of communal research is going on. I sometimes wistfully flick through the *iSpot* pictures, suggesting names or adding my agreement to others, and as usual I wonder why so many people other than me are seeing things like oil beetles. In fact, this very minute, 16:57 hours on Saturday, 18 March 2017, while writing about the site, another damn picture of the violet oil beetle (*Meloe violaceus*) pops up there to taunt me. I hate it. I love it really.

Interactions with other coleopterists, whether through *iSpot*, field meetings, recording schemes, exchange of specimens or identification guides, or the simple dissemination of records and information, brings the study of the Coleoptera within everybody's grasp. Everyone gains. Everyone learns.

* * *

Beetles – so diverse, so numerous, so wonderful – are everywhere, and are amenable to study in so many ways. Don't get bogged down in the detail. Don't be put off by the seeming complexities of identification keys. Don't imagine for a minute that there is no contribution to be made by the beginner. Just get out there and begin – seek, find and marvel. That is all.

Appendix: Online Resources

THIS IS THE MODERN WORLD, so it requires online resources. Web addresses are transient, and change regularly; below are a few that were current in July 2017.

Beetle Fauna of Germany
kerbtier.de
Many useful photo galleries.

Beetle News
amentsoc.org/publications/beetle-news
Downloadable newsletter with several useful introductory identification guides. Currently hosted by the Amateur Entomologists' Society.

Biodiversity Heritage Library
biodiversitylibrary.org
Searchable library of scanned and digitised natural history books, mostly old and out of copyright, but also including many more modern specialist journals.

British Entomological and Natural History Society
benhs.org.uk
Britain's foremost field entomological society. Organises a range of events, including field meetings, identification workshops and an annual exhibition.

The Coleopterist
coleopterist.org.uk
Website for the specialist *Coleoptera* journal. Useful introduction, with links to recording schemes, county and regional recorders, biological record centres, a biographical dictionary of coleopterists and the online newsgroup.

iSpot
ispotnature.org
Public site to upload photos for identification advice.

Koleopterologie
koleopterologie.de
Impressive gallery of German species.

Mark Telfer's Website
markgtelfer.co.uk/beetles
Excellent online keys and identification help for a large number of British beetle families.

Mike's Beetle Keys
sites.google.com/site/mikesinsectkeys/Home
Numerous photo-led identification keys, created by Mike Hackworth.

NBN Atlas
nbnatlas.org
Huge collection of biodiversity information and records, shared by the National Biodiversity Network, including useful, but always rather putative, distribution maps.

Royal Entomological Society
royensoc.co.uk
Click on the 'Publications' link, then on 'Handbooks', for a list of new handbooks available and out-of-print volumes to download for free.

Tomorrow's Biodiversity
tombio.uk
Hosted by the Field Studies Council (FSC), a project to provide identification guides. Search through the 'ID Signpost' tab for beetle PDFs, FSC publications and links to many other external sites.

UK Beetle Recording
coleoptera.org.uk
List and links for UK recording schemes covering the Coleoptera.

Watford Coleoptera Group
thewcg.org.uk
Galleries, information and identification help.

Glossary

Acuminate Pointed, usually sharply so.

Aedeagus Hardened part of internal male genitalia. Sometimes spelled edeagus.

Antennomere One segment of an antenna.

Apodous Legless, referring to larvae.

Aposematic Warningly marked, usually brightly coloured and strongly patterned.

Apposition eye Where each eye facet focuses light onto just one light-sensitive rhabdom beneath, usually in day-active species.

Arrhenotoky Haplodiploidy.

Bifurcate Split or divided into two branches or forks.

Bilobed Divided into two distinct lobes, usually of some or all of the tarsal segments.

Bipectinate Referring to the antennae, each segment having two narrow, protruding lobes, giving a double-comb effect.

Bothrium (pl. bothria) Long sensory (tactile) hair, often set in a depression or puncture.

Campodeiform Narrow, long-legged larval body shape, with distinct antennae and two long tails.

Canthus Ridge, lobe or bar dividing or indenting the eye.

Castaneous Deep reddish-brown colour of chestnuts.

Cercus (pl. cerci) Segmented (or not), feeler-like sensory organ at the tip of the abdomen in some insects, although not beetles (cf. urogomphus).

Cheloniform Flattened and rounded larval body shape, like a tortoise.

Chitin (adj. chitinous) Hard, tough, high-molecular-weight, unbranched amino-polysaccharide in insect cuticle.

Clavate Clubbed, referring to the antennae.

Clypeus Shield-like plate covering the jaws.

Condyle Lobe-and-socket joint by which the jaw or elytron hinges.

Cordate Heart-shaped, usually referring to the pronotum.

Coriaceous Leathery, usually in the texture of the elytra and/or pronotum.

Coxa (pl. coxae) Basal-most segment of the leg, where it attaches to the underside of the thorax.

Cremaster Silk anchor point by which larva attaches its tail end to the substrate for pupation.

Crepuscular Active at dusk.

Cyathiform Cup-shaped, referring to the final antennal or palp segments.

Declivity Sudden inward impression or downward abbreviation, often at the tip of the elytra in some bark beetles.

Decumbent Lying down against the body, referring to pubescence and hairs.

Disc The centre of any flattish body section.

Ectoparasitoid A parasitoid feeding on the outside of its victim rather than internally.

Elateriform Tough, slender, cylindrical larval body form, with short legs.

Elytron (pl. elytra) Hard shell-like or at least leathery wing-case, a pair of which cover the beetle hind wings and abdomen.

Emarginate Cut in, scooped out or indented.

Epipleuron Narrow under-lip along the edge of the elytron.

Eruciform Caterpillar-like, referring to larval body shape.

Exarate Referring to the pupa, whereby the legs, antennae and wings appear free, away from the body.

Exoskeleton The external hard chitinous shell of insects and other arthropods.

Facies Overall general appearance.

Falcate Long, curved and pointed, referring to the jaw.

Femur (pl. femora) Third leg segment, usually the largest, analogous to the thigh.

Ferruginous Yellowish brown or pinkish brown.

Filiform Referring to the antennae, having uniformly matching cylindrical segments.

Flabellate Referring to the antennae, feathery, each segment with a long finger-like extrusion.

Flagellum The third segment onwards of the antenna.

Fossorial Adapted for digging, referring to the leg or body form.

Frass Insect droppings, usually small, hard, roundish pellets.

Frons The forehead – usually the area between the eyes.

Fuliginous Deep black, sooty colour.

Funiculus Intermediate small antennal segments between the scape or pedicel and those enlarged into a terminal club.

Fuscous Dull brown.

Fusiform Spindle-shaped, broad in the middle and narrow at each end.

Gall Swelling, bobble, node or interfered growth of a plant caused by a creature living inside it.

Ganglion (pl. ganglia) Nerve nodule, usually paired, and often repeated in some or all body segments.

Geniculate Referring to the antennae, elbowed, with a long basal segment.

Glabrous Hairless, smooth and usually shining.

Habitus Full-body illustration.

Haemocoele Body cavity filled with haemolymph.

Haemolymph Insect equivalent of blood.

Haplodiploidy Sex determination by chromosome number; usually n = male, 2n = female.

Hastate Spear-shaped, referring to the antennal club.

Hemelytra Partially hardened and coloured front wings of the Hemiptera or 'true' bugs.

Hemimetaboly/hemimetabolism (adj. hemimetabolous) Passing through gradually increasing wing-budded nymphs to fully winged adult without a dormant pupa/chrysalis stage.

Heteromerous Having differing numbers of tarsal segments on different legs, usually five on the front and middle, but four on the rear.

Holometaboly/holometabolism (adj. holometabolous) Having a chrysalis/pupa stage between the wingless larva and adult.

Humerus (pl. humeri) Front outer corner of the elytra, the shoulder.

Hydrofuge Water-repellent.

Hypermetamorphosis With larvae passing through several distinct body forms.

Inquiline An animal invading the living space (nest, gall, gallery) of another, but generally not doing direct harm.

Instar Discrete phase (usually in size) in larval development, between one moult and the next.

Interstices Smooth or convex regions between the linear striae of the elytra.

Johnston's organ Sensory organ in the second antennal segment, measuring movement and sound (as vibrations) via the remainder of the flagellum.

Labium (adj. labial) Lower lip, small plate under the mouthparts.

Labrum Upper lip, small plate covering the mouthparts or the jaws.

Lamellate Referring to the antennae, in which the final segments are expanded into flat plates that stack together into a club.

Larviform Resembling a larva (used in reference to an adult).

Larvule First-instar larva, emerging from the egg.

Luteus Strong yellow, with a hint of red.

Mandibles Jaws, from which comes Mandibulata, a slightly archaic subdivision of insects — those with biting mouthparts.

Maxilla (pl. maxillae, adj. maxilliary) Movable mouthpart under the mandibles, usually softer and more delicate, used for food manipulation.

Melanin Chemical pigment, giving at varying densities colours from yellow and orange to brown and black.

Meso- Middle, usually referring to the second of the three thoracic segments.

Meta- Hind, or rearmost, usually referring to the third of the three thoracic segments.

Metamerism (adj. metameric) Evolution from simple unsegmented body form, through duplication, to multi-segmented body structure.

Microfibrils Bundles of aligned chitin molecules, visible under the electron microscope.

Microsculpture Tiny cracks and lace-like patterns seen down the microscope.

Microtrichia Minute hairs, on some portions of the flight wings.

Mola Inner grinding corner of jaw.

Moniliform Referring to the antennae, having uniformly bead-like segmentation, as a pearl necklace.

Mucro Sharp apical point, sometimes slightly hooked, usually at the end of a leg segment.

Mycangium (pl. mycangiua) Pocket, pore or hollow that carries mutualistic fungal spores, usually in bark beetles.

Mycophagous Feeding on moulds and fungi.

Myrmecophile Living in ant nests.

Neoteny Retention of immature or larval characters into adulthood.

Notum One of the upper plates of the thorax.

Obtect Referring to the pupa, covered by a hard shell, legs, wings and antennae that are attached immovably to the body.

Ocellus (pl. ocelli) Simple light-sensitive dome, usually on the forehead; present in only a few beetle groups.

Ommatidium (pl. ommatidia) Facet that combines with others (sometimes in hundreds or thousands) to form the large compound eyes.

Onisciform Referring to beetle larvae, having segments that are broad and flanged, like those of a woodlouse.

Ostium (pl. ostia) Valved, one-way pores in the dorsal pumping tube of an insect's blood system.

Ovipositor Egg-laying apparatus, a shorter or longer telescopic tube.

Ovoid Egg-shaped, oval.

Paedogenesis (adj. paedogenetic) Creation of viable new eggs or larvae directly by a larva.

Palpus/palp (pl. palpi/palps) Short segmented feeler, attached as a pair to the mouthparts.

Pectinate Referring to the antennae or tarsi, multi-toothed, forming a comb-like array.

Pedicel Second segment of an antenna.

Peduncle Narrow waist between the pronotum and elytra, often with a surmounted scutellum.

Pentamerous Having five segments on the tarsus.

Pharynx Throat.

Pheromone Chemical scent, usually produced in alarm or to attract a mate.

Phoresy Grasping another (often flying) animal to hitch a lift.

Phytophagy (adj. phytophaous) Feeding on plants.

Piceus Reddish black or pitch in colour.

Pilose Covered in (usually longer) hairs.

Plastron Bubble of air covering parts of the body or under the elytra of various water beetles.

Pleuron (pl. pleura) One of the side plates of the thorax.

Plumose Feathery.

Prepupa Sedentary pause in the final larval instar, which usually adopts a stiff, hunched pose, before the final moult produces the pupa.

Pro- Front or fore, usually referring to the first of the three thoracic segments.

Pronotum (adj. pronotal) Large, dominant upper shield-plate of the thorax.

Pubescent Covered in (usually shorter) hairs.

Punctuation Arrangement of dints, pits, pinpricks, dimples or craters decorating various body parts. Sometimes called punctation or puncturation.

Pygidium Last full tergite of the abdomen, sometimes visible peeking out from the end of the elytra.

Pyrophilous Fire-loving, usually of species that breed in burned wood.

Reticulation Wrinkles.

Retinaculum Internal tooth or spike on the middle of the cutting edge of a beetle larva's jaw.

Rhabdom Light-sensitive column array under each of the facets of the compound eye.

Rostrum Long or short snout, with mandibles at the end, found in weevils and a few other beetles.

Rufous Pale reddish, slightly translucent.

Saproxylic Breeding in rotten, fungoid wood.

Scape First, often longer or larger, segment of an antenna.

Scarabaeiform Fat C-shaped larval body form, with short claw-like legs.

Scatoshelling Coating an egg in excrement, using the rear feet, before dropping it.

Sclerotised Hardened and usually smooth or shining.

Scrobe Groove or indentation on the head or snout to receive an antenna, particularly the large basal segment in weevils.

Scutellum (adj. scutellar or scutellary) Small, usually triangular, plate behind the pronotum, between the bases of the elytra.

Securiform Axe- or hatchet-shaped, referring to the final antennal or palp segment.

Sensillum (pl. sensilla) Sensory hair.

Sensorium Small (usually single-segmented) side branch of beetle larval antennae.

Serrate Saw-toothed, referring to the antennae or tarsi, pronotal edge or elytral apex.

Seta (pl. setae) A distinct hair, or very narrow spine.

Spermatheca Often hardened, specifically shaped (usually twisted) sperm-storage organ inside the female abdomen.

Spiracle Hole in the side of an insect segment where an air-breathing tube (trachea) opens.

Stemma (pl. stemmata) Simple, small, domed, light-sensitive area on the head of a beetle larva; usually one to six on each side.

Sternite One of the underside plates of the abdomen.

Sternum (pl. sterna) One of the lower (underside) plates of the thorax.

Stria (pl. striae, adj. striate) Indented groove, or row of punctures, or both, along the elytra.

Stridulation Scraping of one body part past another to produce sound or vibration.

Striole Short stria, or part of a stria.

Style Long, pointed abdomen tip (especially in Mordellidae).

Sub-cortical Occurring under tree bark.

Sub-cylindrical Nearly cylindrical.

Sulphureus Pale but strong yellow, with a hint of green.

Superposition eye Where several facets of the compound eye focus light together onto a light-sensitive rhabdom beneath, usually in low-light, crepuscular or nocturnal species.

Suture (adj. sutural) The usually straight line where the two elytra meet at rest.

Tagmosis Evolutionary combination (and rationalisation) of multiple repeated body segments into aggregate body sections like a head, thorax and abdomen.

Tarsomere Any one of the segments of the tarsus.

Tarsus (p. tarsi, adj. tarsal) Multi-segmented final leg region, analogous to the foot.

Tegmen (pl. tegmina) Short elytron-like wing-case of earwigs and some other insects.

Tergite One of the upperside plates of the abdomen.

Testaceous Orange, reddish or pinkish yellow, slightly translucent, like a snail shell.

Tetramerous Having four segments on the tarsus.

Thanatosis Feigning death using body rigidity and immobility.

Thelytoky Parthenogensis whereby unmated females give birth to all-female offspring.

Tibia (pl. tibiae) Fourth leg segment, often the longest and thinnest, analogous to the shin.

Tibiotarsus Fused tibia and tarsus found in some beetle larval legs.

Trachea (pl. tracheae) Breathing tube, usually paired in each body segment.

Tracheole Blind end of breathing tube (trachea) inside a body segment.

Trichobothrium A bothrium, or thin sensory hair.

Trichome Long hair, usually arising from a pore, on which pheromones or other chemicals are secreted.

Triungulin Minute, active, long-legged first-stage larva of oil beetles etc.

Trochanter The second leg segment, usually small.

Truncate Shortened or cut off before where the end might be expected.

Urogomphus (pl. urogomphi) One of a pair of solid or articulated projections from the 10th abdominal segment of most (but not all) beetle larvae.

Vermiform Worm- or maggot-like larval body form, usually lacking legs.

Vilose Covered in hairs; similar to pubescent, but hairs slightly longer.

Viviparity (adj. viviparous) Giving birth to live larvae, rather than laying eggs.

References

Ahrens, D., Schwarzer, J. & Vogler, A. P.
(2014). The evolution of scarab
beetles tracks the sequential rise of
angiosperms and mammals. *Proceedings
of the Royal Society B*, **281**, 20141470.

Aldrovandi, U. (1602). *De Animalibus Insectis
Libri Septem, cum Singulorum Iconibus ad
Viuum Expressis.* Bononiae, Bologna.

Alexander, K. N. A. (1988). The development
of an index of ecological continuity for
deadwood associated beetles. *Antenna*,
12, 69–70.

**Alexander, K. N. A., Dodd, S. & Denton,
J. S.** (2014). *A Review of the Scarce and
Threatened Beetles of Great Britain.
The Darkling Beetles and their Allies.
Species Status No. 18.* UK Joint Nature
Conservation Committee/Natural
England, Peterborough.

Allaud, C. (1899). *Guide de l'Entomologiste à
Madagascar.* Allaud, Paris.

Allen, A. A. (1951). *Lathridius bifasciatus*
Reitter (Col., Lathridiidae), an
Australian beetle found wild in Britain.
Entomologist's Monthly Magazine, **87**,
114–115.

Allen, A. A. (1956). *Pediacus depressus* in Berks
and Sussex; with a summary of its
British history. *Entomologist's Monthly
Magazine*, **97**, 212.

Allen, A. A. (1969). Notes on some British
serricorn Coleoptera, with adjustments
to the list. 1.– Sternoxia. *Entomologist's
Monthly Magazine*, **104**, 208–216.

Allen, A. A. (1970). *Heterocerus hispidulus* Kies.
(Heteroceridae), a beetle new to Britain.
*Entomologist's Record and Journal of
Variation*, **82**, 232–234.

Allen, A. A. (1972). *Magdalis memnonia* Gyll.
(Col., Curculionidae), a weevil new to
Britain. *Entomologist's Record and Journal
of Variation*, **84**, 22–23.

Allen, A. J. & Hance, D. (2009). *Hololepta
plana* (Sulzer, 1776) (Histeridae) in
Norfolk – new to Britain. *Coleopterist*, **18**,
153–154.

Allen, A. A. & Owen, J. A. (1997). *Philonthus
spinipes* Sharp (Staphylinidae) in Dorset –
new to Britain. *Coleopterist*, **6**, 81–83.

Allen, A. J. & Turner, C. R. (2012). *Conarthrus
praestus* (Boheman in Schönherr, 1838)
(Curculionidae, Cossoninae) in Britain.
Coleopterist, **21**, 21–23.

Anderson, J. M & Coe, M. J. (1974).
Deposition of elephant dung in an arid
tropical environment. *Oecologia*, **14**,
111–125.

Andersson, K. (2012). Pheromone-based
monitoring of *Elater ferrugineus* as
an indicator for species-rich hollow
oak stands. Masters thesis, Swedish

University of Agricultural Sciences, Alnarp.

Aneshansley, D. J., Eisner, T., Widom, J. M. & Widom, B. (1969). Biochemistry at 100 °C: explosive secretory discharge of bombardier beetles (Brachinus). *Science*, **165**, 61–63.

Anon. (1895). *Flea Beetles (Halticae). Leaflet 3*. Board of Agriculture and Fisheries, London.

Anon. (1918–21). *Reports of the Grain Pests (War) Committee*. Harrison & Sons, London.

Anon. (1959). Banana-peeling beetle is dead; museum's insect rare in US. *New York Times*, 9 July, p. 17.

Arenas, L. M., Walter, D. & Stevens, M. (2015). Signal honesty and predation risk among a closely related group of aposematic species. *Scientific Reports*, **5**, 11021.

Arillo, A. & Ortuno, V. M. (2008). Did dinosaurs have any relation with dung-beetles? (The origin of coprophagy). *Journal of Natural History*, **42**, 1405–1408.

Ashe, J. S. & Timm, R. M. (1988). *Chilamblyopinus piceus*, a new genus and species of Amblyopinine (Coleoptera: Staphylinidae) from southern Chile, with a discussion of amblyopinine generic relationships. *Journal of the Kansas Entomological Society*, **61**, 46–57.

Ashe, J. S. & Timm, R. M. (1995). Systematics, distribution, and host specificity of *Amblyopinus* Solsky 1987 (Coleoptera Staphylinindae) in Mexico and Central America. *Tropical Zoology*, **8**, 373–399.

Ashworth, A. C. (1996). The response of arctic carabids (Coleoptera) to climate change based on the fossil record of the Quaternary period. *Annales Zoologici Fennici*, **33**, 125–131.

Atkins, M. D. (1957). An interesting attractant for *Priacma serrata* (Lee), (Cupedidae; Coleoptera). *Canadian Entomologist*, **89**, 214–215.

Atty, D. B. (1983). *Coleoptera of Gloucestershire*. Atty, Cheltenham.

Balfour-Browne, F. (1940, 1950, 1958). *British Water Beetles*. Ray Society, London. 3 vols.

Balfour-Browne, F. (1953). *Coleoptera: Hydradphaga. Handbooks for the Identification of British Insects 4(3)*. Royal Entomological Society, London.

Balfour-Browne, F. (1962). *Water Beetles and Other Things, Half a Century's Work*. Blacklock Farries, Dumfries.

Ball, S. G. (1986). *Terrestrial and Freshwater Invertebrates with Red Data Book, Notable or Habitat Indicator Status. Invertebrate Site Register Internal Report 66*. Nature Conservancy Council, Peterborough.

Barber, H. S. (1913a). Observations on the life history of *Micromalthus debilis* LeC. (Coleoptera). *Proceedings of the Entomological Society of Washington*, **15**, 31–38.

Barber, H. S. (1913b). The remarkable life history of a new family (Micromalthidae) of beetles. *Proceedings of the Biological Society of Washington*, **26**, 185–190.

Barclay, M. V. L. (2003). *Otiorhynchus* (*s. str.*) *armadillo* (Rossi, 1792) and *Otiorhynchus* (*s. str.*) *salicicola* Heyden, 1908 (Curculionidae: Entiminae: Otiorhynchini) – two European vine weevils established in Britain. *Coleopterist*, **12**, 41–56.

Barclay, M. V. L. (2010). A. A. Allen. *Coleopterist*, **19**, 150.

Barclay, M. V. L. (2014). *Bruchidius siliquastri* Delobel, 2007 (Chrysomelidae, Bruchinae) new to Britain. *Coleopterist*, **23**, 41–44.

Barclay, M. V. L. & Notton, D. G. (2015). *Oxythyrea funesta* (Poda) (Scarabaeidae) in London and a summary of recent and historical British records of this species. *Coleopterist*, **24**, 113–116.

Bates, H. W. (1877). On *Ceratorrhina quadrimaculata* (Fabr.), and descriptions of two allied species. *Transactions of the Entomological Society, 1877*, 201–203.

Batten, R. (1986). A review of the British Mordellidae (Coleoptera). *Entomologist's Gazette*, **37**, 225–235.

Beavis, I. C. (1988). *Insects and Other Invertebrates in Classical Antiquity.* Exeter University Press, Exeter.

Bechyne, J. (1956). *Beetles. Open Air Guides.* Trans. C. M. F. von Hayek. Thames & Hudson, London.

Beirne, B. P. (1952). *The Origin and History of the British Fauna.* Methuen, London.

Belmain, S. (2000). The death-watch beetle – accommodated in all the best places. *Pesticide Outlook*, **11**, 233–237.

Belmain, S. R., Simmonds, M. S. J. & Blaney, W. M. (1999). Deathwatch beetle, *Xestobium rufovillosum*, in historical buildings: monitoring the pest and its predators. *Entomologia Experimentalis et Applicata*, **93**, 97–104.

Bense, U. (1995). *Longhorn Beetles; Illustrated Key to Cerambycidae and Vesperidae of Europe.* Margraf Verlag, Weikersheim.

Bercedo, P. & Arnáiz, L. (2010). A new species of the genus *Microsternomorphus* Español, 1977 from England (United Kingdom) (Coleoptera: Ptinidae: Dorcatominae). *Coleopterist*, **19**, 105–110.

Berkov, A., Rodríguez, N. & Centeno, P. (2008). Convergent evolution in the antennae of a cerambycid beetle, *Onychocerus albitarsis*, and the sting of a scorpion. *Naturwissenschaften*, **95**, 257–261.

Bernhardt, P. (2000). Convergent evolution and adaptive radiation of beetle-pollinated angiosperms. *Plant Systematics and Evolution*, **222**, 293–320.

Berry, P. (1993). From cow pat to frying pan: Australian herring (*Arripis georgianus*) feed on an introduced dung beetle (Scarabaeidae). *Western Australian Naturalist*, **19**, 241–242.

Béthoux, O. (2009). The oldest beetle identified. *Journal of Paleontology*, **83**, 931–937.

Beutel, R. G. & Hörnschemeyer, T. (2002). Larval morphology and phylogenetic position of *Micromalthus debilis* LeConte (Coleoptera: Micromalthidae). *Systematic Entomology*, **27**, 169–190.

Beutel, R. G., Friedrich, F., Ge, S.-Q. & Yang, X.-K. (2014). *Insect Morphology and Phylogeny: A Textbook for Students of Entomology.* Gruyter, Berlin.

Bíly, S. (1982). *The Buprestidae (Coleoptera) of Fennoscandia and Denmark. Fauna Entomologica Scandinavica 10.* Scandinavian Science Press, Klampenborg.

Booth, R. (1984). A provisional key to the British species of *Tachyporus* (Coleoptera: Staphylinidae) based on elytral chaetotaxy. *Circaea*, **2**, 15–19.

Booth, R. G. & Galsworthy, A. C. (2015). *Epurea ocularis* Fairmaire, 1849 (Nitidulidae) new to the UK. *Coleopterist*, **24**, 69–71.

Bouchard, P., Bousquet, Y., Davies, A. E. *et al.* (2011). Family group names in Coleoptera (Insecta). *ZooKeys*, **88**, 1–972.

Boussau, B., Walton, Z., Delgado, J. A. *et al.* (2014). Strepsiptera, phylogenomics and the long branch attraction problem. *PLoS ONE*, **9**, e107709.

Böving, A. G. & Craighead, F. C. (1931). *An Illustrated Synopsis of the Principal Larval*

Forms of the Order Coleoptera. Brooklyn Entomological Society, New York.

Bowden, J. & Johnson, C. G. (1976). Migrating and other terrestrial insects at sea. In *Marine Insects* (ed. L. Cheng). New Holland Publishing, Amsterdam.

Bowstead, S. (1999). *A Revision of the Corylophidae (Coleoptera) of the West Palaearctic Region.* Muséum d'Histoire Naturelle, Geneva.

Brackenbury, J. (1992). *Insects in Flight.* Blandford, London.

Brackenbury, J. & Wang, R. (1995). Ballistics and visual targeting in flea-beetles (Alticinae). *Journal of Experimental Biology,* **198**, 1931–1942.

Brakefield, P. M. & de Jong, P. W. (2011). A steep cline in ladybird melanism has declined over 25 years: a genetic response to climate change? *Heredity,* **107**, 574–578.

Brand, J. M., Blum, M. S., Fales, H. M. & Pasteels, J. M. (1973). The chemistry of the defensive secretion of the beetle *Drusilla canaliculata. Journal of Insect Physiology,* **19**, 369–382.

Branham, M. A. & Wenzel, J. W. (2001). The evolution of bioluminescence in cantharoids (Coleoptera: Elateroidea). *Florida Entomologist,* **84**, 565–586.

Branham, M. A. & Wenzel, J. W. (2003). The origin of photic behaviour and the evolution of sexual communication in fireflies (Coleoptera: Lampyridae). *Cladistics,* **19**, 1–22.

Bravo, C., Bautista, L. M., Garcia-Paris, M., Blanco, G. & Alonso, J. C. (2014). Males of a strongly polygynous species consume more poisonous food than females. *PLoS ONE,* **9**, e111057.

Brendell, M. J. D. (1975). *Coleoptera Tenebrionidae. Handbooks for the Identification of British Insects* 5(10). Royal Entomological Society, London.

Britton, E. B. (1956). *Coleoptera Scarabaeoidea. Handbooks for the Identification of British Insects* 5(11). Royal Entomological Society, London.

Brodsky, A. K. (1994). *The Evolution of Insect Flight.* Oxford University Press, Oxford.

Brookes, R. (1763). *The Natural History of Insects, with their Properties and Uses in Medicine. Vol. 4.* Newbery, London.

Brown, P. M. (2011). Decline in native ladybirds in response to the arrival of *Harmonia axyridis*: early evidence from England. *Ecological Entomology,* **36**, 231–240.

Brown, T. (1833). *The Taxidermist's Manual, or, the Art of Collecting, Preparing and Preserving Objects of Natural History, Designed for the Use of Travelers, Conservators of Museums and Private Collectors.* Fullerton, Glasgow.

Buck, F. D. (1948). *Pentarthrum huttoni* Woll. (Col., Curculionindae) and some imported Cossoninae. *Entomologist's Monthly Magazine,* **84**, 153.

Buck, F. D. (1954). *Coleoptera (Lagriidae, Alleculidae, Tetratomidae, Melandryidae, Salpingidae, Pythidae, Mycteridae, Oedemeridae, Mordellidae, Scraptiidae, Pyrochroidae, Rhipiphoridae, Anthicidae, Aderidae and Meloidae). Handbooks for the Identification of British Insects* 5(9). Royal Entomological Society, London.

Buck, F. D. (1957). The British Mycetophagidae and Colydiidae. *Proceedings and Transactions of the South London Entomological and Natural History Society,* **1955**: 53–66.

Buckland, P. I. & Buckland, P. C. (2017). Species as fossils in Quaternary sediments. In *Checklist of Beetles of the British Isles* (A. G. Duff). 3rd edn. Pemberley Books, Iver.

Burrell, J. (1807, 1809). A catalogue of insects found in Norfolk. *Transactions* of the *Entomological Society of London*, 1, 101–240.

Butler, E. A. (1896). *Our Household Insects. An Account of the Insect-pests Found in Dwelling-houses.* Longmans, Green & Co., London.

Callahan, P. S. (1977). Moth and candle: the candle flame as a sexual mimic of the coded infrared wavelengths from a moth sex scent (pheromone). *Applied Optics*, 16, 3089–3097.

Calwer, C. G. (1876). *Käferbuch. Naturgeschichte der Käfer Europas. Zum Handgebrauche für Sammler.* Hoffmann, Stuttgart.

Carlton, C. & Bayless, V. (2011). A case of *Cnestus mutilatus* (Blandford) (Curculionidae: Scolytinae: Xyleborini) females damaging plastic full storage containers in Louisiana, USA. *Coleopterists Bulletin*, 65, 290–291.

Carr, R. (2000). The occurrence of *Nebrioporus canaliculatus* (Lacordaire) (Coleoptera: Dytiscidae) in Britain, with a note on recent taxonomic changes to related British species. *Entomologist's Gazette*, 51, 125–128.

Carroll, A. L., Taylor, S. W., Regniere, J. & Safranyik, L. (2004). Effect of climate change on range expansion by the mountain pine beetle in British Columbia. In *Mountain Pine Beetle Symposium: Challenges and Solutions* (eds T. L. Shore *et al*). Natural Resources Canada, Victoria.

Carter, I. & Owen, J. (1987). *A Numerical List of British Beetles.* Death-watch Press, Cheltenham.

Chapman, J. W., Reynolds, D. R., Smith, A. D., Smith, E. T. & Woiwod, I. P. (2004). An aerial netting study of insects migrating at high altitude over England. *Bulletin of Entomological Research*, 94, 123–136.

Chatzimanolis, S., Grimaldi, D. A., Engel, M. S. & Fraser, N. C. (2012). *Leehermania prorova*, the earliest staphyliniform beetle, from the late Triassic of Virginia (Coleoptera: Staphylinidae). *American Museum Novitates*, 3761, 1–28.

Chinery, M. (1986). *Collins Guide to the Insects of Britain and Western Europe.* HarperCollins, London.

Chu, H. F. (1949). *How to Know the Immature Insects; an Illustrated Key for Identifying the Orders and Families of Many of the Immature Insects with Suggestions for Collecting, Rearing and Studying Them.* W. C. Brown, Dubuque.

Clark, J. T. (1977). Aspects of variation in the stag beetle *Lucanus cervus* (L.) (Coleoptera: Lucanidae). *Systematic Entomology*, 2, 9–16.

Clarke, R. O. S. (1973). Heteroceridae. *Handbooks for the Identification of British Insects. 5(2c).* Royal Entomological Society, London.

Colunga-Garcia, M. & Gage, S. H. (1998). Arrival, establishment, and habitat use of the multicolored Asian lady beetle (Coleoptera: Coccinellidae) in a Michigan landscape. *Environmental Entomology*, 27, 1574–1580.

Colwell, R. K. & Coddington, J. A. (1994). Estimating terrestrial biodiversity through extrapolation. *Philosophical Transactions of the Royal Society of London*, 345, 101–118.

Compton, S. G., Craven, J. C., Key, R. & Key, R. J. D. (2007). Lundy cabbage: past, present, future. In *Lundy Studies: Proceedings of the 60th Anniversary Symposium of the Lundy Field Society* (ed. J. J. George). Lundy Field Society, Lundy.

Comstock, J. H. & Needham, J. G. (1898, 1899). The wings of insects. *American*

Naturalist, **32**, 43–48, 81–89, 231–257, 335–340; **33**, 117–126, 573–582, 845–860.

Coombs, C. W. & Woodroffe, G. E. (1955). A revision of the British species of *Cryptophagus* (Herbst) (Coleoptera: Cryptophagidae). *Transactions of the Royal Entomological Society of London*, **106**, 237–282.

Coope, G. R. (1973). Tibetan species of dung beetle from late Pleistocene deposits in England. *Nature*, **245**, 335–336.

Cooter, J. (1996). Annotated keys to the British Leiodidae (Col., Leiodidae). *Entomologist's Monthly Magazine*, **132**, 205–272.

Cooter, J. & Barclay, M. V. L. (2006). *A Coleopterist's Handbook*. 4th edn. Amateur Entomologist's Society, London.

Cowper, W. (1860). *The Poems of William Cowper Esq., of the Inner Temple, Complete in One Volume*. Crosby & Nichols, Boston.

Cox, H. E. (1874). *A Handbook of the Coleoptera or Beetles, of Great Britain and Ireland*. Janson, London. 2 vols.

Cox, M. L. (2001a). Notes on the natural history, distribution and identification of seed beetles (Bruchidae) of Britain and Ireland. *Coleopterist*, **9**, 113–147.

Cox, M. L. (2001b). The status of the lily beetle, *Lilioceris lilii* (Scopoli, 1763) in Britain (Chrysomelidae: Criocerinae). *Coleopterist*, **10**, 5–20.

Cox, M. L. (2007). *Atlas of the Seed and Leaf Beetles of Britain and Ireland*. Pisces Publications, Newbury.

Cox, M. (2016). Megalopodidae, Orsodacnidae, Chrysomelidae. In *Beetles of Britain and Ireland. Vol. 1: Sphaeriusidae to Silphidae* (A. G. Duff). Duff, West Runton.

Crawshay, L. R. (1903). On the life history of *Drilus flavescens*, Rossi. *Transactions of the Entomological Society of London*, **1903**, 39–51, plates 1–2.

Crotch, G. R. (1863). *A Catalogue of British Coleoptera*. Clay at the University Press, Cambridge.

Crowson, R. A. (1956). *Coleoptera: Introduction and Keys to Families. Handbooks for the Identification of British Insects* 4(1). Royal Entomological Society, London.

Crowson, R. A. (1981). *The Biology of the Coleoptera*. Academic Press, London.

Cunningham, A. (2012). *Charagmus gressorius* (Fabricius, 1792) (Curculionidae) in Devon – new to Britain. *Coleopterist*, **21**, 77–78.

Curtis, J. (1823–40). *British Entomology: Being Illustrations and Descriptions of the Genera of Insects Found in Great Britain and Ireland: Containing Coloured Figures from Nature of the Most Rare and Beautiful Species, and in Many Instances of the Plants upon which They are Found*. Printed for the author, London. 8 vols.

Darby, M. (2012). Ptiliidae Erichson. In *Beetles of Britain and Ireland. Vol. 1: Sphaeriusidae to Silphidae* (A. G. Duff). Duff, West Runton.

Darwin, C. (1839). *Journal of Researches into the Geology and Natural History of the Various Countries Visited by H.M.S.* Beagle *Under the Command of Captain Fitzroy, R.N. from 1832 to 1836*. Henry Colburn, London.

Darwin, C. (1852, 1854). *Living Cirripedia. A Monograph on the Subclass Cirripedia, with Figures of All the Species*. Ray Society, London. 2 vols.

Darwin, C. (1887). Autobiographical chapter. In *The Life and Letters of Charles Darwin* (ed. F. Darwin). John Murray, London.

Dawson, J. F. (1854). *Geodephaga Britannica. A Monograph of the Carnivorous Ground-beetles Indigenous to the British Isles*. Van Voorst, London.

de la Harpe, J., Reich, E., Reich, K. A. & Dowdle, E. B. (1983). Diamphotoxin; the arrow poison of the !Kung bushmen.

Journal of Biological Chemistry, **258**, 11924–11931.

Denny, H. (1825). *Monographia Pselaphidarum et Scydmaenidarum Britanniae: or, an Essay on the British Species of the Genera* Pselaphus, *of Herbst, and* Scydmaenus, *of Latreille, in which those Genera are Subdivided etc.* Wilkin, Norwich.

Denny, H. (1842). *Monographia Anoplurorum Britanniae: or, an Essay on the British Species of Parasitic Insects Belonging to the Order of Anoplura of Leach etc.* Bohn, London.

Denton, J. (2007). *Water Bugs and Water Beetles of Surrey.* Surrey Wildlife Trust, Pirbright.

Deroe, C. & Pasteels, J. M. (1982). Distribution of adult defense glands in chrysomelids (Coleoptera: Chrysomelidae) and its significance in the evolution of defense mechanisms within the family. *Journal of Chemical Ecology*, **8**, 67–82.

Dettner, K. (1987). Chemosystematics and evolution of beetle chemical defenses. *Annual Review of Entomology*, **32**, 17–48.

Dettner, K. (1993). Defensive secretions and exocrine glands in free-living staphylinid beetles – their bearing on phylogeny (Coleoptera: Staphylinidae). *Biochemical Systematics and Ecology*, **21**, 143–162.

Dillwyn, L. W. (1829). *Memoranda Relating to Coleopterous Insects Found in the Neighbourhood of Swansea.* Murray & Rees, Swansea.

Donisthorpe, H. St-J. K. (1927). *The Guests of British Ants; their Habits and Life-histories.* George Routledge, London.

Donovan, E. (1792–1813). *The Natural History of British Insects: Explaining Them in their Several States, with the Periods of their Transformations, their Food, Oeconomy, etc, Together with the History of Such Minute Insects as Require Investigation by the Microscope: The Whole Illustrated by Coloured Figures, Designed and Executed from Living Specimens.* Donovan & Rivington, London. 16 vols.

Donovan, E. (1805). *Instructions for Collecting and Preserving Various Subjects of Natural History; as Quadrupeds, Birds, Reptiles, Fishes, Shells, Corals, Plants & c, together with a Treatise on the Management of Insects in their Several States Selected from the Best Authorities.* Rivington, London.

Döring, T. F., Hiller, A., Wehke, S., Schulte, G. & Broll, G. (2003). Biotic indicators of carabid species richness on organically and conventionally managed arable fields. *Agriculture Ecosystems and Environment*, **98**, 133–139.

Drake, C. M., Lott, D. A., Alexander, K. N. A. & Webb, J. (2007). *Surveying Terrestrial and Freshwater Invertebrates for Conservation Evaluation.* Natural England, Peterborough.

Drury, D. (1770–82). *Illustrations of Natural History; Wherein are Exhibited Upwards of Two Hundred and Forty Figures of Exotic Insects.* Published for the author, London. 3 vols.

Dudley, H. J., Harrison, J. K., Ballance, J. O. et al. (1971). *Scientific Experiments for a Manned Mars Mission.* National Aeronautics and Space Administration, Washington.

Dudley, R. (1998). Atmospheric oxygen, giant Paleozoic insects and evolution of aerial locomotor performance. *Journal of Experimental Biology*, **201**, 1043–1050.

Duff, A. G. (2007). Identification: longhorn beetles. *British Wildlife*, **18**, 406–414, **19**, 35–43.

Duff, A. G. (2008). *Checklist of Beetles of the British Isles.* A. G. Duff, Wells.

Duff, A. G. (2012a). *Checklist of Beetles of the British Isles*. 2nd edn. Pemberley Books, Iver.

Duff, A. G. (2012b.) *Beetles of Britain and Ireland. Vol. 1: Sphaeriusidae to Silphidae*. Duff, West Runton.

Duff, A. G. (2014). *Tanysphyrus ater* Blatchley, 1928 (Erirhinidae) new to the British Isles. *Coleopterist*, **23**, 49–50.

Duff, A. G. (2016). *Beetles of Britain and Ireland. Vol. 4: Cerambycidae to Curculionidae*. Duff, West Runton.

Duff, A. G. (2017). *Checklist of Beetles of the British Isles*. 3rd edn. Pemberley Books, Iver.

Duff, A. G., Campbell-Palmer, R. & Needham, R. (2013). The beaver beetle *Platypsyllus castoris* Ritsema (Leiodidae: Platypsyllinae) apparently established on reintroduced beavers in Scotland, new to Britain. *Coleopterist*, **22**, 9–19.

Duff, A. G., Mahler, N. B. & Leech, A. R. (2015). Europe's smallest beetle, *Baranowskiella ehnstromi* Sörensson, 1997 (Ptiliidae), new to Britain. *Coleopterist*, **24**, 155–161.

Duffy, E. A. J. (1952). Coleoptera (Cerambycidae). *Handbooks for the Identification of British Insects* 5(12). Royal Entomological Society, London.

Duffy, E. A. J. (1953). Coleoptera (Scolytidae and Platypodidae). *Handbooks for the Identification of British Insects* 5(15). Royal Entomological Society, London.

Duftschmidt, K. (1804–25). *Fauna Austriae, oder Beschreibung der Östereichischen Insecten: Für Angehende Freunde der Entomologie*. Academischen Kunst, Musik und Buchhandlung, Linz.

Dyar, H. G. (1890). The number of molts of lepidopterous larvae. *Psyche*, **5**, 420–422.

Easton, A. M. (1967). The Coleoptera of a dead fox (*Vulpes vulpes* (L.)); including two species new to Britain. *Entomologist's Monthly Magazine*, **102 (1966)**, 205–210.

Edgecombe, G. D. & Legg, D. A. (2014). Origins and early evolution of arthropods. *Palaeontology*, **57**, 457–468.

Edwards, P. B. (2007). *Introduced Dung Beetles in Australia 1967–2007 – Current Status and Future Directions*. Dung Beetles for Landcare Farming Community, Maleny.

Eisner, T. & Aneshansley, D. J. (1999). Spray aiming in the bombardier beetle: photographic evidence. *Proceedings of the National Academy of Sciences of the USA*, **96**, 9705–9709.

Eisner, T. & Aneshansley, D. J. (2000). Defence by foot adhesion in a beetle (*Hemisphaerota cyanea*). *Proceedings of the National Academy of Sciences of the USA*, **97**, 6568–6573.

Eisner, T. & Eisner, M. (1992). Operation and defensive role of 'gin traps' in a coccinellid pupa (*Cycloneda sanguinea*). *Psyche*, **99**, 265–273.

Eisner, T. & Eisner, M. (2000). Defensive use of a fecal thatch by a beetle larva (*Hemisphaerota cyanea*). *Proceedings of the National Academy of Sciences of the USA*, **97**, 2632–2636.

Emlen, D. J. (2008). The evolution of animal weapons. *Annual Review of Ecology, Evolution and Systematics*, **39**, 387–413.

Erwin, T. L. (1982). Tropical forests: their richness in Coleoptera and other arthropod species. *Coleopterists Bulletin*, **36**, 74–75.

Esperk, T. Tammaru, T. & Nylin, S. (2007). Intraspecific variability in number of larval instars in insects. *Journal of Economic Entomology*, **100**, 627–645.

Evans, A. G. (2014). *Beetles of Eastern North America*. Princeton University Press, Princeton.

Evans, M. E. G. (1972). The jump of the click beetle (Coleoptera: Elateridae) – a preliminary study. *Journal of Zoology*, **167**, 319–336.

Evans, M. E. G. (1973). The jump of the click beetle (Coleoptera: Elateridae) – energetics and mechanics. *Journal of Zoology*, **169**, 181–194.

Evans, M. E. G. (1977). Locomotion in the Coleoptera Adephaga, especially Carabidae. *Journal of Zoology*, **181**, 189–226.

Exell, A. W. (1991). *The History of the Ladybird: With Some Diversions on This and That.* Blockley Antiquarian Society, Blockley.

Fabre, J. H. (1921). *Fabre's Book of Insects.* Retold by Mrs Rodolph Stawell, illustrated by E. J. Detmold. Tudor Publishing, New York.

Fabricius, J. C. (1775). *Systema Entomologiae, Sistens Insectorum Classes, Ordines, Genera, Species, Adiectis Synonymis, Locis, Descriptionibus, Observationibus.* Kortii, Flensburg.

Fabricius, J. C. (1776). *Genera Insectorum Eorumque Characteres Naturales Secundum Numerum, Figurum, Situm et Proportionem Omnium Patrium Oris Adiecta Mantissa Specierum Nuper Detectarum.* Bartschii, Chilonii.

Fabricius, J. C. (1792–99). *Entomologia Systematica Emendata et Aucta, Secundum Classes, Ordines, Genera, Species, Adjectis Synonymis, Locis, Observationibus, Descriptionibus.* Proft, Hafniae. 4 vols and a supplement.

Fabricius, J. C. (1801). *Systema Eleutheratorum Secundum Ordines, Genera, Species, Adiectis Synonymis, Locis, Observationibus, Descriptionibus.* Academi Novi, Kiliae.

Fairmaire, L. (1902). *Histoire Naturelle de la France, 8e Partie, Coléoptères.* Maison Emile Deyrolle, Paris.

Farhadi, R., Allahyari, H. & Chi, H. (2011). Life table and predation capacity of *Hippodamia variegata* (Coleoptera: Coccinellidae) feeding on *Aphis fabae* (Hemiptera: Aphididae). *Biological Control*, **59**, 83–89.

Farrell, B. (1998). 'Inordinate fondness' explained: why are there so many beetles? *Science*, **281**, 555–559.

Farrow, R. A. & Lewis, E. S. (1971). *Omophron limbatum* (F.) (Col., Carabidae) an addition (or restoration?) to the British list. *Entomologist's Monthly Magazine*, **106**, 219–221.

Feer, F. & Pincebourde, S. (2005). Diel flight activity and ecological segregation within an assemblage of tropical forest dung and carrion beetles. *Journal of Tropical Ecology*, **21**, 21–30.

Fender, K. M. (1951). The Malthini of North America (Coleoptera: Cantharidae). *American Midland Naturalist*, **46**, 513–629.

Finkelman, S., Navarro, S., Rindner, M. & Dias, R. (2006). Effect of low pressure on the survival of *Trogoderma granarium* Everts, *Lasioderma serricorne* (F.) and *Oryzaephilus surinamensis* (L.) at 30 °C. *Journal of Stored Products Research*, **42**, 23–30.

Fleming, L. (2015). *The Entokil man. The Life of Harold Maxwell-Lefroy.* Dexter Haven, London.

Foddai, D. & Minelli, A. (1994). Fossil arthropods from a full-glacial site in northeastern Italy. *Quaternary Research*, **41**, 336–342.

Forbes, W. T. M. (1924). How a beetle folds its wings. *Psyche*, **31**, 254–258.

Forbes, W. T. M. (1926). The wing folding patterns of the Coleoptera. *Journal of the New York Entomological Society*, **34**, 42–68, 91–239.

Forsythe, T. G. (1983). Locomotion in ground beetles (Coleoptera Carabidae):

an interpretation of leg structure in functional terms. *Journal of Zoology*, **200**, 493–507.

Forsythe, T. G. (2000). *Ground Beetles. Naturalists' Handbooks 8*. Richmond Publishing, Slough.

Foster, A. P., Morris, M. G. & Whitehead, P. F. (2001). *Ixapion variegatum* (Wencker, 1864) (Col., Apionidae) new to the British Isles, with observations on its European and conservation status. *Entomologist's Monthly Magazine*, **137**, 95–105.

Foster, G. N. (2008). Whirligigs in Britain and Ireland. *British Wildlife*, **20**, 28–35.

Foster, G. N. (2010). *A Review of the Scarce and Threatened Coleoptera of Great Britain. Part 3: Water Beetles of Great Britain. Species Status No. 1*. UK Joint Nature Conservation Committee, Peterborough.

Foster, G. N. & Eyre, M. D. (1992). *Classification and Ranking of Water Beetle Communities. UK Nature Conservation No. 1*. Joint Nature Conservation Committee, Peterborough.

Foster, G. N. & Friday, L. E. (2011). *Keys to Adults of the Water Beetles of Britain and Ireland (Part 1) (Coleoptera: Hydradephaga: Gyrinidae, Haliplidae, Paelobiidae, Noteridae and Dytiscidae). Handbooks for the Identification of British Insects 4(5)*. Royal Entomological Society, St Albans.

Foster, G. N., Bilton, D. T. & Friday, L. E. (2014). *Keys to the Adults of the Water Beetles of Britain and Ireland (Part 2) (Coleoptera: Polyphaga: Hydrophiloidea – Both Aquatic and Terrestrial Species). Handbooks for the Identification of British Insects 4(5b)*. Royal Entomological Society, St Albans.

Fowler, W. W. (1887–91). *The Coleoptera of the British Islands. A Descriptive Account of the Families, Genera, and Species Indigenous to Great Britain and Ireland, with Notes as to Localities, Habitats, etc*. L. Reeve & Co., London. 5 vols [1887, vol. 1; 1888, vol. 2; 1889, vol. 3; 1890, vol. 4; 1891, vol. 5].

Fowler, W. W. & Donisthorpe, H. St. J. (1913). *The Coleoptera of the British Islands. A Descriptive Account of the Families, Genera, and Species Indigenous to Great Britain and Ireland, with Notes as to Localities, Habitats, etc. Vol. 6, supplement*. L. Reeve & Co., London.

Fowles, A. P., Alexander, K. N. A. & Key, R. S. (1999). The saproxylic quality index: evaluating wooded habitats for the conservation of dead-wood Coleoptera. *Coleopterist*, **8**, 121–141.

Fraedrich, S. W., Harrington, T. C., Rabaglia, R. J. *et al.* (2008). A fungal symbiont of the redbay ambrosia beetle causes a lethal wilt in redbay and other Lauraceae in the southeastern United States. *Plant Disease*, **92**, 215–224.

Frantsevich, L. Dai, Z., Wang, W. Y. & Zhang, Y. (2005). Geometry of elytra opening and closing in some beetles (Coleoptera: Polyphaga). *Journal of Experimental Biology*, **208**, 3145–3158.

Fraser, N. C., Grimaldi, D. A. & Olsen, P. E. (1996). A Triassic lagerstätte from eastern North America. *Nature*, **380**, 615–619.

Freude, H., Harde, K. W. & Lohse, G. A. (1964–83). *Die Käfer Mitteleuropas*. Goeke & Evers, Krefeld. 11 vols [1964, vol. 4; 1965, vol. 1; 1966, vol. 9; 1967, vol. 7; 1969, vol. 8; 1971, vol. 3; 1974, vol. 5; 1976, vol. 2; 1979, vol. 6; 1981, vol. 10; 1983, vol. 11]. Revised editions forthcoming.

Friday, L. E. (1988). A key to the adults of British water beetles. *Field Studies*, **7**, 1–151.

Friedrich, M., Chen, R., Daines, B. *et al.* (2011). Phototransduction and clock gene expression in the troglobiont beetle *Ptomophagus hirtus* of Mammoth Cave. *Journal of Experimental Biology*, **214**, 3532–3541.

Furniss, M. M. & Furniss, R. L. (1972). Scolytids (Coleoptera) on snowfields above timberline in Oregon and Washington. *Canadian Entomologist*, **104**, 1471–1478.

Furth, D. G. (1988). The jumping apparatus of flea beetles (Alticinae) – the metafemoral spring. In *The Biology of Chrysomelidae* (eds P. Jolivet, E. Petitpierre & T. H. Hsiao). Kluwer Academic, Dordrecht.

Furth, D. G. & Suzuki, K. (1992). The independent evolution of the metafemoral spring in Coleoptera. *Systematic Entomology*, **17**, 341–349.

Furth, D. G. & Suzuki, K. (2008). The metatibial extensor tendons in Coleoptera. *Systematic Entomology*, **15**, 443–448.

Gardner, A. E. (1969). *Eucinetus meridionalis* Lap., (Coleoptera: Eucinetidae), a family and species new to Britain. *Entomologist's Gazette*, **20**, 59–63.

Gaston, K. J. (1991). The magnitude of global insect species richness. *Conservation Biology*, **5**, 283–296.

Gemminger, M. & von Harold, E. (1868–76). *Catalogus Coleoptorum Hucusque Descriptorum Synonymicus et Systematicus*. Gummi, Munich. 12 vols.

Geoffroy, E. L. (1762). *Histoire Abrégée des Insectes qui se Trouvent aux Environs de Paris; dans Laquelle ces Animaux sont Rangés Suivant un Ordre Méthodique*. Durand, Paris.

Gerstmeier, R. (1998). *Checkered Beetles: Illustrated Keys to the Cleridae and Thanerocleridae of the Western Palaearctic*. Margraf Verlag, Weikersheim.

Gibson, C. W. D. (1998). *Brownfield: Red Data. The Values Artificial Habitats Have for Uncommon Invertebrates*. English Nature, Peterborough.

Gilbert, P. (2005). *The Entomological Club and Verrall Supper: A History (1826–2004)*. The Entomological Club, London.

Giorgi, J. A., Vandenberg, N. J., McHugh, J. V. *et al.* (2009). The evolution of food preferences in Coccinellidae. *Biological Control*, **51**, 215–231.

Gloyna, K., Thieme, T., Gorb, S. & Voigt, D. (2014). New results on sexual differences in tarsal adhesive setae of *Diabrotica virgifera virgifera* Leconte (Coleoptera, Chrysomelidae, Galerucinae). *European Journal of Environmental Sciences*, **4**, 97–101.

Gønget, H. (1997). *The Brentidae (Coleoptera) of Northern Europe. Fauna Entomologica Scandinavica 34*. Brill, Leiden.

Gorham, H. S. (1901). On a species of *Stenolophus* apparently new to Britain and to science. *Annals of Scottish Natural History*, **1901**, 24–25.

Griffin, F. J. (1940). The first entomological societies. An early chapter in entomological history in England. *Proceedings of the Royal Entomological Society of London (A)*, **15**, 49–68.

Grimaldi, D. & Engel, M. S. (2005). *Evolution of the Insects*. Cambridge University Press, Cambridge.

Gullan, P. J. & Cranston, P. S. (2014). *The Insects: An Outline of Entomology*. 5th edn. Wiley-Blackwell, London.

Haas, F. & Beutel, R. G. (2001). Wing folding and the functional morphology of the wing base in Coleoptera. *Zoology*, **104**, 123–141.

Haas, F. & Wootton, R. J. (1996). Two basic mechanisms in insect wing folding. *Proceedings of the Royal Society of London B*, **263**, 1651–1658.

Haas, F., Gorb, S. & Blickhan, R. (2000). The function of resilin in beetle wings.

Proceedings of the Royal Society of London B, **267**, 1375–1381.

Haas, F., Waloszek, D. & Hartenberger, R. (2003). *Devonohexapodus bocksbergensis, a new marine hexapod from the lower Devonian Hunsrük slates and the origin of Atelocerata and Hexapoda. Organisms Diversity and Evolution*, **3**, 39–54.

Halstead, A. J. (1989). On the move? *The Garden*, **114**, 321–323.

Halstead, D. G. H. (1963). *Coleoptera: Histeroidea. Handbooks for the Identification of British Insects 4(10).* Royal Entomological Society, London.

Hammond, P. M. (1974). Changes in the British coleopterous fauna. In *The Changing Flora and Fauna of Britain. Systematics Association Special Volume No. 6* (ed. D. L. Hawksworth). Academic Press, London.

Hammond, P. M. (1979). Wing-folding mechanisms of beetles, with special reference to investigations of adephagan phylogeny. In *Carabid Beetles, their Evolution, Natural History, and Classification. Proceedings of the First International Symposium of Carabidology* (eds T. L. Erwin, G. E. Ball, D. R. Whitehead & A. Halpern). W. Junk, The Hague.

Hammond, P. M. (1985). Dimorphism of wings, wing-folding and wing-toiletry devices in the ladybird, *Rhyzobius litura* (F.) (Coleoptera: Coccinellidae), with a discussion of inter-population variation in this and other wing-dimorphic beetle species. *Biological Journal of the Linnean Society*, **24**, 15–33.

Hammond, P. (1990). The diversity of Coleoptera. [Short report of lecture given to the British Entomological and Natural History Society, 14 February 1990.] *British Journal of Entomology and Natural History*, **3**, 126.

Hammond, P. M. (1992). Species inventory. In *Global Biodiversity. Status of the Earth's Living Resources. A Report Compiled by the World Conservation Monitoring Centre* (ed. B. Groombridge). Chapman and Hall, London.

Hammond, P. M. & Barham, C. S. (1982). *Laricobius erichsoni* Rosenhaur (Coleoptera: Derodontidae), a species and superfamily new to Britain. *Entomologist's Gazette*, **33**, 35–40.

Hammond, P. M. & Harvey, P. R. (2008). *Scybalicus oblongiusculus* (Dejean) (Carabidae) apparently thriving on brownfield sites in the East Thames Corridor. *Coleopterist*, **17**, 13–16.

Hance, T. (1990). Relationships between crop types, carabid phenology and aphid predation in agroecosystems. In *The Role of Ground Beetles in Ecological and Environmental Studies* (ed. N. E. Stork). Intercept, Andover.

Hance, D. (2007). *Astrapaeus ulmi* (Rossi, 1790) (Staphylinidae) in Britain. *Coleopterist*, **16**, 1–2.

Hansen, M. (1987). *The Hydrophiloidea (Coleoptera) of Fennoscandia and Denmark. Fauna Entomologica Scandinavica 18.* Brill, Leiden.

Hanski, I. & Cambefort, Y. (eds) (1991). *Dung Beetle Ecology*. Princeton University Press, Princeton.

Harde, K. W. (1984). *A Field Guide in Colour to Beetles* (ed. P. M. Hammond). Octopus Books, London.

Hardersen, S., Macagnio, A. L. M., Sacchi, R. & Toni, I. (2011). Seasonal constraints on the mandible allometry of *Lucanus cervus* (Coleoptera: Lucanidae). *European Journal of Entomology*, **108**, 461–468.

Harding, P. T. & Alexander, K. N. A. (1994). The use of saproxylic invertebrates in the selection and evaluation of areas of relic forest in pasture woodland.

British Journal of Entomology and Natural History, **7 (supplement)**, 21–26.

Harding, P. T. & Rose, F. (1986). *Pasture-woodlands in Lowland Britain: A Review of their Importance for Wildlife Conservation.* Institute of Terrestrial Ecology, Abbots Ripton.

Hardy, T. (1891). *Tess of the d'Urbervilles: A Pure Woman Faithfully Presented.* Osgoode, McIlvaine & Co., London.

Harmon, J. P., Stephens, E. & Losey, J. (2007). The decline of native coccinellids (Coleoptera: Coccinellidae) in the United States and Canada. *Journal of Insect Conservation*, **11**, 85–94.

Harrison, J. F., Kaiser, A. & Vandenbrooks, J. M. (2010). Atmospheric oxygen level and the evolution of insect body size. *Proceedings of the Royal Society B: Biological Sciences*, **277**, 1937–1946.

Harrison, S. (2016). *Oxythyrea funesta* (Poda) (Scarabaeidae) established in Britain. *Coleopterist*, **25**, 1–3.

Harvey, D. J., Harvey, H., Harvey, R. P. *et al.* (2017). Use of novel attraction compounds increases monitoring success of a rare beetle, *Elater ferrugineus. Insect Conservation and Diversity*, **10**, 161–170.

Harvey, M. C., Telfer, M. G. & Howard, A. P. (2014). Noble chafer *Gnorimus nobilis* (Linnaeus) (Scarabaeidae) discovered on the Buckinghamshire/ Bedfordshire border. *Coleopterist*, **23**, 25–27.

Harvey, P. H. & Godfray, C. J. (2001). A horn for an eye. *Science*, **291**, 1505–1506.

Háva, J. (2011). *Beetles of the Family Dermestidae of the Czech and Slovak Republics.* Academia, Prague.

Hays, R. (1929). The beetle experiment. *Amazing Stories*, **June**.

Hengeveld, R. (1985). Dynamics of Dutch beetle species during the twentieth century (Coleoptera: Carabidae). *Journal of Biogeography*, **12**, 389–411.

Herbst, J. F. W. (1785–1806). *Natursystem aller Bekannten In- und Ausländischen Insekten: Als Eine Fortsetzung der von Büffonschen Naturgeschichte.* With C. G. Jablonsky. Pauli, Berlin. 10 vols.

Hewitt, G. M. (1999). Post-glacial recolonization of European biota. *Biological Journal of the Linnean Society*, **68**, 87–112.

Hickin, N. E. (1975). *The Insect Factor in Wood Decay. An Account of Wood-boring Insects with Particular Reference to Timber Indoors.* Associated Business Programmes, London.

Hinton, H. E. (1941). The Lathridiidae of economic importance. *Bulletin of Entomological Research*, **32**, 191–247.

Hinton, H. E. (1943). Natural reservoirs of some beetles of the family Dermestidae known to infest stored products, with notes on those found in spider webs. *Proceedings of the Royal Entomological Society of London (A)*, **18**, 33–42.

Hinton, H. E. (1945). *A Monograph of the Beetles Associated with Stored Products.* Vol. 1. British Museum, London.

Hinton, H. E. (1946). The 'gin-traps' of some beetle pupae; a protective device which appears to be unknown. *Transactions of the Royal Entomological Society of London*, **97**, 473–496.

Hintpinzer, U. & Bauer, T. (1986). The antennal setal trap of the ground beetle *Loricera pilicornis*: a specialization for feeding on Collembola. *Journal of Zoology Series A*, **208**, 615–630.

Hodge, P. J. (2010). *Agrilus cuprescens* (Ménétirés, 1832) and *A. cyanescens* Ratzburg, 1837 (Buprestidae) established in Britain. *Coleopterist*, **18**, 85–88.

Hodge, P. J. (2011). *Pseudoperapion brevirostre* (Herbst, 1797) (Apionidae) new to the British Isles. *Coleopterist*, **20**, 111–115.

Hodge, P. J. & Jones, R. A. (1995). *New British Beetles: Species Not in Joy's Practical Handbook.* British Entomological and Natural History Society, Hurst.

Hodge, P. J., Telfer, M. G., Lane, S. A. & Skirrow, M. B. (2016). *Amara (Bradytus) majuscula* (Chaudoir, 1850) (Carabidae) new to Britain from East Norfolk, West Norfolk and Worcestershire. *Coleopterist*, **25**, 99–105.

Hogue, C. L. (1993). *Latin American Insects and Entomology.* University of California Press, Berkeley.

Holland, D. G. (1972). *A Key to the Larvae, Pupae and Adults of the British Species of Elminthidae.* Freshwater Biological Association, Ambleside.

Holland, J. M. & Luff, M. L. (2000). The effects of agricultural practices on Carabidae in temperate agroecosystems. *Integrated Pest Management Review*, **5**, 109–129.

Holmen, M. (1987). *The Aquatic Adephaga (Coleoptera) of Fennoscandia and Denmark. I. Gyrinidae, Haliplidae, Hygrobiidae and Noteridae. Fauna Entomologica Scandinavica 20.* Brill, Leiden.

Holmes, O. W. (1872). *The Poet at the Breakfast Table.* Houghton Mifflin, Boston.

Hope, F. W. (1837–40). *The Coleopterist's Manual, Containing the Lamellicorn Beetles of Linnaeus and Fabricius.* Bohn, London. 3 vols.

Horridge, G. A. (1956). The flight of very small insects. *Nature*, **178**, 45–46.

Howard, L. O. (1930). *A History of Applied Entomology (Somewhat Anecdotal).* Smithsonian Institute, Washington, DC.

Howard, S. (2002). *A Guide to Monitoring the Ecological Quality of Ponds and Canals Using PSYM.* Pond Action, Oxford.

Howden, H. (1952). A new name for *Geotrupes (Peltotrupes) chalybaeus* LeConte with a description of the larvae and its biology. *Coleopterists Bulletin*, **6**, 41–48.

Howe, R. W. (1957). A laboratory study of the cigarette beetle, *Lasioderma serricorne* (F.) (Col., Anobiidae) with a critical review of the literature on its biology. *Bulletin of Entomological Research*, **48**, 9–56.

Hubble, D. S. (2012). *Keys to the Adults of Seed and Leaf Beetles of Britain and Ireland (Coleoptera: Orsodacnidae, Megalopodidae and Chrysomelidae).* Field Studies Council, Telford.

Hubble, D. S. (2014). *A Review of the Scarce and Threatened Beetles of Great Britain. The Leaf Beetles and their Allies: Chrysomelidae, Megalopodidae, Orsodacnidae. Species Status No. 19.* UK Joint Nature Conservation Committee/ Natural England, Peterborough.

Hudson Beare, T. (1930). *A Catalogue of the Recorded Coleoptera of the British Isles.* Janson, London.

Humboldt, A. von. (1852). *Personal Narrative of Travels to the Equinoctial Regions of America During the Years 1799–1804.* Trans. T. Ross. George Routledge, London. 3 vols.

Hyman, P. S. & Parsons, M. S. (1992). *A Review of the Scarce and Threatened Coleoptera of Great Britain. Part 1.* UK Nature Conservation No. 3. UK Joint Nature Conservation Committee, Peterborough.

Hyman, P. S. & Parsons, M. S. (1994). *A Review of the Scarce and Threatened Coleoptera of Great Britain. Part 2.* UK Nature Conservation No. 12. UK Joint Nature Conservation Committee, Peterborough.

Ingpen, A. (1839). *Instructions for Collecting, Rearing and Preserving British and Foreign*

Insects: Also for Collecting and Preserving Crustacea and Shells. William Smith, London.

International Union for Conservation of Nature (2013). *Guidelines for Using the IUCN Red List Categories and Criteria. Version 10.* IUCN Species Survival Commission, Gland.

Invertebrate Link (Joint Committee for the Conservation of British Insects) (2002). A code of conduct for collecting insects and other invertebrates. *British Journal of Entomology and Natural History,* **15**, 1–6.

Irmler, U. & Hoernes, U. (2003). Assignment and evaluation of ground beetle (Coleoptera: Carabidae) assemblages to sites on different scales in a grassland landscape. *Biodiversity and Conservation,* **12**, 1405–1419.

James, M. J. (1973). *The New Aurelians. A Centenary History of the British Entomological and Natural History Society.* BENHS, London.

Janson, O. E. (1921). *Stenopelmus rufinasus* Gyll. an addition to the list of British Coleoptera. *Entomologist's Monthly Magazine,* **57**, 225–226.

Jessop, L. (1986). *Dung Beetles and Chafers. Coleoptera: Scarabaeoidea. Handbooks for the Identification of British Insects 5(11).* Royal Entomological Society, London.

Jobe, J. (2007). *Rhopalapion longirostre* (Olivier) (Apionidae); an earlier British introduction. *Coleopterist,* **16**, 38.

Johnson, C. (1966). *Coleoptera Clambidae. Handbooks for the Identification of British Insects 4(6a).* Royal Entomological Society, London.

Johnson, C. & Booth, R. G. (2004). *Luperomorpha xanthodera* (Fairmaire): a new British flea beetle (Chrysomelidae) on garden centre roses. *Coleopterist,* **13**, 81–86.

Jolivet, P. & Verma, K. K. (2002). *Biology of Leaf Beetles.* Intercept, Andover.

Jones, R. A. (1979). *Hydnobius perrisi* Farim. (Col., Leiodidae) at Rye, Sussex. *Entomologist's Monthly Magazine,* **113**, 248.

Jones, R. A. (1993). Sycamore: an underrated pabulum for insects and some beetles associated with it. *Entomologist's Record and Journal of Variation,* **105**, 1–10.

Jones, R. A. (1994). How big is that bug? Idle meanderings on how we measure insects. *Bulletin of the Amateur Entomologist's Society,* **53**, 217–221.

Jones, R. A. (2002). The beetles and other invertebrates of Sydenham Hill and Dulwich Woods – indicators of ancient woodland. *London Naturalist,* **81**, 87–106.

Jones, R. A. (2003). The beetles of Downham Woodland Walk – little more than a farm track, or ancient woodland? *London Naturalist,* **82**, 161–172.

Jones, R. A. (2006). *Rhopalapion longirostre* (Olivier, 1807) (Apionidae) finally discovered in Britain. *Coleopterist,* **15**, 93–97.

Jones, R. A. (2009). *Larinus turbinatus* Gyllenhal, 1835 (Curculionidae) discovered in Britain. *Coleopterist,* **19**, 49–51.

Jones, R. A. (2013). Aerial plankton and a mass drowning of insects in northern France. *British Journal of Entomology and Natural History,* **26**, 9–15.

Jones, R. A. (2015). *House Guests; House Pests – a Natural History of Animals in the Home.* Bloomsbury, London.

Jones, R. A. (2017). *Call of Nature: The Secret Life of Dung.* Pelagic Publishing, Exeter.

Joy, N. H. (1906a). Coleoptera from old birds' nests. *Entomologist's Monthly Magazine,* **42**, 39–40.

Joy, N. H. (1906b). Coleoptera occurring in the nests of mammals and birds.

Entomologist's Monthly Magazine, **42**, 198–202, 237–243.

Joy, N. H. (1908). A further note on the Coleoptera inhabiting moles' nests. *Entomologist's Monthly Magazine*, **44**, 246–249.

Joy, N. H. (1930). Coleoptera in birds' nests including a species of *Microglossa* new to Britain. *Entomologist's Monthly Magazine*, **66**, 41–42.

Joy, N. H. (1932). *A Practical Handbook of British Beetles*. Witherby, London. 2 vols.

Joy, N. H. (1933). *British Beetles, their Homes and Habitats*. London: Warne.

Joy, N. H. (1936). *How to Know Birds*. Witherby, London.

Junk, W. & Schenkling, S. (eds) (1910–40). *Coleopterorum Catalogus*. Junk, Berlin. Multiple volumes.

Kamoun, S. & Hogenhout, S. A. (1996). Flightlessness and rapid terrestrial locomotion in tiger beetles of the *Cicindela* L. subgenus *Rivacindela* van Nidek from saline habitats of Australia (Coleoptera: Cicindelidae). *Coleopterists Bulletin*, **50**, 221–230.

Karlsson Green, K., Kovalev, A., Svensson, E. I. & Gorb, S. N. (2013). Male clasping ability, female polymorphism and sexual conflict: fine-scale elytral morphology as a sexually antagonistic adaptation in female diving beetles. *Journal of the Royal Society Interface*, **10**, 20130409.

Kathirithamby, J. (1989). Review of the order Strepsiptera. *Systematic Entomology*, **14**, 41–92.

Kenward, H. K. (1975). The biological and archaeological implications of the beetle *Aglenus brunneus* Gyllenhal in ancient faunas. *Journal of Archaeological Science*, **2**, 63–69.

Kenward, H. & Whitehouse, N. (2010). Insects. In *Extinctions and Invasions: A Social History of British Fauna* (T. O'Connor & N. Sykes). Oxbow Books, Oxford.

King, G. A. (2013). Establishing a foothold or six: insect tales of trade and migration. In *Mobility, Transition and Change in Prehistory and Classical Antiquity. Proceedings of the Graduate Archaeology Organization Conference on the Fourth and Fifth of April 2008 at Hertford College, Oxford, UK* (eds P. R. Preston & K. Schörle). Archaeopress, Oxford.

Kirby, W. (1802). *Monographia Apum Angliae; or, an Attempt to Divide into their Natural Genera and Families Such Species of the Linnean Genus* Apis *as Have Been Discovered in England… etc.* J. Raw, Ipswich.

Kirby, W. (1808). The genus *Apion* of Herbst's Natursystem considered, its characters laid down, and many of the species described. *Transactions of the Linnean Society*, **9**, 1–80.

Kirby, W. (1811). Descriptions of seven new species of *Apion. Transactions of the Linnean Society of London*, **10**, 347–357.

Kirby, W. (1835). *On the Power Wisdom and Goodness of God as Manifested in the Creation of Animals and in their History, Habits and Instincts*. Pickering, London.

Kirby, W. & Spence, W. (1815–26). *Introduction to Entomology, or, Elements of the Natural History of Insects: Comprising an Account of Noxious and Useful Insects, of their Metamorphoses, Food, Stratagems, Habitations etc.* Longman, Brown, Green and Longman, London. 4 vols [1815, vol. 1; 1817, vol. 2; 1826, vol. 3; 1826; vol. 4].

Kirejtshuk, A. G., Poschmann, M., Prokop, J., Garrouste, R. & Nel, A. (2013). Evolution of the elytral venation and structural adaptations in the oldest

Palaeozoic beetles (Insecta: Coleoptera: Tshekardocoleidae). *Journal of Systematic Palaeontology*, **12**, 575–600.

Kirkendall, L. R., Kent, D. S. & Raffa, K. F. (1997). Interactions among males, females and offspring in bark ambrosia beetles: the significance of living in tunnels for the evolution of social behaviour. In *The Evolution of Social Behaviour in Insects and Arachnids* (eds J. C. Choe & B. J. Crespi). Cambridge University Press, Cambridge.

Kirk-Spriggs, A. H. (1996). *Pollen Beetles. Coleoptera: Kateretidae and Nitidulidae: Meligethinae. Handbooks for the Identification of British Insects 5(6a)*. Royal Entomological Society, London.

Klausnitzer, B. (1991–2001). *Die Larven der Käfer Mitteleuropas*. Goeke & Evers, Krefeld. 6 vols.

Klemperer, H. G. (1982a). Normal and atypical nesting behaviour of *Copris lunaris* (L.): comparison with related species (Coleoptera: Scarabaeidae). *Ecological Entomology*, **7**, 69–83.

Klemperer, H. G. (1982b). Parental behaviour in *Copris lunaris* (Coleoptera: Scarabaeidae): care and defence of brood balls and nest. *Ecological Entomology*, **7**, 155–167.

Kloet, G. S. & Hincks, W. D. (1945). *A Check List of British Insects*. Buncle & Co., Arboath.

Knell, R. (2011). Male contest competition and the evolution of weapons. In *Ecology and Evolution of Dung Beetles* (eds L. W. Simmons & T. J. Ridsdill-Smith). Wiley-Blackwell, Oxford.

Knell, R. J., Pomfret, J. C. & Tomkins, J. L. (2004). The limits of elaboration: curved allometries reveal the constraints on mandible size in stag beetles. *Proceedings of the Royal Society of London B*, **271**, 523–528.

Koch, R. L. (2003). The multicoloured Asian lady beetle, *Harmonia axyridis*: a review of its biology, uses in biological control, and non-target impacts. *Journal of Insect Science*, **3**, 1–16.

Koskela, H. (1979). Patterns of diel flight activity in dung-inhabiting beetles: an ecological analysis. *Oikos*, **33**, 419–439.

Kühl, G. & Rust, J. (2009). *Devonohexapodus bocksbergensis* is a synonym of *Wingertshellicus backesi* (Euarthropoda) – no evidence for marine hexapods in the Devonian Hunsrük Sea. *Organisms Diversity and Evolution*, **9**, 215–231.

Kukalová-Peck, J. (1978). Origin and evolution of insect wings and their relation to metamorphosis, as documented by the fossil record. *Journal of Morphology*, **156**, 53–126.

Kukalová-Peck, J. (1983). Origin of the insect wing and articulation from the arthropodian leg. *Canadian Journal of Zoology*, **71**, 1618–1669.

Kukalová-Peck, J. & Beutel, R. G. (2012). Is the Carboniferous *Adiphlebia lacoana* really the 'oldest beetle'? Critical reassessment and description of a new Permian beetle family. *European Journal of Entomology*, **109**, 633–645.

Kukalová-Peck, J. & Lawrence, J. E. (1993). Evolution of the hind wing in Coleoptera. *Canadian Entomology*, **125**, 181–258.

Lacordaire, T. (1854–76). *Histoire Naturelle des Insects. Genera des Coléoptères ou Exposé Méthodique et Critique de Tous les Genres Proposés Jusqu'ici dans cet Ordre d'Insectes*. Roret, Paris.

Lane, S. A. & Mendel, H. (2016). *Chrysomela vigintipunctata* (Scopoli, 1763) (Chrysomelidae) new to Britain. *Coleopterist*, **25**, 57–62.

Langley, S. P. & Very, F. W. (1890). On the cheapest form of light. *American Journal of Science, Series 3*, **40**, 97–113.

Latreille, M. & Dejean, P. F. M. A. (1822–24). *Histoire Naturelle et Iconographie des Insectes Coléoptères d'Europe.* Crevot, Paris.

Lawrence, J. F. & Newton, A. F. (1982). Evolution and classification of beetles. *Annual Review of Systematic Entomology*, **13**, 261–290.

Lawrence, J. F., Slipinski, A., Seago, A. E., Thayer, M. K., Newton, A. F. & Marvaldi, A. E. (2011). Phylogeny of the Coleoptera based on morphological characters of adults and larvae. *Annales Zoologici*, **61**, 1–217.

Leather, S. R. (2015). An entomological classic – the pooter or insect aspirator. *British Journal of Entomology and Natural History*, **28**, 52–54.

Leclerc, G.-L. (1749–89). *Histoire Naturelle, Génerale et Particulière, avec la Description du Cabinet du Roi.* Imprimerie Royale, Paris. 36 vols.

Lefkovitch, L. P. (1958). Unusual antennal characters in some Laemophloeinae (Coleoptera: Cucujidae) and their taxonomic importance. *Proceedings of the Royal Entomological Society of London (B)*, **27**, 93–100.

Lefkovitch, L. P. (1959). A revision of the European Laemophloinae (Coleoptera: Cucujidae). *Transactions of the Royal Entomological Society of London*, **111**, 95–118.

Leseigneur, L. (1972). *Coléoptères Elateridae de la Faune de France Continentale et de Corse. Supplément au Bulletin Mensuel de la Societé Linnéenne de Lyon.* Societé Linnéenne de Lyon, Lyon.

Lettsom, J. C. (1774). *The Naturalist's and Traveller's Companion, Containing Instructions for Collecting and Preserving Objects of Natural History and for Promoting Enquiries After Human Knowledge in General.* Dilly, London.

Levey, B. (1977). *Coleoptera Buprestidae. Handbooks for the Identification of British Insects 5(1b).* Royal Entomological Society, London.

Levey, B. (2009). *British Scraptiidae. Handbooks for the Identification of British Insects 5(18).* Royal Entomological Society, London.

Lewis, R. (1979). *The Devil's Coach Horse.* Hamlyn, London.

Linder, C., Lorenzini, F. & Kehrli, P. (2009). Potential impact of processed *Harmonia axyridis* on the taste of 'chasselas' and 'pinot noir' wines. *Vitis*, **48**, 101–102.

Lindroth, C. H. (1935). The boreo-British Coleoptera. A study of the faunistical connections between the British Isles and Scandinavia. *Zoogeografica*, **2**, 579–634.

Lindroth, C. H. (1972). Taxonomic notes on certain British ground-beetles (Col., Carabidae). *Entomologist's Monthly Magazine*, **96**, 44–47.

Lindroth, C. H. (1973). Systematics specializes between Fabricius and Darwin: 1800–1859. In *History of Entomology* (eds R. E. Smith, T. E. Mittler & C. N. Smith). Annual Reviews, Palo Alto.

Lindroth, C. H. (1974). *Coleoptera: Carabidae. Handbooks for the Identification of British Insects 4(2).* Royal Entomological Society, London.

Lindroth, C. H. (1985–86). *The Carabidae (Coleoptera) of Fennoscandia and Denmark. Fauna Entomologica Scandinavica 15, parts 1 & 2.* Brill, Leiden. 2 vols.

Linnaeus, C. (1758). *Systema Naturae per Regna Tria Naturae, Secundum Classes, Ordines, Genera, Species, cum*

Characteribus, Differentiis, Synonymis, Locis. 10th edn. [1st edn. 1735.] Alurentii Salvii, Stockholm.

Linssen, E. F. (1959). *Beetles of the British Isles.* Frederick Warne, London. 2 vols.

Linz, D. M., Hu, A. W., Sitvarin, M. I. & Tomoyasu, Y. (2016). Functional value of elytra under various stresses in the red flour beetle, *Tribolium castaneum. Scientific Reports,* **6**, 34813.

Lloyd, J. E. (1983). Bioluminescence and communication in insects. *Annual Review of Entomology,* **28**, 131–160.

Löbl, I., Smetana, A. & Löbl, D. (eds) (2003–16). *Catalogue of Palaearctic Coleoptera.* Apollo Books, Stenstrup, and Brill, Leiden. Multiple volumes.

Lohse, G. A. & Lucht, W. H. (1989–94). *Die Käfer Mitteleuropas. Supplementband.* Goeke & Evers, Krefeld. 3 vols [1989, vol. 12; 1992, vol. 13; 1994, vol. 14].

Lott, D. A. (2009). *The Staphylinidae (Rove Beetles) of Britain and Ireland. Part 5: Scaphidiinae, Piestinae, Oxytelinae. Handbooks for the Identification of British Insects* 12(5). Royal Entomological Society, St Albans.

Lott, D. A. (2010). A new species of *Quedius* Stephens subgenus *Microsaurus* Dejean (Coleoptera: Staphylinidae) from Scotland and Ireland. *Coleopterist,* **19**, 135–141.

Lott, D. A. & Anderson, R. (2011). *The Staphylinidae (Rove Beetles) of Britain and Ireland. Parts 7 and 8: Oxyporinae, Steninae, Euaesthetinae, Pseudopsinae, Paederinae and Staphylininae. Handbooks for the Identification of British Insects* 12(7). Royal Entomological Society, St Albans.

Lucht, W. M. (1987). *Die Käfer Mitteleuropas. Katalog.* Goeke & Evers, Krefeld.

Lückmann, J. & Assmann, T. (2005). Reproductive biology and strategies of nine meloid beetles from Central Europe (Coleoptera: Meloidae). *Journal of Natural History,* **39**, 4101–4125.

Lückmann, J. & Niehuis, M. (2009) *Die Ölkäfer in Rheinland-Pfalz und im Saarland.* Gesellschaft für Naturschutz und Ornithologie Rheinland-Pfalz, Mainz.

Luff, M. L. (1998). *Provisional Atlas of the Ground Beetles (Coleoptera, Carabidae) of Britain.* Biological Records Centre, Abbots Ripton.

Luff, M. L. (2007). *The Carabidae (Ground Beetles) of Britain and Ireland. Handbooks for the Identification of British Insects* 4(2). 2nd edn. Royal Entomological Society, London.

Lyneborg, L. (1977). *Beetles in Colour.* Trans. G. Vevers. Blandford Press, Poole.

MacKechnie Jarvis, C. (1976). A history of the British Coleoptera. *Proceedings of the British Entomological and Natural History Society,* **1976**, 91–112.

McKenna, D. D., Wild, A. L., Kanda, K. *et al.* (2015). The beetle tree of life reveals that Coleoptera survived end-Permian mass extinction to diversify during the Cretaceous terrestrial revolution. *Systematic Entomology,* **40**, 835–880.

Majerus, M. E. N. (1994). *Ladybirds. Collins New Naturalist Library.* Harper Collins, London.

Majerus, M. E. N. & Kearns, P. (1989). *Ladybirds. Naturalists' Handbooks* 10. Richmond Publishing, Slough.

Majerus, M. E. N., Majerus, T. O., Bertrand, D. & Walker, L. E. (1997). The geographic distribution of ladybirds (Coleoptera: Coccinellidae) in Britain (1984–1994). *Entomologist's Monthly Magazine,* **133**, 181–203.

Majerus, M. E. N., Strawson, V. & Roy, H. E. (2006). The potential impacts of the arrival of the harlequin ladybird, *Harmonia axyridis* (Pallas) (Coleoptera:

Coccinellidae), in Britain. *Ecological Entomology*, **31**, 207–215.

Majerus, M. E. N., Roy, H. E. & Brown, P. M. J. (2016). *A Natural History of Ladybirds*. Cambridge University Press, Cambridge.

Mann, D. J. (2006). *Ptilodactyla exotica* Chapin, 1927 (Coleoptera: Ptilodactylinae) established breeding under glass in Britain, with a brief discussion on the family Ptilodactylidae. *Entomologist's Monthly Magazine*, **142**, 67–79.

Marples, N. M. (1993). Toxicity assays of ladybirds using natural predators. *Chemoecology*, **4**, 33–38.

Marples, N. M., Brakefield, P. M. & Cowie, R. J. (1989). Differences between the 7-spot and 2-spot ladybird beetles (Coccinellidae) in their toxic effects on a bird predator. *Ecological Entomology*, **14**, 79–84.

Marples, N. M., van Veelen, M. & Brakefield, P. M. (1994). The relative importance of colour, taste and smell in the protection of an aposematic insect *Coccinella septempunctata*. *Animal Behaviour*, **48**, 967–974.

Marsh, R. (1897). *The Beetle, a Mystery*. Skeffington & Co., London.

Marsham, T. (1802). *Entomologia Britannica, Sistens Insecta Britanniae Indigena Secundum Linneum Disposita*. White, London.

Marske, K. A. & Ivie, M. A. (2003). Beetle fauna of the United States and Canada. *Coleopterists Bulletin*, **57**, 495–503.

Martyr, P. (1555). *The Decades of the Newe Worlde or West India, Conteyning the Navigations and Conquests of the Spanyardes with the Particular Description of the Moste Ryche and Large Landes and Lands Lately Founde in the Weste Ocean Perteyning to the Inheritaunce of the Kinges*

of *Spayne*. Trans. R. Eden. William Powell, London.

Maruyama, M. (2012). A new genus and species of flightless, microphthalmic Corythoderini (Coleoptera: Scarabaeidae: Aphodiinae) from Cambodia, associated with *Macrotermes* termites. *Zootaxa*, **3555**, 83–88.

Marvell, A. (1681). *Miscellaneous Poems*. Robert Boulter, London.

Matthews, A. H. (1872). *Trichopterygia Illustrata et Descripta. A Monograph of the Trichopterygia*. London: Janson.

Mendel, H. (1991). *Bembidion argenteolum* Ahrens (Coleoptera: Carabidae) in the British Isles. *British Journal of Entomology and Natural History*, **4**, 139–141.

Mendel, H. & Barclay, M. V. L. (2008). *Semanotus russicus* (Fabricius, 1776) (Cerambycidae) breeding in Britain. *Coleopterist*, **17**, 1–4.

Mendel, H. & Morris, M. G. (2014). *Simo hirticornis* (Herbst, 1795) (Curculionidae) new to Britain; yet another alien from Brooklands, Surrey (VC17). *Coleopterist*, **23**, 97–100.

Menzies, I. S. & Cox, M. L. (1996). Notes on the natural history, distribution and identification of British reed beetles. *British Journal of Entomology and Natural History*, **9**, 137–162.

Miller, K. B. (2003). The phylogeny of diving beetles (Coleoptera: Dytiscidae) and the evolution of sexual conflict. *Biological Journal of the Linnean Society*, **79**, 359–388.

Milne, A. A. (1927). *Now We Are Six*. Methuen, London.

Miquel, M. E. (2005). Selective mortality of stag beetles in Orpington, Kent. *British Journal of Entomology and Natural History*, **18**, 267–269.

Mocquerys, M. S. (1880). *Tératologie Entomologique. Recuile de Coléoptères Anormaux*. Léon Deshays, Rouen.

Moczec, A. P. & Nijhout, H. F. (2004). Trade-offs during the development of primary and secondary sexual traits in a horned beetle. *American Naturalist*, **163**, 184–191.

Moed, L., Shwayder, T. A. & Chang, M. W. (2001). Cantharidin revisited. A blistering defence of an ancient medicine. *Archives of Dermatology*, **137**, 1357–1360.

Moore, B. & Brown, W. (1989). Graded levels of chemical defence in mimics of lycid beetles of the genus *Metrorrhynchus* (Coleoptera). *Journal of the Australian Entomological Society*, **28**, 229–233.

Moret, P., Aráuz, M. D. L. A, Gobbi, M. & Barragan, A. (2016). Climate warming effects in the tropical Andes: first evidence for upslope shifts of Carabidae (Coleoptera) in Ecuador. *Insect Conservation and Diversity*, **9**, 342–350.

Morris, M. G. (1990). *Orthocerous Weevils. Coleoptera: Curculionoidea (Nemonychidae, Anthribidae, Urodontidae, Attelabidae and Apionidae). Handbooks for the Identification of British Insects 5(16).* Royal Entomological Society, London.

Morris, M. G. (1997). *Broad-nosed weevils. Coleoptera: Curculionidae (Entiminae). Handbooks for the Identification of British Insects 5(17a).* Royal Entomological Society, London.

Morris, M. G. (2002). *True Weevils (Part 1). Coleoptera: Curculionoidea (Subfamilies Raymondionyminae to Smicronychinae). Handbooks for the Identification of British Insects 5(17b).* Royal Entomological Society, London.

Morris, M. G. (2008). *True Weevils (Part 2). (Coleoptera: Curculionidae, Ceutorhynchinae). Handbooks for the Identification of British Insects 5(17c).* Royal Entomological Society, St Albans.

Morris, M. G. (2012). *True Weevils (Part 3). (Coleoptera: Curculioninae, Baridinae, Orobitinae). Handbooks for the Identification of British Insects 5(17d).* Royal Entomological Society, St Albans.

Mouffett, T. (1634). *Insectorum Sive Minimorum Animalium Theatrum, Olim ab Edoardo Wottono, Conrado Gesnero, Thomasve Pennio Inchoatum.* Cotes, London.

Mouffett, T. (1658). *The Theatre of Insects: Or Lesser Living Creatures as Bees, Flies, Caterpillars, Spiders, Worms, & Co, a Most Elaborate Work.* Cotes, London.

Muggleton, J. (1977). Ladybirds on the move. *Country Life*, **161**, 601.

Nadein, K. & Betz, O. (2016). Jumping mechanisms and performance in beetles. 1. Flea beetles (Coleoptera: Chrysomelidae: Alticini). *Journal of Experimental Biology*, **219**, 2015–2027.

Neave, S. A. (1933). *The History of the Entomological Society of London, 1833–1933.* Royal Entomological Society, London.

Nel, A., Roques, P., Nel, P., Prokin, A. A. et al. (2013). The earliest known holometabolous insects. *Nature*, **503**, 257–261.

Neumann, P. & Elzen, P. J. (2004). The biology of the small hive beetle (*Aethina tumida*, Coleoptera: Nitidulidae): gaps in our knowledge of an invasive species. *Apidologie*, **35**, 229–247.

Niemitz, C. (1972). Biochemische, verhaltensphysiologische und morphologische Untersuchingen an *Necrophorus vespillo*. *Forma et Functio*, **5**, 209–230.

Nilsson, A. N. & Holmen, M. (1995). *The Aquatic Adephaga (Coleoptera) of Fennoscandia and Denmark. II. Dytiscidae. Fauna Entomologica Scandinavica 32.* Brill, Leiden.

Novák, V. (2014). *Beetles of the Family Tenebrionidae of Central Europe*. Prague: Academia.

Oakley, J. N. (2004). The resurgence of *Zabrus tenebroides* (Goeze) (Carabidae) in southern England. *Coleopterist*, **13**, 99–102.

Ogilvie, B. W. (2012). Attending to insects: Francis Willughby and John Ray. *Notes and Records of the Royal Society*, **66**, 357–372.

Olivier, G.-A. (1789–1808). *Entomologie, ou Histoire Naturelle des Insectes, avec leurs Caractères Génériques et Spécifiques, leur Description, leur Synonymie, et leur Figure Eluminée. Coléoptères*. Baudouin, Paris. 6 vols.

Ormerod, E. (1881). *A Manual of Injurious Insects with Methods of Prevention and Remedy for their Attacks to Food Crops, Forest Trees, and Fruit, and with a Short Introduction to Entomology*. Sonnenschein & Allen, London.

Orwell, G. (1949). *Nineteen Eighty-four, a Novel*. Secker & Warburg, London.

Ostojá-Starzewski, J. C. (2005). The western corn rootworm *Diabrotica virgifera virgifera* LeConte (Col., Chrysomelidae) in Britain: distribution, description and biology. *Entomologist's Monthly Magazine*, **141**, 175–182.

Päivinen, J., Ahlroth, P. & Kaitala, V. (2002). Ant-associated beetles of Fennoscandia and Denmark. *Entomologica Fennica*, **13**, 20–40.

Parry, J. A. (2012). *Scymnus interruptus* (Goeze, 1777) (Coccinellidae) now established in Britain. *Coleopterist*, **21**, 47–49.

Pasteels, J. M., Rowell-Rahier, M., Braekman, J. C., Daloze, D. & Duffet, S. (1989). Evolution of exocrine chemical defense in leaf beetles (Coleoptera: Chrysomelidae). *Cellular and Molecular Life Sciences*, **45**, 295–300.

Paykull, G. von (1798–1800). *Fauna Suecica. Insecta*. Edman, Uppsala. 3 vols.

Paykull, G. von (1811). *Monographia Histeroidum*. Stenhammar & Palmblad, Uppsala.

Peacock, E. R. (1977). *Coleoptera Rhizophagidae. Handbooks for the Identification of British Insects 5(5a)*. Royal Entomological Society, London.

Peacock, E. R. (1993). *Adults and Larvae of Hide, Larder and Carpet Beetles and their Relatives (Coleoptera Dermestidae) and of Derodontid Beetles (Coleoptera Derodontidae). Handbooks for the Identification of British Insects 5(3)*. Royal Entomological Society, London.

Pearce, E. J. (1957). *Coleoptera (Pselaphidae). Handbooks for the Identification of British Insects 4(9)*. Royal Entomological Society, London.

Pearce, E. J. (1975). A revised annotated list of the British Pselaphidae (Coleoptera). *Entomologist's Monthly Magazine*, **110**, 13–26.

Peris, D., Pérez-de la Fuente, R., Peñalver, E., Delclòs, X., Barrón, E. & Labandeira, C. C. (2017). False blister beetles and the expansion of gymnosperm-insect pollination modes before angiosperm dominance. *Current Biology*, **27**, 897–904.

Peschke, K. & Eisner, T. (1987). Defensive secretion of the tenebrionid beetle, *Blaps mucronata*: physical and chemical determinants of effectiveness. *Journal of Comparative Physiology A*, **161**, 377–388.

Petrie, W. M. F. (1917). *Scarabs and Cylinders with Names, Illustrated by the Egyptian Collection in University College, London*. University College, London.

Pickering, G., Lin, J., Riesen, R., Reynolds, A., Brindle, I. & Soleas, G. (2004). Influence of *Harmonia axyridis* on the sensory properties of white and red wine. *American Journal of Enology and Viticulture*, **55**, 153–159.

Pike, T. W. (2015). Interference coloration as an anti-predator defence. *Biology Letters*, **11**, 20150159.

Plant, C. W., Morris, M. G. & Heal, N. F. (2006). *Pachyrhinus lethierryi* (Desbrochers, 1875) (Curculionidae) new to Britain and evidently established in south-east England. *Coleopterist*, **15**, 59–65.

Poe, E. A. (1843). The gold-bug. *Dollar Newspaper* [Philadelphia], 28 June, pp. 1, 4.

Polilov, A. A. (2015). How small is the smallest? New record and remeasuring of *Scydosella musawasensis* Hall, 1999 (Coleoptera, Ptiliidae), the smallest known free-living insect. *Zookeys*, **526**, 61–64.

Pollock, D. A. & Normark, B. B. (2002). The life cycle of *Micromalthus debilis* LeConte (1878) (Coleoptera: Archostemata: Micromalthidae): historical review of evolutionary perspective. *Journal of Zoological Systematics and Evolutionary Research*, **40**, 105–112.

Pope, R. D. (1953). *Coleoptera Coccinellidae and Sphindidae. Handbooks for the Identification of British Insects 5(7)*. Royal Entomological Society, London.

Potter, B. (1910). *The Tale of Mrs Tittlemouse.* Warne, London.

Poulton, E. B. (1933). 'The Society of Entomologists of London for the Study of Insects': 1780–1782. *Proceedings of the Royal Entomological Society of London*, **8**, 97–104.

Pringle, J. A. (1938). A contribution to the knowledge of *Micromalthus debilis* LeC. (Coleoptera). *Transactions of the Royal Entomological Society of London*, **87**, 271–286.

Prischmann, D. A. & Sheppard, C. A. (2002). Love bugs? A world view of insects as aphrodisiacs, with special reference to Spanish fly. *American Entomologist*, **48**, 208–220.

Raven, C. E. (1947). *English Naturalists from Neckham to Ray: A Study of the Making of the Modern World.* Cambridge University Press, Cambridge.

Ray, J. (1691). *The Wisdom of God Manifested in the Works of Creation. In Two Parts viz the Heavenly Bodies, Elements, Meteors, Fossils, Vegetables, Animals etc.* Innys and Manby, London.

Ray, J. (1705). *Methodus Insectorum: Seu Insecta in Methodum Aliqualem Digesta.* Smith & Walford, London.

Ray, J. (1710). *Historia Insectorum.* Churchill, London.

Reitter, E. (1908–16). *Fauna Germanica. Die Käfer des Deutschen Reiches. Nach der Analysischen Methode Bearbeitet.* Lutz, Stuttgart. 5 vols.

Rennie, J. (1832). *A Conspectus of the Butterflies and Moths Found in Britain; with their English and Systematic Names, Times of Appearance, Sizes, Colours; their Caterpillars, and Various Localities.* Orr, London.

Ribak, G. & Weihs, D. (2011). Jumping without legs: the jump of the click-beetles (Elateridae) is morphologically constrained. *PLoS ONE*, **6**, e20871.

Richardson, J. (2014). *Stictoleptura cordigera* (Fuessly, 1775) (Cerambycidae) new to Britain. *Coleopterist*, **23**, 88–89.

Robinson, M. A. (2000). Coleopteran evidence for the elm decline, neolithic activity in woodland, clearance and the use of the landscape. In *Plants in Neolithic Britain and Beyond* (ed. A. S. Fairbairn). Oxbow Books, Oxford.

Ross, A. (1998). *Amber. The Natural Time Capsule.* Natural History Museum Publishing, London.

Roy, H. E., Frost, P. M. J. & Poland, R. L. (2011). *Ladybirds (Coccinellidae) of Britain and Ireland*. Field Studies Council, Shrewsbury.

Roy, H. E., Dariaens, T., Isaac, N. *et al.* (2012). Invasive alien predator causes rapid declines of native European ladybirds. *Diversity and Distributions*, **18**, 717–725.

Roy, H. E., Brown, P. M. J., Comont, R., Poland, R. I. & Sloggett, J. J. (2013). *Ladybirds. Naturalists' Handbooks No. 10*. Pelagic Publishing, Exeter.

Rye, E. C. (1866). *British Beetles: An Introduction to the Study of our Indigenous Coleoptera*. L. Reeve, London.

Rye, E. C. (1870). Coleoptera. New British species, corrections of nomenclature, etc., noticed since the publication of the *Entomologist's Annual*, 1869. *Entomologist's Annual*, **1870**, 31–120.

Sage, B. (2014). *Quedius lyszkowskii* Lott (Staphylinidae) new to Norfolk and the first record for England. *Coleopterist*, **23**, 51–52.

Sage, B. L. & French, P. (2015). *Paropsisterna selmani* Reid & de Little (Chrysomelidae) in Britain. *Entomologist's Monthly Magazine*, **138**, 77–80.

Sahlberg, C. R. (1817–34). *Insecta Fennica, Dissertationibus Academicis*. Frenckel, Helsingfosiae.

Salisbury, A. (2008). The biology of the lily beetle, *Lilioceris lilii* (Scopoli) (Coleoptera: Chrysomelidae). An extract from 'Impact, host range and chemical ecology of the lily beetle, *Lilioceris lilii*', Doctor of Philosophy thesis, Imperial College London, London.

Salisbury, A. & Booth, R. G. (2004). *Rodolia cardinalis* (Mulsant), the vedalia ladybird (Coleoptera: Coccinellidae) feeding on *Icerya purchasi* Maskell, cottony cushion scale (Hemiptera: Margarodidae) in London gardens. *British Journal of*

Entomology and Natural History, **17**, 103–104.

Saul-Gershenz, L. S. & Millar, J. G. (2006). Phoretic nest parasites use sexual deception to obtain transport to their host's nest. *Proceedings of the National Academy of Sciences of the USA*, **103**, 14039–14044.

Schmidt, F. (1832). *Leptodirus hohenwartii* und *Elater grafii*. In *Faunus Zeitschrift für Zoologie und Vergleichende Anatomie. Vol. 1* (ed. J. Gistl). Lindauer, Munich.

Schmitz, A., Sehrbrock, A. & Schmitz, H. (2007). The analysis of the mechanosensory origin of the infrared sensilla in *Melanophila acuminata* (Coleoptera; Buprestidae) adduces new insight into the transduction mechanism. *Arthropod Structure and Development*, **36**, 291–303.

Scholtz, C. H., Davis, A. & Kryger, U. (eds) (2009). *Evolutionary Biology and Conservation of Dung Beetles*. Pensoft Publishers, Sofia.

Schultz, T. D. (1986). Role of structural colors in predator avoidance by tiger beetles of the genus Cicindela (Coleoptera: Cicindelidae). *Bulletin of the Entomological Society of America*, **32**, 142–146.

Scopoli, G. A. (1763). *Entomologia Carniolica Exhibens Insecta Carnioliae Indigena et Distributa in Ordines, Genera, Species, Varietates Methodo Linnaeana*. Trattner, Vienna.

Scott, A. C. (1938). Paedogenesis in the Coleoptera. *Zeitschrift für Morphologie und Ökologie der Tiere*, **33**, 633–653.

Scott, A. C. (1941). Reversal of sex production in *Micromalthus*. *Biological Bulletin*, **81**, 420–431.

Seago, A. E., Brady, P., Vigneron, J.-P. & Schultz, T. D. (2009). Gold bugs and beyond: a review of iridescence and structural colour mechanisms in beetles (Coleoptera). *Journal of the Royal Society Interface*, **6**, S165–S184.

Seidl, R., Rammer, W., Jäger, D. & Lexer, M. J. (2008). Impact of bark beetle (*Ips typographus* L.) disturbance on timber production and carbon sequestration in different management strategies under climate change. *Forest Ecology and Management*, **256**, 209–220.

Seybold, S. & Vanderwel, D. (2003). Biosynthesis and endocrine regulation of pheromone production in the Coleoptera. In *Insect Pheromones: Biochemistry and Molecular Biology. The Biosynthesis and Detection of Pheromones and Plant Volatiles* (eds G. Blomquist & R. Vogt). Academic Press, San Diego.

Sharp, D. (1869). A revision of the British species of *Homalota*. *Transactions of the Entomological Society of London*, **17**, 91–272.

Sharp, D. (1871). *Catalogue of British Coleoptera*. Janson, London.

Sharp, D. (1882). On aquatic carnivorous Coleoptera or Dytiscidae. *Scientific Transactions of the Royal Dublin Society*, **2**, 1–1033.

Sharp, D. (1909). *The Cambridge Natural History. Insects, Part 2*. Macmillan, London.

Sharp, D. & Fowler, W. W. (1893). *Catalogue of British Coleoptera*. L. Reeve, London.

Sharp, W. E. (1899). Some speculations on the derivation of our British Coleoptera. *Proceedings and Transactions of the Liverpool Biological Society*, **13**, 163–184.

Shirt, D. B. (ed.) (1987). *British Red Data Books: 2. Insects*. Nature Conservancy Council, Peterborough.

Sierra, J. R., Woggon, W.-D. & Schmid, H. (1976). Transfer of cantharidin during copulation from the adult male to the female *Lytta vesicatoria* ('Spanish flies'). *Experientia*, **32**, 142–144.

Simmons, L. W. & Emlen, D. J. (2006). Evolutionary trade-off between weapons and testes. *Proceedings of the National Academy of Sciences of the USA*, **103**, 16346–16351.

Simmons, L. W. & Ridsdill-Smith, T. J. (eds) (2011a). *Ecology and Evolution of Dung Beetles*. Wiley-Blackwell, Oxford.

Simmons, L. W. & Ridsdill-Smith, T. J. (2011b). Reproductive competition and its impact on the evolution and ecology of dung beetles. In *Ecology and Evolution of Dung Beetles* (eds L. W. Simmons & T. J. Ridsdill-Smith). Wiley-Blackwell, Oxford.

Sloggett, J. J. & Majerus, M. E. N. (2003). Adaptations of *Coccinella magnifica*, a myrmecophilous coccinellid, to aggression by wood ants (*Formica rufa* group). II. Larval behaviour and ladybird oviposition location. *European Journal of Entomology*, **100**, 337–344.

Smart, J. (1940). *British Museum (Natural History) Instructions for Collectors, No. 4A. Insects*. British Museum, London.

Smith, D. & Kenward, H. (2011). Roman grain pests in Britain: implications for grain supply and agricultural production. *Britannia*, **42**, 243–262.

Smith, D. N. (1962). Prolonged larval development in *Buprestis aurulenta* L. (Coleoptera: Buprestidae). A review with new cases. *Canadian Entomologist*, **94**, 586–593.

Smith, H. H. (1884). Antennae of a beetle used as defensive weapons. *American Naturalist*, **18**, 727–728.

Smith, R. E., Mittler, T. E. & Smith, C. N. (eds) (1973). *History of Entomology*. Annual Reviews, Palo Alto.

Smith, R. L. & Olson, C. A. (1982). Confused flour beetle and other Coleoptera in stored marijuana. *Pan-Pacific Entomologist*, **58**, 79–80.

Smith, S. G. (1971). Parthenogenesis and polyploidy in beetles. *American Zoologist*, **11**, 341–349.

Smith, S. G. & Virkki, N. (1978). *Animal Cytogenetics. Vol. 3, Insects 5. Coleoptera*. Gebrüder Borntraeger, Berlin.

Sprecher, E. & Taroni, G. (2004). *Lucanus cervus Depictus*. Giorgio Taroni, Como.

Spry, W. & Shuckard, W. E. (1840). *The British Coleoptera Delineated, Consisting of Figures of All the Genera of British Beetles*. Crofts, London.

Stanley-Smith, F. (1955). The history of the 'South London' society. *Proceedings of the South London Entomological and Natural History Society*, **1953–54**, 55–75.

Steiner, K. (1998). The evolution of beetle pollination in a South African orchid. *American Journal of Botany*, **85**, 1180–1193.

Stephens, J. F. (1827–34). *Illustrations of British Entomology; or, a Synopsis of Indigenous Insects Containing their Generic and Specific Distinctions with an Account of their Metamorphoses, Times of Appearance, Localities, Food, and Economy, as Far as Practicable*. Baldwin & Cradock, London. 5 vols [Mandibulata: Coleoptera].

Stephens, J. F. (1834). Mode of killing insects. *Entomologist's Magazine*, **2**, 436–437.

Stephens, J. F. (1839). *A Manual of British Coleoptera, or Beetles, Containing a Brief Description of All the Species of Beetles Hitherto Ascertained to Inhabit Great Britain and Ireland, Together with a Notice of their Chief Localities, Times and Places of Appearances, etc.* Longman, Orme, Brown, Green, and Longmans, London.

Stork, N. E. (1988). Insect diversity: facts, fiction and speculation. *Biological Journal of the Linnean Society of London*, **35**, 321–337.

Stork, N. E. (1993). How many species are there? *Biodiversity and Conservation*, **2**, 215–232.

Stork, N. E., McBroom, J., Gely, C. & Hamilton, A. J. (2015). New approaches narrow global species estimates for beetles, insects, and terrestrial arthropods. *Proceedings of the National Academy of Sciences of the USA*, **112**, 7519–7523.

Straw, N. A., Fielding, N. J., Tilbury, C., Williams, D. T. & Inward, D. (2015). Host plant selection and resource utilisation by Asian longhorn beetle, *Anoplophora glabripennis* (Coleoptera: Cerambycidae) in southern England. *Forestry*, **88**, 84–95.

Stubbs, A. E. (1982). Conservation and the future for the field entomologist. Presidential address part 2. *Proceedings of the British Entomological and Natural History Society*, **15**, 55–67.

Sturm, J. (1805–57). *Deutschlands Insecten, Käfer*. Verfasser, Nuremburg. 23 vols.

Sun, J. & Bhushan, B. (2012). Structure and mechanical properties of beetle wings: a review. *Royal Society of Chemistry Advances*, **2**, 12606–12623.

Svacha, P. & Danilevsky, M. L. (1986). Cerambycoid larvae of Europe and Soviet Union (Coleoptera, Cerambycoidea), part 1. *Acta Universitatis Carolinae – Biologica*, **30**, 1–186.

Svensson, G. P., Larsson, M. C. & Hedin, J. (2004). Attraction of the larval predator *Elater ferrugineus* to the sex pheromone of its prey, *Osmoderma eremita*, and its implication for conservation biology. *Journal of Chemical Ecology*, **30**, 353–363.

Swammerdam, J. (1669). *Historia Insectorum Generalis*. Meinardus van Dreunen, Utrecht.

Swammerdam, J. (1737–38). *Biblia Naturae.* Isaak Severinus, Boudewijn van der Aa, Pieter van der Aa, Leyden. 2 vols.

Swammerdam, J. (1858). *The Book of Nature or, the History of Insects.* Seyffert, London.

Symondson, W. O. C. (2004). Coleoptera (Carabidae, Staphylinidae, Lampyridae, Drilidae and Silphidae) as predators of terrestrial gastropods. In *Natural Enemies of Terrestrial Molluscs* (ed. G. M. Barker). CAB International, Oxford.

Taylor, J. E. (1876). *Notes on Collecting and Preserving Natural History Objects.* Hardwicke & Bogue, London.

Telfer, M. G. (2001). *Bembidion coeruleum* Serville (Carabidae) new to Britain and other notable carabid records from Dungeness, Kent. *Coleopterist,* **10**, 1–4.

Telfer, M. G. (2012). *Quedius lucidulus* Erichson, 1839 (Staphylinidae) new to Britain. *Coleopterist,* **21**, 129–131.

Telfer, M. G. (2013). *Olibrus norvegicus* Munster, 1901 (Phalacridae) new for Britain. *Coleopterist,* **22**, 25–26.

Telfer, M. G. (2015). *Carpelimus nitidus* (Baudi di Selve, 1848) (Staphylinidae): another beetle new to Britain from Dungeness. *Coleopterist,* **24**, 100–105.

Telfer, M. G. (2016). *A Review of the Beetles of Great Britain: Ground Beetles (Carabidae): Special Status No. 25.* Natural England, Peterborough.

Telnov, D. (2010). Ant-like flower beetles (Coleoptera: Anthicidae) of the UK, Ireland and Channel Islands. *British Journal of Entomology and Natural History,* **23**, 99–117.

Tennyson, A. (1847). *The Princess.* Moxon, London.

Thomas, M. B. (1990). The role of man-made grassy habitats in enhancing carabid populations in arable land. In *The Role of Ground Beetles in Ecological and Environmental Studies* (ed. N. E. Stork). Intercept, Andover.

Thompson, R. T. (1958). *Coleoptera Phalacridae. Handbooks for the Identification of British Insects* 5(5b). Royal Entomological Society, London.

Thompson, R. T. (1994). *Rhynchaenus erythropus* (Germar) (Curculionidae) not (yet) British. *Coleopterist,* **2**, 68–69.

Tolasch, T., von Fragstein, M. & Steidle, J. L. M. (2007). Sex pheromone of *Elater ferrugineus* L. (Coleoptera: Elateridae). *Journal of Chemical Ecology,* **33**, 2156–2166.

Tottenham, C. E. (1930). *Saprosites mendax* Blackb. in Sussex. *Entomologist's Monthly Magazine,* **66**, 231.

Tottenham, C. E. (1954). *Coleoptera: Staphylinidae (Piestinae to Euaesthetinae). Handbooks for the Identification of British Insects* 4(8a). Royal Entomological Society, London.

Toussaint, E. F. A., Sedel, M., Arriaga-Varela, E., Hájek, J., Král, D. & Fikácek, M. (2016). The peril of dating beetles. *Systematic Entomology,* **42**, 1–10.

Trautner, J. & Geigenmüller, K. (1987). *Tiger Beetles, Ground Beetles: Illustrated Key to the Cicindellidae and Carabidae of Europe.* Josef Margraf, Aichtal.

Tribe, G. D. & Burger, B. V. (2011). Olfactory ecology. In *Ecology and Evolution of Dung Beetles* (eds L. W. Simmons & T. J. Ridsdill-Smith). Wiley-Blackwell, Oxford.

Tschinkel, W. R. (1975). A comparative study of the chemical defensive system of tenebrionid beetles. Defensive behaviour and ancillary features. *Annals of the Entomological Society of America,* **68**(3), 439–453.

Turner, C. R. (2011). *Pentarthrum elumbe* (Boheman) (Curculionidae, Cossoninae) in Britain. *Coleopterist,* **20**, 20–22.

Turner, C. R. (2014). *Stenotrupis marshalli* Zimmerman 1938 (Culculionidae: Cossoninae) in Britain. *Coleopterist*, **23**, 101–104.

Tuxen, S. L. (1973). Entomology systematizes and describes: 1700–1815. In *History of Entomology* (eds R. E. Smith, T. E. Mittler & C. N. Smith). Annual Reviews, Palo Alto.

Twinn, D. C. (2009). *Tillus elongatus* (Linnaeus) (Cleridae) – over a century of error. *Coleopterist*, **18**, 81–83.

Tylden, W. (1865). Curious habit of *Notoxus monocerus*. *Entomologist's Monthly Magazine*, **2**, 118–119.

Tyler, J. & Trice, E. (2001). A description of a possible defensive organ in the larva of the European glow-worm *Lampyris noctiluca* (Linnaeus) (Lampyridae). *Coleopterist*, **10**, 75–78.

Unwin, D. M. (1984). A key to the families of British Coleoptera (and Strepsiptera). *Field Studies*, **6**, 149–197.

van Emden, F. (1939–49). Larvae of British beetles. *Entomologist's Monthly Magazine*, **75**, 257–273; **76**, 7–13; **77**, 117–127, 181–192; **78**, 73–79, 206–226, 253–272; **79**, 209–223, 259–270; **81**, 13–37; **83**, 154–171; **85**, 265–283.

Vega, F. E. & Hofstetter, R. W. (eds) (2015). *Bark Beetles: Biology and Ecology of Native and Invasive Species*. Elsevier, Amsterdam.

Vereecken, N. J. & Mahé, G. (2007). Larval aggregations of the blister beetle *Stenoria analis* (Schaum) (Coleoptera: Meloidae) sexually deceive patrolling males of their host, the solitary bee *Colletes hederae* Schmidt & Westrich (Hymenoptera: Colletidae). *Annales de la Société Entomologique de France (NS)*, **43**, 493–496.

Wallin, H., Nylander, U. & Kvamme, T. (2009). Two sibling species of *Leiopus* Audinet-Serville, 1834 (Coleoptera: Cerambycidae) from Europe: *L. nebulosus* (Linnaeus, 1758) and *L. linnei* sp. nov. *Zootaxa*, **2010**, 31–45.

Walters, J. M., Barclay, M. V. L. & Geiser, M. F. (2016). *Xylotes griseus* (Fabricius, 1775) (Cerambycidae; Lamiinae), the New Zealand fig longhorn, breeding in Devon, new to Britain and Europe. *Coleopterist*, **25**, 53–56.

Ward, W. (1987). Scarab typology and archaeological context. *American Journal of Archaeology*, **91**, 507–532.

Waterhouse, C. O. (1903). Description of a new coleopterous insect belonging to the Curculionidae. *Annals and Magazine of Natural History (7th series)*, **11**, 229–230.

Waterhouse, G. R. (1858). *Catalogue of British Coleoptera*. Taylor & Francis, London.

Wessel, A. (2006). Stridulation in the Coleoptera – an overview. In *Insect Sounds and Communication: Physiology, Behaviour, Ecology and Evolution* (eds S. Drosopoulos & M. F. Claridge). Taylor & Francis, Boca Raton.

Westwood, J. O. (1839, 1840). *Introduction to the Modern Classification of Insects; Founded on the Natural Habits and Corresponding Organisation of the Different Families*. Orme, Brown, Green and Longmans, London. 2 vols.

Wheater, C. P. & Cook, P. A. (2003). *Studying Invertebrates. Naturalists' Handbooks 28*. Richmond Publishing, Slough.

White, M. J. D. (1973). *Animal Cytology and Evolution*. Cambridge University Press, Cambridge.

White, R. E. (1983). *Beetles. A Field Guide to the Beetles of North America*. Peterson Field Guide Series. Houghton Mifflin, Boston.

Whitehouse, N. J. (2006). The Holocene British and Irish forest fossil beetle fauna: implications for forest history, biodiversity and faunal colonisation. *Quaternary Science Reviews*, **25**, 1755–1789.

Williams, D. W. & Liebhold, A. M. (2002). Climate change and the outbreak ranges of two North American bark beetles. *Agricultural and Forest Entomology*, **4**, 87–99.

Wollaston, T. V. (1854). *Insecta Maderensia, Being an Account of the Insects of the Islands of the Madeiran Group*. Van Voorst, London.

Wood, T. (1884). A new species of *Cis*. *Entomologist's Monthly Magazine*, **21**, 130–131.

Wootton, R. J. (1992). Functional morphology of insect wings. *Annual Review of Entomology*, **37**, 113–140.

Wordsworth, W. (1904). *The Complete Works of William Wordsworth, in Ten Volumes. Vol. IV, 1801–1805*. Cosimo, New York.

Wright, S. (2000). The Asian longhorn beetle *Anoplophora glabripennis* Motschulsky (Cerambycidae) – could it become established in Britain? *Coleopterist*, **9**, 94–95.

Xu, Z., Lenaghan, S. C., Reese, B. E., Jia, X. & Zhang, M. (2012). Experimental studies and dynamics modelling analysis of the swimming and diving of whirligig beetles (Coleoptera: Gyrinidae). *PLoS Computational Biology*, **8**, e1002792.

Yin, Z.-W., Cai, C.-Y., Huang, D.-Y. & Li, L.-Z. (2017). Specialized adaptations for springtail predation in mesozoic beetles. *Scientific Reports*, **7**, 98.

Zahradnik, J. (1977). *A Field Guide in Colour to Insects*. Octopus Books, London.

Zahradnik, P. (2013). *Beetles of the Family Ptinidae of Central Europe*. Academia, Prague.

Zwolfer, H. & Bennett, F. D. (1969). *Ludovix fasciatus* Gyll. (Col., Curculioninae), an entomophagous weevil. *Entomologist's Monthly Magazine*, **105**, 122–123.

Index

SPECIES INDEX

Page numbers in **bold** include illustrations.

GENERAL INDEX

Page numbers in **bold** include illustrations.

Buffon, Georges-Louis Leclerc, Count 157
Bug's Life, A (Disney) 291
Buprestidae *see* jewel beetles
Burrell, Rev. John 363–4
burrowing beetles 175, 177, 188
burying (sexton) beetles (*Nicrophorus*) 96, **97**,
    105, 109, 110, 136, 142, 196–7, **213**
Bushy, London 171
butterflies/moths (Lepidoptera) 3, 23, 69, 141,
    313, 319, 327, 354
Byrrhidae (pill beetles) 44, 185, **223**, **273**
Byturidae **233**

caeca 25
Cairngorms 168, 353
Caledonian pine forests 168, 171–2, 332
California 56, 286–7, 305
Calliphoridae (blue-/green-bottle flies) 21
Camber Sands, Sussex 167, 186, 195
Cambrian 143, 312
Cambridge Ladybird Survey 350
Cambridge, University of 370, 371
camouflage **54**, **55**, 60, 61
campodeiform (larval type) 72, **73**, 74–5, 81,
    99, 107
Canada 197, 200, 237, 331
Canada balsam 398
cannabis 108
cannibalism 103, 226
Cantharidae (soldier beetles) 119, **120**, **228**,
    278, 319, 358
    larvae 67, **101**
cantharidin ('Spanish fly') 121–2, 244–5
captive breeding 226
Carabidae *see* ground beetles
Carboniferous 313, 314, 315
cardinal beetles (Pyrochroidae) 9, **68**, 94, 124,
    125, **142**, **246**, **275**
carotenoids 54
carpet beetles (*Anthrenus*) 1, 97–8, 108, 191,
    229, 298
carrion 96–7, 195–7, 207, 210, 213, 229, 233, 237
carrion beetles (Silphidae) 24, 61, 74, 96–7, 98,
    158, 195–7, **213**, **267**, **273**
Cassidinae (Chrysomelidae subfamily)
    (tortoise beetles) 38, 46, 61, **62**, 87, **252**
castaneous (colour) 20
catalepsy (catatonic rigidity) 135
Catalogue of British Coleoptera (Waterhouse) 369

Catalogue of Palaearctic Coleoptera (Löbl *et al.*) 202
cave beetles 9, 80n, 107, 185–6
cellar beetles 42, **61**, 83, 107, 190, 191–2, 242,
    243, 344
    see also darkling beetles
cellulose 19
centipedes/millipedes (Myriapoda) 8, 311
Cerambycidae *see* longhorn beetles
cerambycoid form 76
cerci 27, 29, 69
Cerylonidae **238**
Cetoniinae (Scarabaeidae subfamily) 141, 220
Ceutorhynchinae (Curculionidae subfamily)
    258
chafers (Scarabaeidae, subfamilies
    Melolonthinae, Rutelinae and Cetoniinae)
    1, **22**, **48**, **154**, **163**, **293**, **341**, **345**, 356, 357
    antennae 47, **48**
    in art 294, **295**
    bee mimicry 57, 153, 180
    cetoniine 21, **345**
    coloration/patterning 59–60, 220
    elytra 16, 220
    etymology 3
    feeding 90, 182, 220–1
    flight 141, 142, 145, 220
    guts 24
    introductions 340, **345**, 352
    larvae 220
    as pests 220, 294
    as pets 285
    range/distribution 163
    stridulation/sound 136, 294
chalk downland 163, 179–80, 290
    see also South Downs
cheloniform 73
chemical defences 58, 60, 61, 121–6, 130–1, 244–5
chemical scents *see* pheromones
chemoreceptors 10–11, 24, 47–9, 153
chemosynthesis 123, 130–1
Chilocorinae (Coccinellidae subfamily) 81, **82**
chitin 18–19, 145, 395
Chobham, Surrey 347
Cholevidae (Leiodidae subfamily) 212
Christian symbolism 287, 292–3
Christie, Laurie 161n
chromosomes 77
Chrysididae (ruby-tailed wasps) 21
Chrysomelidae *see* leaf beetles

The New Naturalist Library

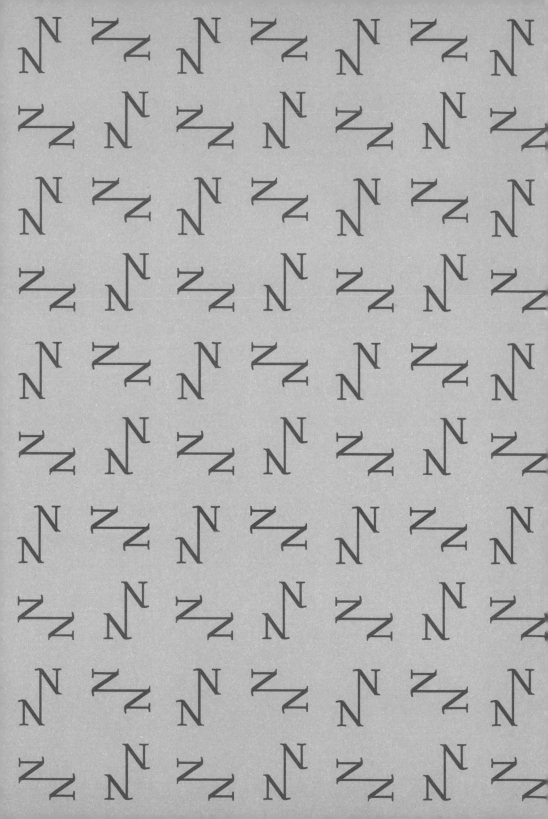